ADVANCES IN SURFACE ACOUSTIC WAVE TECHNOLOGY, SYSTEMS AND APPLICATIONS (Vol.1)

SELECTED TOPICS IN ELECTRONICS AND SYSTEMS

Editor-in-Chief: **M. S. Shur**

Published

Vol. 1: Current Trends in Integrated Optoelectronics
ed. *T. P. Lee*

Vol. 2: Current Trends in Heterojunction Bipolar Transistors
ed. *M. F. Chang*

Vol. 3: Current Trends in Vertical Cavity Surface Emitting Lasers
ed. *T. P. Lee*

Vol. 4: Compound Semiconductor Electronics: The Age of Maturity
ed. *M. Shur*

Vol. 5: High Performance Design Automation for Multichip Modules and Packages
ed. *J. Cho and co-ed. P. D. Franzon*

Vol. 6: Low Power VLSI Design and Technology
eds. *G. Yeap and F. Najm*

Vol. 7: Current Trends in Optical Amplifiers and Their Applications
ed. *T. P. Lee*

Vol. 8: Current Research and Developments in Optical Fiber Communications in China
eds. *Q.-M. Wang and T. P. Lee*

Vol. 9: Signal Compression: Coding of Speech, Audio, Text, Image and Video
ed. *N. Jayant*

Vol. 10: Emerging Optoelectronic Technologies and Applications
ed. *Y.-H. Lo*

Vol. 11: High Speed Semiconductor Lasers
ed. *S. A. Gurevich*

Vol. 12: Current Research on Optical Materials, Devices and Systems in Taiwan
eds. *S. Chi and T. P. Lee*

Vol. 13: High Speed Circuits for Lightwave Communications
ed. *K.-C. Wang*

Vol. 14: Quantum-Based Electronics and Devices
eds. *M. Dutta and M. A. Stroscio*

Vol. 15: Silicon and Beyond
eds. *M. S. Shur and T. A. Fjeldly*

Vol. 16: Advances in Semiconductor Lasers and Applications to Optoelectronics
eds. *M. Dutta and M. A. Stroscio*

Vol. 17: Frontiers in Electronics: From Materials to Systems
eds. *Y. S. Park, S. Luryi, M. S. Shur, J. M. Xu and A. Zaslavsky*

Vol. 18: Sensitive Skin
eds. *V. Lumelsky, Michael S. Shur and S. Wagner*

Forthcoming

Vol. 20: Advances in Surface Acoustic Wave Technology, Systems and Applications (Two volumes), volume 2
eds. *C. C. W. Ruppel and T. A. Fjeldly*

ns in Electronics and Systems – Vol. 19

ADVANCES IN SURFACE ACOUSTIC WAVE TECHNOLOGY, SYSTEMS AND APPLICATIONS (Vol.1)

Editors

Clemens C. W. Ruppel
Siemens AG, Germany

Tor A. Fjeldly
Norwegian University of Science and Technology, Norway

World Scientific

Published by

World Scientific Publishing Co. Pte. Ltd.
P O Box 128, Farrer Road, Singapore 912805
USA office: Suite 1B, 1060 Main Street, River Edge, NJ 07661
UK office: 57 Shelton Street, Covent Garden, London WC2H 9HE

British Library Cataloguing-in-Publication Data
A catalogue record for this book is available from the British Library.

ADVANCES IN SURFACE ACOUSTIC WAVE TECHNOLOGY, SYSTEMS
AND APPLICATIONS (Vol. 1)

Copyright © 2000 by World Scientific Publishing Co. Pte. Ltd.

All rights reserved. This book, or parts thereof, may not be reproduced in any form or by any means, electronic or mechanical, including photocopying, recording or any information storage and retrieval system now known or to be invented, without written permission from the Publisher.

For photocopying of material in this volume, please pay a copying fee through the Copyright Clearance Center, Inc., 222 Rosewood Drive, Danvers, MA 01923, USA. In this case permission to photocopy is not required from the publisher.

ISBN 981-02-4414-2

Printed in Singapore.

PREFACE

ADVANCES IN SURFACE ACOUSTIC WAVE TECHNOLOGY, SYSTEMS AND APPLICATIONS

CLEMENS C. W. RUPPEL
SIEMENS AG, ZT MS 1, Otto-Hahn-Ring 6
81730 Munich/Germany

TOR A. FJELDLY
UniK - Center for Technology at Kjeller, Norwegian University of
Science and Technology, N-2027 Kjeller, Norway.

Surface acoustic wave (SAW) devices are recognized for their versatility and efficiency in controlling and processing electrical signals. Basically, we may think of a SAW device as consisting of a solid substrate with an input and an output transducer. The input transducer converts the incoming signal by the inverse piezoelectric effect into acoustic waves that propagate along the planar surface of the solid. At the output transducer, the surface acoustic waves are reconverted to an electrical signal. Hence, a fundamental property of the SAW device is to act as a signal delay line. The relatively slow propagation velocity of the surface acoustic waves of typically 3500 m/s allows delays of several microseconds on a small chip. However, the versatility of the SAW technology lies in the great flexibility in configuring the transducers, the substrate, and the path of the propagating surface acoustic wave. Since the introduction of the first SAW devices in the mid-1960's, this flexibility has given room for a great deal of ingenuity in the design of different types of devices. This has resulted in a multitude of device concepts for a wide range of signal processing functions, such as delay lines, filters, resonators, pulse compressors, convolvers, and many more. As a consequence, the production volume has risen to millions of devices produced every day, as the SAW technology has found its way into mass markets such as TV receivers, pagers, keyless entry systems, and cellular phones. For such high-volume applications, the unit price of packaged SAW band-pass filters is in the range of up to a few US dollars. At the other end of the scale, we find specialized high performance signal processing SAW devices for satellite communication and military applications, such as radar and electronic warfare, that may run into thousands of dollars per unit.

In two issues of IJHSES, we present an overview of recent advances in SAW technology, systems and applications by some of the foremost researchers and engineers contributing to this exciting field today. Here follows a survey of the seven contributions included in the first issue.

The present overview would not be complete without a historic account of the development that has taken place over the last 35 years. Hence, in Chapter 1, David P. Morgan walks us through the extensive efforts that have led to the present state of affairs in SAW technology. He has emphasized issues related to wave propagation, materials and devices, leading up to the modern low-loss filters used in today's cellular phones. By deferring from a very detailed presentation of the operation of the devices, this chapter also presents itself as a valuable tutorial on the basic design techniques and functionality of SAW devices.

SAW devices derive much of their versatility from the use of and flexibility in shaping the metallic thin films transducers. They provide input and output directional control and can be shaped to perform a multitude of signal processing functions. Other thin films of dielectric, piezoelectric, or semiconductor materials can selectively modify the propagation characteristics of the surface waves, permit the use of non-piezoelectric substrates, and enable the integration of SAWs with active semiconductor electronics. Still other films may provide additional enhancements of the SAW devices. In Chapter 2, Fred S. Hickernell leads us through the many fascinating aspects of SAW thin film technology. In Chapter 3, Eric L. Adler goes a step further by considering in detail the acoustic propagation characteristics of surface as well as bulk acoustic waves in anisotropic multilayer media. Layered geometries are advantageous for enhancing the performance in many device types, but lead to complicated propagation characteristics. He presents an analytical method that offers a systematic and simplified approach to handling such complicated structures in terms of a transmission matrix formalism that maps the variables between neighboring layers. The presentation includes sufficient detail to make it both a tutorial introduction to this field, as well as a presentation of a state-of the-art analysis of the propagation of acoustic waves in layered structures.

The basic functionality of SAW devices is derived from the conversion from electrical to acoustic and back to electrical signals, and the interaction of the acoustical signal with a signal processing grating structure in the propagation path. Hence, details of the interaction between the acoustic signal and the metal grating structure is of primary concern in the analysis and design of SAW devices. In Chapter 4, Ken-Ya Hashimoto, Tatsuya Omori, and Masatsune Yamaguchi review numerical techniques used for the analysis of excitation and propagation of surface acoustic waves under periodic metallic grating structures. Emphasis is placed on the use of a combination of the finite element method (FEM) and the spectral domain analysis (SDA).

Much of the advances in SAW technology over the last decade or so relate to devices for high-frequency applications, in the low GHz frequency range. The applications are in areas such cellular phones, microwave test and measurement equipment and in various military applications. This development has been based on significant advances in design, fabrication, materials, as well as the type of acoustic waves utilized. One such advance is the adoption of surface transverse waves (STWs) for the realization of high-quality GHz resonators, as discussed by Ivan D. Avramov in Chapter 5. Besides a review of STW propagation, this chapter includes details on important design aspects and applications of STW resonators.

In recent years, the application of SAW devices has seen an impressive expansion because of the growth in mobile communication systems, in particular the large expansion of the cellular phones in the private sector. Today, SAW filters are produced in large quantities for such systems. A key component in radio tranceivers of mobile communication systems is the antenna duplexer, which protects the receiver from the transmitted signals. In two chapters, the pioneering development of new antenna filters with SAW resonators, replacing the previous ceramic resonator filters, is presented by some of the foremost contributors to this technology: Mitsutaka Hikita (Chapter 6) and Yoshio Satoh and Osamu Ikata (Chapter 7).

The second issue on Advances in Surface Acoustic Wave Technology, Systems and Application includes additional contributions on SAW materials, device analysis, propagation modes, visualization of modes, interaction with electrons and light, and SAW sensors.

Special Issue EDITORS

Clemens C.W. Ruppel was born in Munich, Germany, in 1952. In 1978 he received the Diploma in mathematics from the Ludwig-Maximilians University of Munich, Germany. Afterwards he has participated in research projects, solving mathematical problems related to bio chemistry and power plant safety, at the university. In 1981 he joined the micro-acoustics research group at Siemens AG as a doctorate student. In 1986 he received his Ph.D. degree for works on the design of surface acoustic waves (SAW) filters from the Technical University of Vienna, Austria. In 1984, he became member of the micro-acoustics group at the Corporate Research and Development of Siemens AG in Munich. In 1990, he became Group Manager. He was responsible for the development of software for the simulation and synthesis of SAW filters. Since 1991, he has been a member of the Technical Program Committee of the IEEE Ultrasonics Symposium, and since 1997 of the IEEE Frequency Control Symposium. In 2000 he has become an elected committee member of the IEEE UFFC AdCom. He has been a voting member of IEEE 802.11. His research interests include all SAW related subjects, especially the design of bandpass filters, dispersive transducers, low-loss filters, and mathematical procedures and algorithms needed for the design and simulation of SAW devices. He is author/co-author of approximately 50 papers (including 7 invited papers) on the design and simulation of SAW filters, and sensors based on SAW devices. In his leisure time he likes to play guitar in a rock band, and enjoys cooking and dining.

Tor A. Fjeldly received the M. Sc. degree in physics from the Norwegian Institute of Technology, 1967, and the Ph.D. degree from Brown University, Providence, RI, in 1972. From 1972 to 1994, he was with Max-Planck-Institute for Solid State Physics in Stuttgart, Germany. From 1974 to 1983, he worked as a Senior Scientist at the SINTEF research organization in Norway. Since 1983, he has been on the faculty of the Norwegian University of Science and Technology (NTNU), where he is a Professor of Electrical Engineering. He is presently with NTNU's Center for Technology at Kjeller, Norway. He was Head of the Department of Physical Electronics at NTNU, and he also served as an Associate Dean of the Faculty of Electrical Engineering and Telecommunication. From 1990 to 1997, he held the position of Visiting Professor at the Department of Electrical Engineering, University of Virginia, Charlottesville, VA, and from 1997 he has been Visiting Professor at the Electrical, Computer and Systems Engineering Department, Rensselaer Polytechnic Institute, Troy, NY. His research interests have included fundamental studies of semiconductors and other solids, development of solid-state chemical sensors, electron transport in semiconductors, modeling and simulation of semiconductor devices, and circuit simulation. He has written about 150 scientific papers, several book chapters, and is a co-author of the books *Semiconductor Device Modeling for VLSI* (Englewood Cliffs, NJ: Prentice Hall, 1993) and *Introduction to Device Modeling and Circuit Simulation* (New York, NY: Wiley & Sons, 1998). He is also a co-developer of the circuit simulator *AIM-Spice*. Since 1998, he has been a Co-Editor-in-Chief of the *International Journal of High Speed Electronics and Systems*, Singapore. Dr. Fjeldly is a Fellow of IEEE and a member of the Norwegian Academy of Technical Sciences, the American Physical Society, the European Physical Society and the Norwegian Society of Chartered Engineers.

Clemens. C. W. Ruppel

Tor A. Fjeldly

CONTENTS

Preface   v

A History of Surface Acoustic Wave Devices   1
   D. P. Morgan

Thin-Films for SAW Devices   51
   F. S. Hickernell

Bulk and Surface Acoustic Waves in Anisotropic Solids   101
   E. L. Adler

Analysis of SAW Excitation and Propagation under Periodic Metallic Grating Structures   133
   K.-Y. Hashimoto, T. Omori, and M. Yamaguchi

High-Performance Surface Transverse Wave Resonators in the Lower GHz Frequency Range   183
   I. D. Avramov

SAW Antenna Duplexers for Mobile Communication   241
   M. Hikita

Ladder Type SAW Filter and its Application to High Power SAW Devices   273
   Y. Satoh and O. Ikata

A HISTORY OF SURFACE ACOUSTIC WAVE DEVICES

David P. Morgan
Impulse Consulting, Northampton NN3 3BG, U.K.

This paper gives a historical account of the development of Rayleigh-wave, or surface-acoustic-wave (SAW), devices for applications in electronics. The subject was spurred on initially by the requirements of pulse compression radar, and became a practical reality with the planar interdigital transducer, dating from 1965. The accessibility of the propagation path gave rise to substantial versatility, and a huge variety of devices were developed. Passive SAW devices are now ubiquitous, with applications ranging from professional radar and communications systems to consumer areas such as TV, pagers and mobile phones.

The paper describes the extensive work, particularly in the 1970s, to investigate SAW propagation in crystalline media, including piezoelectric coupling, diffraction and temperature effects. This led to identification of many suitable materials. Concurrently, many devices began development, including pulse compression filters, bandpass filters, resonators, oscillators, convolvers and matched filters for spread spectrum communications. In the 1970s, many of these became established in professional systems, and the SAW bandpass filter became a standard component for domestic TV.

In the 1980s and 90s, SAW responded to the new call for low-loss filters, particularly for mobile phones. With losses as low as 2 dB required (and subsequently achieved) at RF frequencies around 900 MHz, a raft of new technologies was developed. Additionally, for IF filters special techniques were evolved to reduce the physical size needed for narrow bandwidths. Such devices are now manufactured in very large quantities. In order to satisfy these needs, new types of surface wave, particularly transverse leaky waves, were investigated, and materials using such waves now have their place alongside more traditional materials

1. Introduction

Acoustic waves have been used in electronics for many years, notably in quartz resonators which provide high Q-values as a result of the low acoustic losses. Also, delay lines, exploiting the low acoustic velocities, give a long delay in a small space. In 1965, the first *surface* acoustic wave (SAW) devices were made, introducing exceptional versatility because the propagation path was accessible to components for generating, receiving or modifying the waves. In the subsequent 35 years there has been an explosion in the development of the subject and a huge range of devices has emerged, including delay lines, bandpass filters, resonators, oscillators and matched filters, all of these having a variety of forms. These are now ubiquitous in applications ranging from professional radar and communications systems to consumer areas such as TV, pagers and mobile phones. World-wide production stands at hundreds of millions annually. Despite this the devices are not generally well known, perhaps because they are esoteric components whose presence in systems is not readily evident to the uninitiated.

This paper is a historical review of the subject, without details of the operation of the devices. The coverage is limited to electronics applications and therefore excludes other areas in which SAW's are found, namely non-destructive evaluation, seismology, acousto-optics, acoustic microscopes and sensors (though sensors are described by Scholl and Schmidt elsewhere in this issue). The account is biased towards devices, as opposed to phenomena or physics. At the beginning of the references, some books and special issues are listed as Refs. 1-19 and recent reviews are given in Refs. 20-22.

We begin by describing initial discoveries which showed that the subject might be practical and useful (Sec. 2, up to 1970). Section 3 describes the period 1970-85, in which the subject developed into practicality and began to demonstrate useful functions in real systems. The period after 1985 has been dominated by demands for high-performance low-loss devices which present special problems, invoking a variety of special solutions. Hence it is convenient to defer this to a new section, Sec.4, which begins with an assessment of the position around 1985. Dates given in the headings are approximate only.

2. Beginnings

Seismology and non-destructive evaluation

The existence of the basic type of surface acoustic wave, in an isotropic material, was first demonstrated by Lord Rayleigh (J. Strutt) in his 1885 paper,[23] and hence the wave is often called a Rayleigh wave. This straight-crested wave propagates along the plane surface of a half-space, with the particle motion in the sagittal plane (the plane containing the surface normal and propagation direction), and the amplitude decreases with distance from the surface. Rayleigh was interested in the seismic signals observed following a ground shock. He showed that a late component, following the expected signals due to bulk longitudinal and transverse waves, could be explained by the existence of the slower surface wave. The signal could also be relatively strong owing to the wave spreading in two dimensions rather than three.

Subsequently, there was substantial work by other geophysicists with seismic interests. Love's remarkable treatise,[24] originally published in 1911, includes a study of *shear* surface waves, with motion perpendicular to the sagittal plane. This wave, called the Love wave, can exist when a halfspace is covered with a layer of material with lower bulk shear wave velocity. Love also showed that a Rayleigh-type wave, with sagittal particle motion, could exist in a layered system. Work on the layered Rayleigh wave at the Earthquake Research Institute, Tokyo, in the 1920s showed that a series of higher modes could exist in addition to the fundamental.[25] The first higher mode, known as the Sezawa wave, has been used in SAW devices. The many other developments in seismic surface waves are not generally of much relevance to SAW engineers, but the interested reader will find more information in, for example, Brekhovskikh and Godin[26] and Ewing et al.[27]

Fig. 1. Wedge (left) and comb (right) transducers. In each case, a bulk wave generated by a piezoelectric plate is converted to surface waves.

As for bulk acoustic waves, surface waves were found to be useful for non-destructive evaluation, for example detection of cracks near the surface. In this context Viktorov's book[1] was an important source. At that time, in the 1960s, the commonest methods for generating surface waves were the wedge and the comb, shown in Fig.1. In both cases a bulk wave is generated by a piezoelectric plate transducer and subsequently converted into Rayleigh waves (in both directions in the case of the comb). At this time there were also many other methods, and 24 were described in White's review.[28]

Radar

The interest in surface waves for electronics applications is of course comparatively recent, arising originally from radar requirements. The extensive use of radar in World War II was accompanied by much classified research. After the war, work on pulse compression was revealed in a classic paper by Klauder et al.[29] It was shown that the range capability of a radar can in principle be substantially improved if the radiated pulse is lengthened without changing its power level, and preferably without changing its bandwidth since this determines the resolution. It was envisaged that this would be done by transmitting a chirp pulse, that is, one whose frequency varies with time. In the receiver, there would be a 'matched filter' to optimize the signal-to-noise ratio, basically a dispersive delay line such that the various frequencies of the received signal are delayed by different amounts, arriving at the output at the same time. This system was well understood theoretically – it merely required some means of implementation!

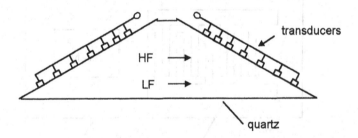

Fig. 2. Wedge delay line (Mortley and others)

The long pulse lengths, tens of microseconds, implied that the filter would need a technology giving substantial signal delays, and 'conventional' methods such as dispersive L-C circuits or cables would be very bulky. It was natural to consider acoustic waves, with velocities some 10^5 times smaller than those of electromagnetic waves. The many possibilities were reviewed by Court[30] in 1969. In particular, several types of dispersive acoustic waves were investigated, including magnetoelastic waves, Lamb waves used in 'strip delay lines',[31] and Love waves.[32] However, for the device response to be dispersive, it is not necessary for the wave itself to be dispersive. Non-dispersive acoustic waves were used in the wedge, or diffraction, delay line sketched in Fig.2. Here a set of transducers is fabricated on each of the inclined faces of a block of crystal quartz, with varied spacing. The transducers generate waves traveling horizontally in the figure; at high frequencies the waves are generated most strongly where the transducers are closer together, so the acoustic path length, and hence the delay, varies with frequency. Mortley[33] demonstrated a device of this type, using interleaved electrodes as the 'transducers', and similar devices were produced by others.[30] Knowing that surface waves can exist on a half-space, it is not too large a step to imagine this device collapsed on to a plane surface, Fig.3, so that it is now planar and the transducers generate surface waves instead of bulk waves. This suggestion was made independently by Rowen and Mortley in two patents in 1963,[34] and these were the first publications on planar SAW transducers. The surface wave device was a substantial advance since the waves can only propagate in one direction, simplifying its behaviour, and the fabrication is much simpler.

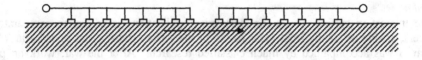

Fig. 3. Planar SAW pulse compression filter (Mortley, Rowen, 1963)

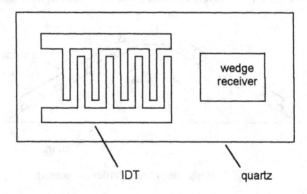

Fig. 4. Interdigital SAW transducer (White and Voltmer, 1965)

The first SAW devices
The first experimental realization of such ideas was that of White and Voltmer[35] in 1965. They demonstrated the uniform (constant-pitch) interdigital transducer (IDT), generating and receiving the waves on a crystal quartz substrate and exploiting piezoelectricity to couple electric to elastic fields. The IDT, Fig.4, consisted simply of interleaved metal electrodes, connected alternately to two bus bars. To behave like a half-space, the substrate only needed to be a few wavelengths thick because the wave has a small penetration depth.

Shortly afterwards, in 1969, Tancrell et al[36] published the first results for a dispersive interdigital SAW device, with both transducers dispersive as in Fig.5. The device had a lithium niobate substrate, and had a center frequency of 60 MHz, a bandwidth of 20 MHz and a time dispersion of about 1 μsec. This paper also suggested that the electrode overlaps could be varied, a technique later called 'apodization', to provide weighting. It was already known[29] that weighting could in principle reduce the 'time-sidelobes' of the radar output pulse, which might otherwise be falsely interpreted as extra targets. Hartemann and Dieulesaint[37] were the first to demonstrate this in a practical SAW device, producing 'compressed' waveforms obtained by applying a frequency-swept pulse to the device, and demonstrating reduction of the time-sidelobes. This is a vital factor in pulse-compression radars.

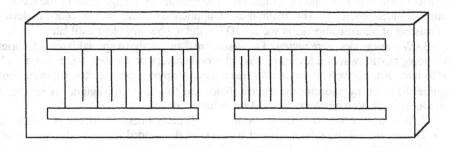

Fig. 5. Interdigital pulse compression filter (Tancrell *et al*, 1969)

These developments were the origin of modern SAW devices. In place of the clumsy wedge or comb transducers, the IDT was a structure that could be fabricated easily by photolithography. A crucial factor was the use of a piezoelectric material, such as quartz or lithium niobate, in which electric and elastic fields are coupled, a phenomenon discovered by P. and J. Curie in 1880. Thus the propagation medium, the substrate, also plays a part in the transduction process, converting electric signals into acoustic waves and, at the receiver, vice versa. At this stage (1969), the basic elements for SAW were in place:
(a) Suitable crystalline piezoelectric materials – quartz and lithium niobate – were available, both having been investigated previously for bulk acoustic wave applications. In particular, quartz had been available for many decades, and its artificial growth, though difficult, was well established.
(b) Lithographic techniques for fabrication were already established.
(c) The interdigital transducer, which could be made by the above techniques, was established as an effective component, though it was not very well understood at the time.

It was realized that this technology could have a substantial future, since the two-dimensional nature of the device made all points in the propagation path accessible, unlike a bulk wave device. This would confer substantial versatility, since almost arbitrary structures, highly complex if required, could be produced with high precision. Indeed, much of the history of SAW can be seen as an exploration of the functions obtainable by variously-shaped structures on the substrate surface, all fabricated by the same basic process. This line of thinking was reinforced by the concurrent development of integrated circuits, in which the use of a planar technology was enabling a remarkable increase of device complexity, a process still continuing today.

Other early developments (to 1969)

Court's paper[30] noted above appears in a 1969 special issue of IEEE Transactions on Microwave Theory and Techniques, which gives a useful snapshot of SAW development up to that time.[15] As might be expected for a new technology involving waves, there was a notable influence of microwave ideas, reflected in both the analysis techniques and the applications envisaged. Thus, waveguides had been shown to be effective, amplification of SAW's using proximity semiconductors was demonstrated, and miniature resonators might be feasible if an efficient SAW reflector could be found. A micro-miniature receiver circuit, with components some 10^5 times smaller than conventional, might be feasible. Much of this thinking turned out to be impractical though some of the ideas are still of much interest. The main present application areas, which concern devices consisting of planar components such as IDT's, did not become clear until later.

SAW waveguides were reviewed by Ash *et al*[38] in the above special issue. A variety of 'topographic' waveguides, using shaped protuberances or grooves, were shown to be effective. Alternatively, it was shown that a metal strip on a piezoelectric substrate could guide the wave by shorting the electric field, and this '$\Delta v/v$ waveguide' is simpler to produce; it has been much used in SAW non-linear convolvers.

To extend the delays obtainable, 'wrap-around' delay lines were developed, in which the ends of the substrate were ground into a smooth rounded shape so that the SAW's could circulate around it. Potentially, these devices could employ waveguides to eliminate diffraction and proximity-coupled semiconductor amplifiers to overcome propagation loss. However, they have not received practical usage.

3. Development to 1985

3.1 *Propagation and materials* (1970 - 1985)

To be effective, a prime requirement for SAW technology is the availability of one or more substrate materials on which the wave propagation is sufficiently well behaved. Extensive studies were therefore undertaken to establish the practicality of the technology and to determine the limitations imposed by the behaviour of potential substrate materials. The basic needs are listed as follows:

(a) Piezoelectricity
(b) Existence of a surface wave, with adequate piezoelectric coupling
(c) Low diffraction effects
(d) Low temperature coefficient
(e) Low attenuation
(f) Low dispersion
(g) Low non-linear effects
(h) Propagation not much affected by SAW components such as IDT's
(i) Minimal degradation due to excitation of unwanted bulk waves.

A remarkable conclusion to be drawn from the early studies is that it is possible to choose a material such that *most of the above requirements are easily and simultaneously satisfied* – the propagation of SAW's is, for many practical purposes, ideal. A number of 'standard' substrates (materials with specified orientations), with well-characterized SAW properties suitable for device applications, became established. Thus, the SAW designer can often, as a result of these early studies, embark on his task with the knowledge that he will not need to compensate for SAW attenuation, dispersion or diffraction, and that non-linear effects will be negligible at the power levels to be used. However, temperature effects usually need to be considered because a material with many attractive properties may have inadequate temperature stability for some applications, an example being lithium niobate.

In practice, a piezoelectric material, which is necessarily anisotropic, is usually either a crystal or a ceramic. Ceramics such as PZT are used at low frequencies, but beyond 50 MHz or so the SAW attenuation becomes unacceptable. There have been a great number of publications assessing SAW propagation in anisotropic materials, an extensive topic because the solutions depend not only on the material considered but also on its orientation (the surface normal direction and the wave propagation direction, relative to the crystal lattice). Initially, it seemed that there was no SAW solution for some orientations, but later this was found to be untrue; problems had arisen due to the very complex nature of the numerical methods necessary. In 1968, Campbell and Jones[39] explained the numerical method for finding SAW solutions, using elastic and piezoelectric constants obtained from measurements on the bulk material. They also pointed out that the piezoelectric coupling for SAW's can be characterized by evaluating two velocities, one for the SAW with a vacuum above the surface (v_0), and the other when the surface is covered by a metal film which shorts out the tangential electric field but is considered to be too thin to have any mechanical effect (v_m). The fractional difference, $\Delta v/v \equiv (v_0 - v_m)/v_0$, is directly related to the amplitude of surface waves generated by an IDT, as shown rigorously by later theory. In particular, the results showed[39] that the coupling is strong, with $\Delta v/v = 2.4$ %, for Y-cut lithium niobate with SAW propagation in the Z-direction. This case, known by convention as Y-Z lithium niobate, is one of the most popular choices.

Diffraction of SAW's, which can be strongly influenced by anisotropy, was first considered theoretically in 1971 by Kharusi and Farnell[40]. They used the 'angular spectrum of plane waves' method (ASoW), previously established for optics but here applied for the first time to the anisotropic case. The angular dependence of the SAW velocity gives sufficient information to allow for anisotropy. Various authors, notably Slobodnik,[46,47] investigated this and several associated topics. Beam steering can occur, so that the SAW energy flow direction is not normal to the wavefronts and the beam propgation direction is not normal to the electrodes. In many anisotropic materials, the SAW velocity is approximately a parabolic function of propagation angle, in which case the diffraction pattern becomes the same as that for an isotropic material except for scaling along the main axis; the diffraction spreading can be less or more than for the isotropic case. Non-parabolic materials are more difficult to analyze, though the ASoW method still applies. A particular example of this is Y-Z lithium niobate, in which the anisotropy is such that diffraction spreading is substantially reduced – this 'minimal diffraction' behaviour is another advantage of this material[41]. Somewhat later, Seifert et al[42] surveyed the many methods for analysis of diffraction in practical devices, and design methods to compensate for it.

The temperature coefficient of SAW velocity is obtainable by calculating the velocity from elastic constants appropriate for two or more temperatures. However, the temperature coefficient of delay (TCD), which also involves the thermal expansion of the material, is the appropriate parameter for a SAW device. For quartz there is an orientation such that the velocity and expansion effects cancel, giving a TCD of zero at one temperature, as might be expected in view of the AT and BT cuts for bulk waves. The delay is a parabolic function of temperature. The orientation is a 42.75° rotated Y cut, known as the ST-cut, with propagation along X. This case, identified in 1970 by Schulz et al,[43] is essential for many temperature-stable devices such as resonators and oscillators. Later, it was shown[44] that the 'turn-over temperature', at which the TCD is zero, can conveniently be adjusted by changing the cut angle. The presence of transducers also affects the turn-over temperature, by an amount dependent on film thickness.[45]

Attenuation was shown to be low in many crystalline materials provided the surface is well polished. Experiments[46,47] bore out the theoretical predictions that the attenuation (in dB) has an f^2 term due to the material viscosity,[48,49] and an f term due to generation of bulk waves in the air.[49,50] The former is usually dominant, though the total attenuation is often small – for example Y-Z lithium niobate gives about 1 dB/μsec at 1 GHz.

Dispersion is theoretically non-existent, though very small amounts have been found in practice, probably due to surface treatment of the material during preparation. Non-linear effects are also usually insignificant. Early works such as Ref. 51 showed that the former affects the latter. The non-linearity gives rise to harmonic-frequency waves which combine to reconstruct the fundamental, but the extent to which this happens can be limited by dispersion (and can be used to measure dispersion). However, this work[51] relates to a free surface and does not apply quantitatively to practical devices, where the (small) dispersion is mainly associated with SAW components such as IDT's. Usually, such components are made from aluminium film. This choice was arrived at in the 1960s by examining a variety of film materials, comparing their resistivity and their effects (attenuation, reflections and velocity shifts) on the SAW's.[52] With this choice, the components cause little disturbance of the propagation except when intended, as in the case of reflectors for example.

It will be seen from the above that many of the requirements can be investigated using numerical calculations of SAW velocities. A great deal of such work was done in the 1970s to find suitable materials, and it continues to the present day. Extensive theoretical calculations were done by, for example, Farnell[53] and Slobodnik,[46,47] and in 1970-74 the latter published extensive catalogues of SAW velocities in many materials.[54-56] In addition, much experimental work was done to evaluate for example attenuation, diffraction and non-linear effects,[47] and for this purpose optical probes were widely used.[47,57] Paying attention particularly to piezoelectric coupling, diffraction effects and temperature stability, a variety of promising material orientations were identified, though not all have been exploited because of practical factors such as cost, size and availability.

Main materials

A popular alternative to Y-Z lithium niobate is the 128°Y-X orientation, which was shown by Shibayama et al[58] to give much less excitation of unwanted bulk waves. Lithium tantalate has intermediate piezoelectric coupling and temperature coefficient, and the X-112°Y orientation gives relatively good temperature stability and low bulk-wave excitation.[59] Zinc oxide films have been investigated since the beginning.[60] Although effective for bulk waves, the fabrication requirements for SAW applications are more demanding, and it was some time before they became common (using glass substrates) in economical TV filters.[61] Recent results demonstrate impressive high-frequency performance, using sapphire substrates.[62] Another promising material, introduced by Whatmore in 1981, is lithium tetraborate.[63] This material has piezoelectric coupling better than that of quartz and a TCD much less than that of lithium niobate. In fact, the

TCD is zero at one temperature as for ST-X quartz, though the stability is not so good. The above materials are commonest for practical devices, but many others have been investigated.

The fundamental work on SAW solutions also yielded a new type of surface wave, called the Bleustein-Gulyaev-Shimizu wave after its three independent discoverers.[64-66] This wave is transverse, with particle displacement normal to the sagittal plane, and exists in a piezoelectric material with a diad axis in this direction. The wave ceases to be guided along the surface if the material is not piezoelectric.

Section 4.2 below discusses further SAW materials, and includes a summary in Table 2.

3.2 Pulse compression filters

As explained in Sec.2 above, pulse compression radar was the main driving force behind the initial development of SAW, and the first SAW pulse compression filter was that of Tancrell et al[36] in 1969, illustrated in Fig.5. In addition to demonstrating a dispersive response, the device was used to compress a 1 μsec chirp pulse to 50 nsec, simulating a pulse compression radar. Conversely, it was also shown that a 1 μsec chirp pulse could be generated by applying an unmodulated 50 nsec pulse to the device. This technique, called 'passive generation', was to become a popular method of chirp generation because of the accuracy provided by SAW devices. Soon afterwards, comprehensive theoretical modeling was done,[67] as described below. To determine the electrode positions it was shown[36] that for a chirp pulse $\cos[\phi(t)]$, in which the phase $\phi(t)$ must be a non-linear function, the electrodes can be located at places corresponding to uniform phase increments, such that $\phi(t) = n\pi$.

In the 1970s these devices were refined and design methods were improved, including in particular compensation for diffraction. The high-quality reproducible performance of the devices made pulse compression radar a reality. Time-bandwidth products were typically 50 to 500, and the sidelobe rejection of the compressed pulse was typically 25 to 40 dB. The chirps were linear, that is, the frequency of the pulse varied linearly with time. Single-electrode transducers were used initially, but later it was common to use double-electrode transducers to suppress electrode reflections. The substrate was often quartz, chosen for temperature stability, but for some chirp characteristics this would lead to high insertion loss, in which case lithium niobate or lithium tantalate could be used. Quartz and lithium niobate were available as long crystals, and chirp lengths up to about 30 μs could be produced, requiring crystals around 12 cm long. When SAW devices were used for passive generation (in the transmitter) as well as pulse compression (in the receiver), the two devices (the 'expander' and 'compressor') were often made as a matched pair mounted in the same package, as this helped to reduce errors due to differences of fabrication or temperature. However, at the present time the generation is often done digitally if the bandwidth is not too large (up to say 20 MHz), as this can be more accurate.

Another development was the use of non-linear chirps as a weighting technique, that is, the rate of frequency sweep in the chirp signal was made to vary with time, with the result that a corresponding amplitude variation was introduced in the spectrum. This enables the sidelobes of the compressed pulse to be reduced while maintaining a flat time-domain amplitude, as normally required in a radar transmitter; in signal processing terms, the receiver is better matched to the transmitter and the output signal-to-noise ratio is somewhat better than in the linear-chirp case. This principle was proposed in 1964,[68] but could not be implemented until SAW devices became available. In 1976, Newton[69] explored the various design trade-offs, and device results were described by Butler.[70] The non-liner chirp system is more affected by Doppler shifts than a linear chirp, but intermediate waveforms can be designed if necessary. In this and other topics, SAW engineers were the main instigators of radar signal development.

The Plessey AR3D radar (1979) used non-linear chirps in a novel way, in connection with electronic scanning of the radar beam, and had a large non-linear chirp SAW filter in the receiver.

To improve the performance of interdigital devices, the transducers can be slanted so that the wave only needs to pass through relatively few electrodes. This reduces degradation due to electrode reflections and other perturbations.[71] Although double-electrode transducers were soon available to ameliorate these problems, the slanted geometry remained of interest for high frequencies, because it could use relatively wide $\lambda/4$ electrodes. The device has been developed continually, for example by Potter,[72] and excellent results were obtained recently at 1.5 GHz by Jen and Hartmann.[73]

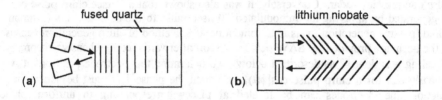

Fig. 6. Pulse compression filters using groove arrays. (**a**): single reflection (Sittig and Coquin, 1968, TB = 540), (**b**): RAC using two reflections (Williamson and Smith, 1972, TB = 1500).

Reflective arrays

Interdigital devices become inconvenient if large time-bandwidth products (TB > 1000) are needed. This is because only part of a chirp transducer is active at any one frequency – the remainder is effectively a capacitor connected across the active part, increasing the device insertion loss. For large TB products, the Reflective Array Compressor (RAC) was introduced in 1972 by Williamson and Smith.[74] Here, as shown in Fig.6b, the SAW is launched by a short uniform IDT, reflected through 90° by an array of grooves, reflected again through 90° by another groove array, and finally reaches another short uniform IDT at the output. The groove pitch is graded so that the path length varies with frequency. The device resembles an earlier device, the IMCON, in which *bulk* waves were reflected by grooves.[75] The IMCON, dating from 1970, used the non-dispersive SH wave propagating in a parallel-sided metal plate. Another precursor was the 1968 demonstration by Sittig and Coquin[76] of a chirp device using reflection of SAW's at a graded-periodicity grooved grating, using wedge transducers and giving TB = 540. This device, shown in Fig.6a, is probably the first published SAW device to use a reflective array. The RAC normally has a lithium niobate substrate because the short wide-band transducers need strong piezoelectric coupling to be efficient. As for the interdigital devices, the long crystals available make long chirp lengths feasible. In the RAC the chirp length is effectively doubled because the SAW transits the device twice, and chirp lengths up to 100 μs were produced, using crystal lengths of around 20 cm.

The RAC enabled TB products as large as 16,000 to be produced.[77] Also, a shaped metal film could be added between the groove arrays to compensate for measured phase errors, and good sidelobe levels, down to − 45 dB, were achieved. However, it was necessary to vary the groove depths to control the amplitude, making the fabrication time-consuming and therefore expensive.

Signal processing using chirp filters

If we consider a radar using linear chirps, it can be expected that a Doppler shift, causing a small shift of the signal frequency, will cause the output pulse to be displaced in time. In view of this, it is not surprising that chirp filters can be used to obtain the spectrum of an applied signal. The method is to multiply the input signal by a linear chirp and apply the result to a chirp filter with opposite dispersion. The output, as a function of time, then gives a scaled version of the spectrum of the input. This was noted by Klauder (Ref.29, p.761) and described in detail by Darlington.[78] The system is closely related to Fourier transformation by optical lenses. If the input signal is taken to be CW, the system can be regarded as a method for frequency measurement, and is then called a compressive receiver, with applications including signal measurements for electronic intelligence. The earliest SAW example appears to be that of Alsup et al[79] in 1973, and it was subsequently developed by for example Atzeni et al,[80] Jack et al[81] and Moule.[82] For large TB products, many systems have used RAC's, notably those of Gerard[77] and Williamson.[83] A more recent example is that of Li et al.[84] There are also chirp systems providing variable delay, variable bandwidth bandpass filtering, or chirp filtering with variable time dispersion.[6]

3.3 Bandpass filters (transversal)

Following the development of apodization in pulse compression filters, described above, it was realized that apodization could also be applied to a transducer with constant electrode pitch in order to control its frequency response, thus enabling it to provide bandpass filtering, as illustrated in Fig.7. This was demonstrated by Hartemann and Dieulesaint[85] in 1969. Soon afterwards it was realized that the technology was well suited to IF filters for domestic TV receivers, the first example being that of Chauvin et al in 1971, with a 33 MHz center frequency and about 6 MHz bandwidth.[86] Analysis using both the delta-function model and equivalent networks, allowing for apodization, was given by Tancrell and Holland.[67]

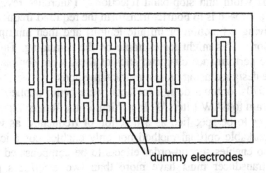

Fig. 7. Basic bandpass filter using apodized transducer

Transversal bandpass filter

The apodized IDT, with constant electrode pitch, is a foundational SAW component. Following the delta-function model, it can be regarded as a sequence of equally-spaced sources whose amplitudes (proportional to the electrode lengths or overlaps) can be chosen at will. Now, diffraction and dispersion are often small, and often the electrodes do not disturb incident waves very much. It follows that the spatial shape of the transducer, given by the apodization, is essentially a scaled version of its impulse response. In turn, the latter is, inevitably, the inverse Fourier transform of the frequency response. Thus a transducer can be designed simply by Fourier transformation of the required frequency response, using the resulting time-domain function to determine the electrode weights. This property was demonstrated by the above authors.[85,86] It expresses the enormous degree of flexibility available in SAW devices, because *any* frequency response whatsoever can be synthesized using this principle. There are, of course, practical limitations, but typically many hundred wavelengths are available, giving many degrees of freedom. In practice the device has two transducers, arranged so that the overall response is simply the product of the two transducer responses. Apodization introduces spaces between each shortened electrode and the opposite bus bar, and these are usually filled with 'dummy electrodes' (1971) to improve the uniformity of SAW velocity.[87]

The above design principle can be expressed more rigorously by noting that the frequency response of a source with position corresponding to time t_0 is proportional to $\exp(-j\omega t_0)$. The transducer, consisting of a regular sequence of these sources, thus has a frequency response of the form

$$H(\omega) = E(\omega)\sum_n A_n \exp(-jn\omega\tau) \tag{1}$$

where τ is the source spacing in time units and A_n are the source amplitudes, which can be taken to be the electrode lengths or overlaps. The term $E(\omega)$ is the 'element factor', representing the frequency response of each source; this was not known initially, but it could be assumed to be approximately constant. With E constant, Eq.(1) is, rigorously, the response of a *transversal filter*, or finite impulse response (FIR) filter, which had earlier been much studied by designers of digital filters. SAW designers could therefore take advantage of methods for designing digital filters, to optimize parameters such as in-band ripple, skirt width and stop-band rejection. Tancrell[88] reviewed the position in 1974. A simple approach is to Fourier transform the required frequency response into the time domain (giving a function of infinite length) and then multiply by a finite-length weighting function. This might for example be the Hamming, Taylor or Kaiser-Bessel function, and the performance characteristics of these were compared. However, many early filters were designed using more empirical methods.

In the mid-1970s, optimal design methods were being applied to digital transversal filters,[89] and in turn to SAW filters.[88] In particular, the Remez algorithm[89,90] enables an arbitrary *tolerance* to be specified as a function of frequency, as well as the amplitude itself. This remarkable optimal method not only enables complicated responses to be designed but also enables second-order effects to be compensated. To fully exploit its flexibility, the transducer must have more than two electrodes per center-frequency wavelength, and typically a double-electrode transducer will be used. Fortunately, this also eliminates electrode reflections, which could cause severe distortion. An alternative and equally flexible design method is linear programming, applicable to a wide range of mathematical optimization problems and applied to SAW filter design a little later.[91]

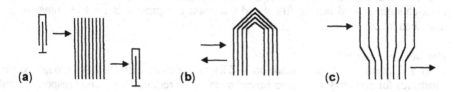

Fig. 8. Multi-strip coupler (MSC) (Marshall and Paige, 1971). (a) basic coupler; (b) mirror; (c) one type of beam compressor (Maerfeld and Farnell, 1973).

Multi-strip coupler

Many bandpass filters also incorporate a multi-strip coupler (MSC). This component, consisting of an array of unconnected electrodes parallel to the SAW wavefronts, was introduced in 1971 by Marshall and Paige.[92,93] The two transducers in the filter are placed in separate, parallel, tracks, and the MSC transfers the SAW's from one track to the other, as shown in Fig.8a. In doing this, bulk waves are not transferred, thus eliminating an unwanted signal which occurs particularly in Y-Z lithium niobate substrates. The MSC also allows the device to be designed with both transducers apodized, giving more flexibility. Several other MSC components were introduced including, for example, an efficient surface-wave reflector made by bending an msc into

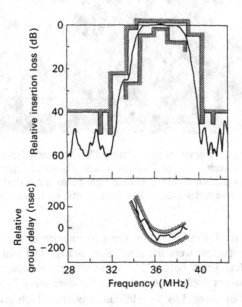

Fig.9. Frequency response of a TV IF filter on Y-Z lithium niobate, using a MSC. The complicated specification, shown by shaded lines, illustrates the versatility of SAW technology. Reprinted from Ref. 6, Fig. 8.12, with permission from Elsevier Science.

a 'U' shape as in Fig.8b. A beam compressor, converting a wide SAW beam into a narrow one with little power loss, can be made by using different strip pitches in the two tracks[94], as in Fig.8c. The basic MSC is often used in bandpass filters, while an MSC track-changer is used in ring filters and the beam compressor is used in convolvers, as described later.

Product development

Early TV filters were usually made on Y-Z lithium niobate with an MSC, though ceramic substrates for economy were also investigated. The required frequency response would normally be a complicated function, often including a frequency-dependent delay as in the example shown in Fig.9. The delay variation mimics the behaviour of the conventional L-C filter, which had to be compensated by pre-distorting the signal at the transmitter. Figure 10 is a scanning electron micrograph of a similar TV filter in operation, showing the SAW propagation graphically.

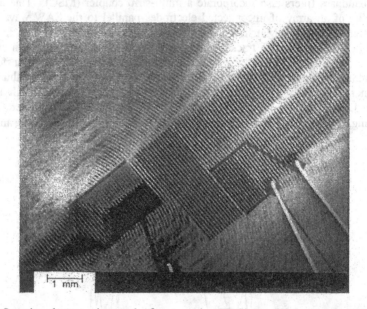

Fig. 10. Scanning electron micrograph of an operating TV filter, using heterodyning to make the SAW wavefronts appear stationary. An apodized transducer at bottom left generates SAW's which are split into two beams by the central MSC, and the right beam is received by a uniform output transducer. A variety of unwanted waves, mostly having little effect, can be seen. Courtesy of C.Ruppel, Siemens.

Compared with its rival, the L-C filter, the performance of the SAW was not the crucial factor – more important were the small size and the fact that trimming was not necessary. It was a few years before TV manufacturers accepted SAW devices in quantity. For this consumer application, cost was the overriding factor, and the device area was soon reduced by using 128°Y-X lithium niobate, which gives low bulk wave excitation and therefore eliminates the need for an MSC. Other substrates now used[95] are X-112°Y lithium tantalate,[59] piezoelectric ceramics and ZnO on glass. A crucial factor in reducing cost is high-volume production (millions of devices annually), made possible by the simple planar structure and semiconductor-based fabrication techniques. Plastic packages are normally used, and TV filters are made to a wide variety of specifications. Such filters also find their way into VCR's and DBS receivers.

Simultaneously, there was extensive work on filters for very exacting requirements, notably vestigial sideband (VSB) filters for TV broadcasting equipment. Typically, these have 6 MHz bandwidth centered at 40 MHz, in-band ripple of ± 0.2 dB in amplitude and ± 20 nsec in delay, and skirt width 500 kHz. To realize this performance, sophisticated designs are needed, usually compensating for diffraction[96,97] and for circuit effects, that is, the distortion associated with the use of finite terminating impedances. Devices meeting the requirements emerged in the late 1970's, for example Kodama's device[98] using X-112°Y lithium tantalate, though most devices used Y-Z lithium niobate with an MSC. These devices are widely used for CATV. Somewhat later, similar filters were developed for digital radio, in which quadrature amplitude modulation (QAM) is used; here each symbol is a pulse whose amplitude and phase can take many values, giving many bits per symbol. The filter requirements for this system are even more demanding than those for VSB. Ganss-Puchstein et al[99] show examples with ripple as low as ± 0.05 dB for 64-QAM, with bandwidth 20 MHz.

As an alternative to apodization, in 1973 Hartmann introduced 'withdrawal weighting', in which selected electrodes are removed from a uniform IDT.[100] This method, much less affected by diffraction, gives excellent results in narrow-band filters. In the 1970s there were studies allowing for the complex electrostatic effects present in the short transducer sections remaining. However, later it was realized that this was unnecessary because the weighting can be done by modifying the electrode polarities rather than physically removing them.

Fig. 11. Bandpass filter for deep space transponder on Voyager spacecraft. From Ref. 22, courtesy F.Hickernell and copyright Academic Press, 1999.

In the 1970s, there was also much work on developing filter banks for applications such as ESM. These could be based on multiple SAW devices,[101] and alternatively Solie[102] developed an efficient MSC-based method, effectively a frequency-selective beam router.

Figure 11 shows a photo of an early bandpass filter developed for the deep space Voyager mission. Filters such as this were subject to very rigorous environmental testing (for temperature changes, vibration etc.), an important factor for SAW's generally if they were to be accepted in practical systems. After lift-off in August 1977, Voyager flew past Jupiter, Saturn, Uranus and, in August 1989, Neptune. In January this year (2000) the spacecraft transmitter was still operational and was tracked as being at 9000 million km from earth; it appears that the SAW's are still working.[103]

3.4 Gratings, resonators and oscillators

In microwave devices, resonances can be exploited to provide a narrow-band response in a relatively small device. SAW resonators are not quite analogous to this because there is no well-localized method of providing efficient reflection. However, in 1970 Ash[104] showed that good reflectivity can be obtained, over a limited bandwidth, by using a regular array of weak reflectors, giving maximum reflectivity when the pitch is $\lambda/2$. Subsequently, these gratings have usually been composed of aluminium strips (for convenience of fabrication) or grooves (for better performance). As shown in Fig. 12, a typical resonator would have two reflective gratings, forming a resonant cavity, with two IDT's in between, one as the input and the other as the output. For temperature stability, quartz would normally be the substrate. In the 1970s there was substantial research on such devices, investigating for example the performance of gratings, optimum positioning of transducers and velocity shifts caused by electrodes or grooves, which could strongly influence the results. In some cases the performance was found to be degraded by the transducers behaving as waveguides, causing unwanted perturbations due to higher modes, and the latter were reduced by incorporating apodization.

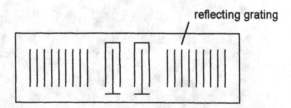

Fig. 12. Two-port resonator. The gratings are composed of (a) grooves or (b) shorted or open-circuit metal strips, with spacing $\lambda/2$.

Analysis of gratings was initially done by modeling as a periodic transmission line having sections with alternating impedances, as explained for example by Li and Melngailis.[105] The impedance changes give rise to reflections, and the values needed can be deduced from measurements. Later, first-order analysis of reflection by one strip or groove, in terms of material constants, was derived by Suzuki et al[106] and independently by Datta and Hunsinger.[107, 108] For a metal strip there are two terms, one arising from electrical effects[106, 108] and significant when $\Delta v/v$ is large, and the other due to mechanical effects[106, 107] and dominant in a weakly-piezoelectric material such as quartz. For a variety of materials, these formulae give reasonable agreement with experimental measurements, notably those of Dunnrowitz[109] and Wright[110]. The first-order analysis also shows that the SAW velocity in a grating is expected to vary linearly with the normalized film thickness h/λ. However, measurements revealed an additional $(h/\lambda)^2$ term, which was attributed to stored energy. In early papers, this was represented by adding susceptances to the transmission-line model.[105] Given the scattering behaviour of a single strip, the grating behaviour can be found by algebraically cascading the transmission line sections, though Suzuki[106] used the coupled-mode (COM) analysis. In either case, there are closed-form formulae for the grating as a whole.

Resonators and oscillators

The resonator, consisting typically of either one or two transducers in the cavity formed by two gratings, was described by Staples *et al*[111] in 1974, and many others began work almost simultaneously. A major application of the resonator is as the controlling element for a SAW oscillator. For modest requirements the resonator can be made in one lithographic step, using metal strips for the gratings as well as transducers. To avoid degradation due to electrode reflections in the transducers, one possibility is to position the transducer electrodes as extensions of the adjacent grating pattern, connecting them alternately to the bus bars so as to form a single-electrode transducer. This 'synchronous' design, common since about 1980, produces a skewed frequency response, but this is still acceptable for an oscillator and it avoids the need for a double-electrode transducer with its narrow electrodes. The insertion loss, without tuning inductors, is typically 5 dB.

For high performance, groove gratings are used and the electrode reflections can be minimized by recessing the electrodes into grooves in the quartz surface. Considerable efforts to optimize the performance have included the development of a quartz package, instead of the commoner metal packages, to minimize stresses due to differential expansion. Q-values in the region of 10^5 are achieved, and the long-term stability can be better than 1 ppm per year.[112] The frequency stability of these devices is comparable to that of bulk wave resonator oscillators when the effects of frequency up-conversion (from typical bulk-wave to SAW frequencies) are taken into account. The SAW devices offer a convenient solution for fundamental mode operation in the range 100 to 2000 MHz. Improved Q-values were achieved later using STW's (Sec. 4.2 below).

A SAW oscillator can also be realized using a delay line, with the output fed back to the input via an amplifier. In fact, the first SAW oscillator, produced in 1969 by Maines *et al*,[113] was of this type; it was put forward as a precise method for determining the substrate temperature coefficient, since frequencies can be measured very accurately. The Q-factor of a delay line is essentially its length in wavelengths, typically a few hundred and therefore much less than that of the resonator. However, this is not necessarily a disadvantage. Analysis of the short-term stability[112, 114] shows that the input power level is also an important factor. For the resonator the high Q-factor magnifies the internal power level, so that the input power must be more constrained in order to avoid damage to the electrodes. In fact, the two devices can give similar performance,[112] though the resonator is of course more compact and needs less input power.

The resonators considered above can also be regarded as narrow-band bandpass filters but their sharply peaked responses, corresponding to an L-C filter with only one pole, are not attractive. For multiple-pole designs, with potential for flatter passbands, a wide variety of methods for coupling single-pole devices were investigated in the 1970s.[115] However, these were not popular for various reasons such as critical fabrication, narrow bandwidth capability or poor stop band rejection. Later, we consider the transverse-coupled resonator, an effective solution to this problem.

3.5 Convolvers and spread-spectrum devices

This Section is concerned with devices for correlation of coded waveforms such as PSK, as used in spread-spectrum communications, and in particular with programmable devices. The popularity of the topic in the 1970s is illustrated by the fact that about a quarter of the 1976 Special Issue of Proceedings IEEE[17] is devoted to it.

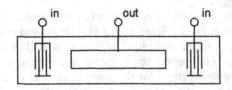

Fig. 13. SAW non-linear convolver (Luukkala and Kino, 1971)

Many of these devices employ non-linearities associated with the SAW propagation. In a basic 'convolver', contra-directed SAW's are generated by transducers at both ends of the substrate, so that they overlap in the space between. The non-linearity gives rise to an electric field proportional to the product of the two SAW amplitudes. The field is sensed and spatially averaged by a uniform electrode on the surface between the two input transducers (this process also rejects the two input signals, which are of course the strongest signals present). For general input waveforms, it can be shown that the output is ideally the convolution of the two input waveforms (apart from a time-contraction by a factor of 2), hence the name 'convolver'. Consequently, this device behaves mathematically like a *linear* filter, but with the remarkable property that the 'impulse response', which is in fact one of the input waveforms, is practically arbitrary. To correlate a coded signal, this 'reference' input needs to be the time-reverse of the signal itself, and the device is particularly suited to spread-spectrum signals for which the reference is easily generated.

Mixing of contra-directed SAW's was first observed in 1969 by Svaasand,[116] using a quartz substrate. The non-linear effect is however much stronger in lithium niobate, and the convolution process was first verified in this material by Quate and Thompson[117] using bulk waves, and by Luukkala and Kino[118] using SAW's with the arrangement of Fig.13. In view of earlier remarks that non-linear effects are usually insignificant in SAW devices, the deliberate use of them here is at first sight surprising. However, the non-linearity is indeed weak, causing little perturbation of the input waves; despite this, the weak output signal can be strong enough to meet the essential requirement that it is not appreciably corrupted by thermal noise.

Improvement of efficiency

The efficiency is nevertheless of considerable concern, and in 1974 new designs emerged with the SAW beamwidth reduced to around 3λ, thus increasing the power density and hence the strength of the non-linear interaction.[119, 120] The narrow sensing electrode also served as a $\Delta v/v$ waveguide, with the added advantage that the beam was well guided. With the addition of ground electrodes on the surface adjacent to the guide, instead of a ground plane beneath the crystal, the device was much more efficient.[119] The bilinearity factor, defined by $C = P_{out} - P_{in1} - P_{in2}$ (with the powers in dBm), was $C \approx -71$ dBm. Many later devices were similar, though usually with SAW's in two waveguides and a subtracting arrangement to reduce spurious signals,[121] and bilinearity factors improved by a few dB's. Special methods were used for generation of narrow SAW beams, namely a multi-strip beam compressor such as that in Fig.8c,[119] waveguide horns,[120] chirp transducers[122] and transducers with curved electrodes.[123] Typical devices would have 100 MHz bandwidth and 16 μsec interaction length, capable of correlating waveforms with time-bandwidth products up to 1600. In 1981 it was shown that the device could be characterized by a two-dimensional frequency response, and this facilitated measurement of the spatial uniformity of the device.[6, 124]

For better efficiency, a variety of devices exploited electronic non-linearity in semiconductors. It was known in the 1960s that SAW's in lithium niobate, for example, could be amplified by a drift current of electrons in an adjacent semiconductor (silicon).[125] To minimize unwanted propagation loss it was necessary to maintain a small gap, of thickness comparable to the wavelength, between the semiconductor and the substrate. In the early 1970s it was shown that, with waves generated at each end of the lithium niobate, non-linearity in the semiconductor gave convolution action with improved efficiency.[126, 127] Bilinearity factors in the region of $C \approx -60$ dBm were obtained.[127, 128] An alternative device used a zinc oxide film on a silicon substrate,[127,129] in which case an air gap is not necessary, though the growth of the ZnO film is a complex matter. Recent devices of this type have used the Sezawa wave (the first higher Rayleigh wave mode) because it can give stronger piezoelectric coupling, and gave very high efficiency with $C = -41$ dBm.[130] These acoustoelectric devices are also capable of other functions including signal storage and optical scanning, as discussed in Kino's review.[127]

Tapped delay lines; integration
Another approach to correlation of waveforms such as PSK is the tapped delay line, in which the taps are short transducers with polarities (0 or 180°) corresponding to the code. These devices originated in the 1960s.[131] It was soon realized that programmable filters could be produced by arranging external circuitry to switch the phase of the output from individual taps.[132, 133] A recent device of this type[134] had amplitude as well as phase programmability, and was demonstrated as an adaptive notch filter to reject CW interference in a communication system.

It was recognized in the early days that these devices were limited by the space and complexity of the bond wires needed for individual taps. To overcome this, integration of the SAW and semiconductor components was envisaged, potentially giving also the substantial versatility of combining these two technologies in one device. This aim has prompted research work for many years. For example, Hagon[135] demonstrated in 1973 a 64-tap filter on aluminium nitride, with switching circuitry in silicon, both materials being in the form of films on the same sapphire substrate. Later, Hickernell[136] demonstrated an integrated programmable 31-tap device in silicon, using MOSFET's as taps (not requiring piezoelectricity), and with a ZnO film for the IDT. Recently, another technology has emerged, the acoustic charge transport (ACT) device in which an unmodulated SAW propagates along a (piezoelectric) GaAs substrate and packets of charge are transported in the potential wells associated with the SAW.[137, 138] Zinc oxide is needed for the SAW transducers, but the charge packets can be sensed by electrodes, much as in a CCD. A fully integrated technology would appear to have much potential, allowing circuits to incorporate the delays obtainable with SAW's, but to date there does not seem to be much practical realization.

3.6 Transducer analysis (1965–1985)

This topic is vital to SAW devices, and indeed often encompasses the analysis of a device. Because of its importance, it has been pervasive throughout the history of SAW. The complexity of IDT behaviour on a piezoelectric substrate is enough to justify the use of several different approaches in parallel. Thus, simple approximate methods are convenient to use and may serve as the basis for the inverse process of transducer design. Complex, comprehensive methods predict details with more accuracy, as needed for sophisticated devices, and give insight into factors limiting performance.

In the 1960s there were many early studies, such as those of Coquin and Tiersten,[139] Joshi and White[140] and Skeie.[141] Emtage[142] had an approach somewhat related to modern COM analysis. These works laid some useful groundwork but their complexity, and the limitation to unweighted single-electrode transducers, made them not very attractive to design engineers. A quite general approach was the normal mode theory of Auld and Kino (Ref. 143; Ref. 11, vol. II, pp.179-185) which, for SAW's, relates the transducer response to the surface charge density. This was applied to analysis of single-electrode transducers, though its validity is much wider. Datta[7, 107, 108] developed a method to analyze reflections of SAW's by metal strips, involving either electrical or mechanical loading, arriving at useful simple formulae.

SAW transducer analysis is plagued by the complication of electrode reflections. The electrode pitch in a single-electrode transducer is $\lambda/2$ (at the center frequency), so that reflections from all the electrodes are in phase. This can cause substantial distortion, as shown in the experimental and theoretical study of Jones et al.[144] The double-electrode transducer, introduced in 1972 by Bristol et al,[145] virtually eliminates this problem by using an electrode pitch of $\lambda/4$, so that electrode reflections cancel in pairs. Subsequently, all devices requiring high performance, for example VSB bandpass filters, have used transducers of this type or similar. Hence, for many purposes a theory that ignores electrode reflections is acceptable.

The simplest theoretical approach is the approximate 'delta-function' model of Tancrell and Holland,[67] which envisages a localized source at each edge of each electrode. A localized source can be represented by a spatial delta function, hence the name. To accommodate apodization, the sources can be considered to be present only where electrodes of different polarity overlap. Excitation is analysed simply by adding the waves generated by the sources, ignoring the presence of other electrodes (thus excluding electrode reflections). This leads to the formula of Eq.(1) in Sec.3.3 above. The method gives excellent value in that a straightforward calculation yields a great deal of information about the response, and consequently it is still popular today. The main limitation is that the transducer impedance is not obtainable, thus excluding evaluation of the device insertion loss and the circuit effect (i.e. distortion associated with terminating impedances).

These deficiencies were overcome by the equivalent circuit models introduced in 1969 by Smith et al[146] – these actually pre-dated the delta-function model, though initially they were applied only to uniform transducers. The IDT was modeled as an array of bulk wave transducers. For each of these an accurate equivalent network derived by Mason was used, and this could be cascaded to model the IDT. For simplicity, the electric field generated by the electrodes was assumed to be either vertical or horizontal, leading to the 'crossed-field model' or 'in-line field model' respectively. Unlike the crossed-field model, the in-line model predicts electrode reflections. However, it was later concluded that the crossed-field model corresponded more closely to reality, with reflections included if necessary by incorporating transmission lines with impedance discontinuities.[147] Although physically unrealistic, this model includes all the basic phenomena present – capacitance, transduction, propagation and reflection – though appropriate parameters need to be supplied from either experimental experience or more basic theory. It has the advantage that it does not demand any knowledge of acoustics, and it is easily extended to cover both apodized and chirped (varied-pitch) transducers, as shown by Tancrell and Holland.[67]

An important parameter for the network model is the piezoelectric coupling constant, k^2. For bulk wave transducers this is unambiguously defined, and by analogy the value for a SAW IDT was expected to be approximately $k^2 \approx 2\,\Delta v/v$.[146] Values in this region were obtained from measurements on various substrates. Most authors now define $k^2 = 2\,\Delta v/v$. From the Q-factor of a simple transducer, the present author concluded (Ref. 6, p.153) that the appropriate value was $k^2 = 2.25\,\Delta v/v$, in agreement with Auld's normal mode theory (Ref. 143; Ref. 11, vol. II, p.185). However, the difference is not very significant practically.

Hartmann's 'impulse model'[148] provided a simpler approach, extending the delta-function method to give, for example, transducer impedances. Like many other theories, this made use of the fact that the acoustic susceptance $B_a(\omega)$ of a transducer is the Hilbert transform of its conductance $G_a(\omega)$, as first noted by Nalamwar and Epstein.[149]

Effective permittivity

A major advance was Ingebrigtsen's introduction[150] of the effective permittivity $\varepsilon_s(\beta)$ in 1969. Taking all relevant quantities to vary as $\exp(\,j\beta x)$, this is defined by

$$\varepsilon_s(\beta) = \frac{\sigma(\beta)}{|\beta|\phi(\beta)} \qquad (2)$$

This is essentially the ratio $\Delta D_n/E_p$, where ΔD_n is the change of normal component of **D** at the surface, equal to the charge density σ, and E_p is the tangential electric field, related to the surface potential ϕ. In a transducer, ϕ gives the electrode voltages and σ gives the current (after spatial integration and time differentiation). The function $\varepsilon_s(\beta)$ expresses *all* electrical and acoustic phenomena related to the electrical variables at the surface, including all types of acoustic waves. If there is a Rayleigh wave solution, $\varepsilon_s(\beta)$ will be zero when $\beta = k_0$, the wavenumber for the wave on a free surface, and $\varepsilon_s(\beta) = \infty$ when $\beta = k_m$, the wavenumber for the wave on a metalized surface. This approach was further developed by, in particular, Milsom *et al*[151]. In the spatial domain, they expressed the relation between the variables as the convolution

$$\phi(x) = G(x) * \sigma(x) = \int_{-\infty}^{\infty} G(x-x')\sigma(x')\,dx' \qquad (3)$$

where $G(x)$ is the frequency-dependent Green's function, which can be calculated numerically from $\varepsilon_s(\beta)$. Taking account of the boundary conditions, this leads to an analysis applicable to arbitrary one-dimensional transducer geometries and arbitrary types of acoustic waves. Milsom showed in particular that it could effectively predict the unwanted bulk wave responses observed in simple Rayleigh-wave devices, though mechanical loading was excluded.

Although inconvenient in some cases because of computational complexity, the above approach also leads to useful approximate results. Milsom showed that $G(x)$, the surface potential due to a line of charge, could be expressed as a sum of electrostatic, SAW and bulk wave terms. The SAW term has the form $\Gamma_s.\exp(-jk_0|x|)$, where Γ_s is a constant. Also, Ingebrigtsen[150] proposed that, for Rayleigh waves, $\varepsilon_s(\beta)$ would be approximately proportional to $(\beta - k_0) / (\beta - k_m)$, giving the pole and zero noted above. From this it can be shown[6] that Γ_s is proportional to $\Delta v/v$, thus justifying the earlier use of this parameter as a measure of the piezoelectric coupling.[39] The Ingebrigtsen approximation was also used by Blotekjaer et al[152] in the analysis of SAW propagation in a regular array of metal strips, using Floquet expansions with the fields expressed as summations of Legendre functions. The width of the predicted stop band is consistent with the electrode reflection coefficient deduced later by Datta.[108]

Electrostatics
Ignoring the bulk-wave Green's function, as is often a good approximation, the remaining SAW and electrostatic terms still lead to some complexity because they involve electrode reflections. However, the latter are often small because, for example, double-electrode transducers are used or $\Delta v/v$ is small. To exclude reflections, the analysis can be cast in an approximate 'quasi-static' form[6, 153]. This shows that the SAW amplitude generated by an IDT is proportional to the Fourier transform of the electrostatic charge density, which also gives the capacitance. In fact, the role of the electrostatic solution had been recognized from earlier work in the 1960s, and an algebraic solution for the single-electrode transducer, involving elliptic integrals, was derived in 1969 by Engan.[154] This was the first analysis to show how the piezoelectric coupling, and the relative levels of the harmonics, depend on the metalization ratio a/p. Later, it was realized that an element factor could be defined in terms of the charge densities on a regular array of electrodes with a voltage applied to one electrode,[155] and Peach[156] derived an algebraic expression for this. Using superposition, this function easily gives the response of a wide variety of transducers (single-electrode, double-electrode, withdrawal weighted, etc.) and accurately quantifies the effects of varying the metalization ratio, including the behaviour of the harmonics. Numerical electrostatic calculations have also been used extensively, initially in 1972 by Hartmann and Secrest[157] in connection with end effects. Smith's work[158] is also notable, and a recent account is given by Biryukov et al.[159]

It should be noted that little of the above analyses include mechanical loading, a complex subject which will be mentioned later. However, diffraction simulations were well advanced (Sec. 3.1 above).

In most of the devices mentioned so far the unwanted triple-transit signal is not very significant, though if required it could be included in the analysis quite conveniently. In contrast, a resonator such as that of Fig. 12 relies heavily on multiple reflections of the waves. Such cases can be analysed conveniently by using a scattering matrix for each component; for a transducer, this relates the four SAW amplitudes (two input, two output) to the two variables at the electric port. This approach was first used by Cross and Schmidt[160] and later by Campbell (Ref. 12, p.305).

3.7 *Low-loss filters* (to 1985)

Throughout the history of SAW, the problem of obtaining good performance with low insertion loss has been a major concern. The problem arises because well-matched IDT's reflect the waves strongly, giving unwanted multiple-transit signals which are usually unacceptable. To avoid this, most early filters (and all high-performance filters) had quite high insertion losses, typically more than 20 dB. Resonator filters afford one solution, but only for very small bandwidths.

As far back as 1969, Smith et al[161] used a tuned auxiliary IDT to reflect the unwanted 'backward wave' from an IDT, giving a device with reduced loss and small ripple. This principle still survives in some present devices in the form of a reflective grating added to an IDT, a form of SPUDT (see below). However, it is effective only over a relatively small frequency band.

In 1972, Lewis[162] demonstrated a three-transducer device, in which two identical transducers were connected to the device input, and a symmetrical output transducer was located between them. The output transducer receives waves with the same amplitude on both sides, and in this situation it does not reflect if it is electrically matched. Thus, reflections are suppressed and the minimum loss is ideally 3 dB. A related device is the ring filter of Fig.14, in which waves generated (in both directions) by the input transducer are transferred to an adjacent track, with propagation directions reversed. A central output transducer receives these waves and, again, gives no reflection if it is electrically matched. Here, the minimum loss is ideally 0 dB. The waves can be transferred between the tracks by a multi-strip (MSC) trackchanger incorporating a 3 dB coupler and and two mirrors.[93] However, the earliest ring filter[163] used a simpler MSC arrangement, and Brown[164] developed further modifications. Recent devices used an adapted multi-strip trackchanger which also improves the stop band suppression of the filter, and achieved a 1 dB insertion loss.[165]

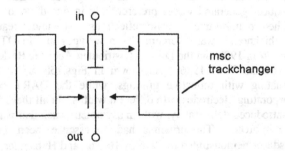

Fig. 14. Ring filter. With identical waves incident from both sides, the output transducer ideally gives no reflections and no loss when matched.

Unidirectional transducers

In the 1970s a variety of multi-phase transducers were investigated, initially Hartmann's three-phase type.[166] This transducer has three electrodes per wavelength, connected sequentially to three bus-bars. A simple L-C circuit is used to apply voltages to the bus-bars with phases incrementing by 120°, with the result that SAW's are generated in one direction only. This is therefore a 'unidirectional transducer', or UDT. A device with two such transducers ideally gives 0 dB insertion loss and no reflections, at the center frequency. A disadvantage is that connections to one of the three sets of electrodes require the use of insulating 'cross-overs', complicating the fabrication. To avoid this, Yamanouchi et al[167] introduced a 'group-type' version using a meander line to provide one set of connections. Impressive results were obtained with such devices (Ref. 6, p.176), but they fell out of favour in the 1980s, probably due to competition from the following.

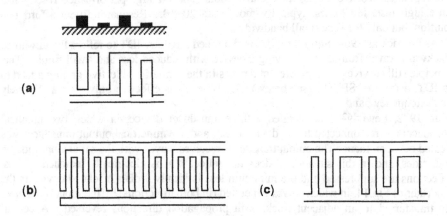

Fig. 15. Single-phase unidirectional transducers (SPUDT's). (a) Using two-thickness metalization, one period shown (Hartmann, 1982); (b) Group-type (Lewis, 1983); (c) DART (Kodama, 1986)

A more practicable solution was the 'single-phase unidirectional transducer', or SPUDT, introduced in 1982 by Hartmann.[168] The initial structure was a double-electrode transducer in which asymmetry was introduced by making alternate electrodes have different film thicknesses, as in Fig.15a. When a voltage was applied, this transducer generated waves preferentially in one direction, and it could be electrically matched to minimize acoustic reflections. The undesireable complication of different film thicknesses was overcome by new types of SPUDT, notably the group type of Lewis[169] in 1983 and the DART (Distributed Acoustic Reflection Transducer) introduced by Kodama[170] in 1986. As shown in Figs.15b and c, the group type uses IDT's alternating with reflective gratings, while the DART consists of asymmetric cells incorporating electrodes with different widths. In all these SPUDT's, internal reflections are introduced deliberately, with an asymmetry such that waves launched in one direction are reinforced. The principle had, in essence, been anticipated much earlier in a transducer demonstrated in 1976 by Hanma and Hunsinger,[173] though this is not popular now because it incorporates $\lambda/16$ electrodes. Another SPUDT is the floating-electrode UDT, or FEUDT, introduced in 1984 and having a variety of forms.[171, 172]

An early experimental example is the group-type SPUDT filter response shown in Fig.16. The insertion loss is 2.7 dB, but with the SPUDT's reversed so that the 'backward' ports face each other (right plot) the center-frequency loss is 31.9 dB. The difference of 29.2 dB equals the sum of the two transducer directivities.

Yet another possibility is a simple single-electrode transducer fabricated on an asymmetric substrate. Demonstrated by Wright[174] in 1985, this simply uses a substrate orientation with asymmetric mechanical properties, such as ST-X+25° quartz, and is called a 'natural SPUDT', or N-SPUDT. Although it might seem surprising that a symmetric transducer could be directional, earlier analyses were general enough to predict that the phase of the electrode reflection coefficient could be affected by the substrate orientation.[106, 107] Thorvaldsson[175] showed that this phenomenon, due to mechanical loading, could account for the N-SPUDT effect. However, this SPUDT is of limited applicability because the 'forward' direction is determined by the substrate, so there is no simple way of producing two N-SPUDT's with their 'forward' ports facing each other. The effect is not normally seen in practice because established SAW materials have symmetry high enough to exclude it.

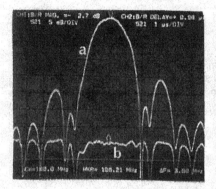 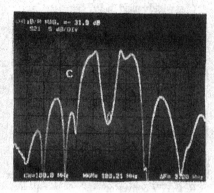

Fig. 16. Responses of group-type SPUDT filters. The devices, each consisting of two SPUDT's, are identical except that the left case has the 'forward' ports adjacent, and the right case has the 'backward' ports adjacent. Vertical scales 5 dB/div., horizontal scales 0.5 MHz/div. centered at 100 MHz. Trace (b) is the group delay, at 1 μsec/div. From IEEE Ultrasonics Symp., Ref. 169, courtesy M.Lewis and copyright 1983 IEEE.

4. Recent Devices (1985–present)

4.1 *Introduction*

Summarising the above, the period 1970-85 saw the establishment of the basic elements of SAW technology – materials, propagation effects, analysis of transducers and gratings, design techniques – and the development of a wide variety of devices meeting exacting performance requirements. The technology had been set up and shown to be effective. Much of this effort (with the notable exception of the TV filter) was in response to military and professional requirements, particularly in radar and communications, with emphasis on optimising the performance. In contrast, the post-1985 period was driven more by consumer requirements, particularly for mobile phones, where cost, insertion loss and size were more important than before. It is therefore convenient to review the position at this point.

Initial development of SAW's was greatly helped by three factors already present – suitable materials (lithium niobate, quartz), photolithography and computing (for device analysis, design, and mask pattern generation). These have always been crucial to SAW development. Optical lithography, originally spurred by integrated circuit requirements, was also basic to other technologies starting about the same time, namely CCD's,

Table 1. Summaries of SAW Device Development.

(A) Selected from Williamson,[177] (1977)

Developed SAW Devices
* Bandpass filter   * CATV filter   Convolver
* Delay line   Filter bank   Fourier transform
* Front-end filter   Memory correlator   Mixer
  Oscillator (*delay line, resonator)
* Pulse compression filter (generation, compression)   * PSK matched filter
  Recirculating delay line   * Tapped delay line (PSK generation, correlation)
* TV broadcast filter   * TV receiver filter   **plus** 28 others
  (* = widely used in 1977)

Government Systems using SAW Devices
No. of systems: 45
Application areas: Radar, comms., satellite comms., spread spectrum, GPS, etc.
Devices: Pulse compression filters (used in 23 systems);   Bandpass filters (19);
         Delay lines (11); Oscillators (4); Programmable correlators (2)

(B) Selected from Hartmann,[178] (1985)

Consumer applications
TV; satellite TV receiver	IF filter
VCR; CATV convertor	IF filter, modulator resonator
Garage door control; medical alert	Transmitter resonator
Cordless phone; pager	Front-end filters

Commercial Applications
CATV/MATV headend	VSB bandpass filter; IF filter
Local area network	Transmit/receive filters
Digital microwave radio	Bandpass filters for shaping QAM
Fibre-optic repeater	Clock recovery filter
RF synthesizers; spectrum analysers	Oscillators; filters

Military Applications
Radar (chirp)	Pulse compression filters.
Radar (monopulse)	IF matched filters; tapped delay line.
Radar (Doppler)	Local oscillator; delay line for clutter reference.
ESM receivers	Filter bank; delay lines for queued analysis; chirp filters for frequency measurement; delay lines for IFM.
ECM: jammers	Delay lines.
ECCM (spread spectrum)	Pulse shaping filters; matched filters (fixed, programmable); convolvers.
Navigation	Front end and IF filters; oscillators.
IFF	Stable oscillators.
Communications (voice, data)	IF filters; oscillators; digital demodulators.

integrated optics and bubble memories. Around 1970, line width capabilities were in the region of 1 μm, limiting SAW devices to about 1 GHz. In present manufacturing, linewidths around 0.3 μm are feasible using non-contacting optical steppers projecting a reduced image. Devices with sub-micron lines were in fact made quite early using electron-beam fabrication,[176] demonstrating device feasibility at high frequencies, though this method is not suitable for volume manufacturing.

The success of SAW devices can be attributed to several key factors:

(a) Crystalline materials are available on which the SAW propagation is almost ideal, and also with adequate piezoelectric coupling and temperature stability.
(b) Long delays, needed for chirp or coded waveforms, are obtainable compactly.
(c) Substantial design versatility is obtained because photolithography enables arbitrary patterns to be etched from a film on the surface. The number of degrees of freedom is of the order of the length of the substrate in wavelengths, typically several hundred, hence the sophistication of bandpass filter designs.
(d) Excellent accuracy and reproducibility arise from the precision of photomask generation machines and the reproducibility of crystalline substrate materials.
(e) Complex designs, exploiting the versatility, are feasible because of sophisticated numerical analysis and design techniques.
(f) Single-stage lithography is well suited to mass production of low-cost devices, with many devices per wafer.

The progress of the subject is illustrated by the reviews of Williamson[177] in 1977 and Hartmann[178] in 1985. Williamson listed 45 distinct species of SAW device, with 10 of them well established in practical systems. Forty-five government systems used SAW devices, particularly matched filters, bandpass filters and delay lines. Prices ranged from $1.50 to $10,000. Hartmann listed 29 types of SAW device *in use* in systems, again with many military applications, including radar, ESM, ECM and ECCM. In addition, the move towards consumer applications was already evident, including pagers, cordless phones and VCR's, as well as TV IF filters. The practical significance of SAW was shown by the fact that many systems designers were already relying on performance standards unobtainable by any other technology. Some selected data from these two papers are shown in Table 1.

In the Soviet Union, and in eastern European countries, there was also substantial SAW development, but awareness of this among western scientists was restricted because of the limited communications between the two blocs. However, most at least of the topics mentioned above were pursued in the east, in particular the TV filter. Useful reviews of work in Siberia and in eastern Europe were given in 1991 by Yakovkin[179] and Buff,[180] respectively.

In the post-1985 period, consumer applications, particularly mobile and cordless phones, demanded low-loss compact filters, including RF filters with losses of 3 dB or less. In addition to the development of many new device types, this requirement has led to the use of novel types of wave. We therefore consider these waves before describing the devices.

This section is mainly concerned with bandpass filters. Other topics, such as pulse compression and convolvers, continued to be active, but they appeared less in the research literature because the main principles were established earlier.

4.2 Transverse waves and new materials (1975 - present)

The devices considered earlier have all used piezoelectric Rayleigh waves, for which the particle displacement is in or close to the sagittal plane and the velocity is less than that of the slowest bulk wave. However, in many materials there are special orientations giving solutions with shear displacement (normal or almost normal to the sagittal plane), and often with higher velocities. Several of these have practical advantages over Rayleigh waves for some situations and have become widely exploited, particularly in response to the strong demand for low-loss RF filters for mobile phones, described below. The topic originated around 1970, but it is convenient to consider it here because the applications are relatively recent.

In 1977, Lewis[181] demonstrated that *bulk* shear waves could be used in rotated Y-cut quartz, with propagation normal to X. These 'surface-skimming bulk waves' (SSBW) were shown to have amplitude proportional to $x^{-1/2}$. This was expected because, as for a shear horizontal wave in an isotropic material, the displacement was practically parallel to the surface and the boundary conditions were predicted to have little effect. Orientations close to BT-cut and AT-cut respectively gave good temperature stability (much better than ST-X quartz) and a high velocity (5100 m/s) attractive for high-frequency devices. To reduce the propagation loss, Auld et al[182] showed (initially, in 1976, using an isotropic material) that a shear wave could be trapped at the surface by means of a grooved grating, thus practically eliminating the $x^{-1/2}$ decay. In later studies using AT-cut quartz, the grating was provided simply by the transducer electrodes or metal strips. The wave became known as a 'surface transverse wave' (STW), and was used in resonators.[183] Avramov's work on resonator design showed that, in addition to enabling the frequencies to be increased, STW resonators could handle power levels higher than their SAW counterparts and give higher Q-values, giving improved short-term stability when used to control oscillators.[184]

On the other hand, *leaky surface waves* were first identified in X-cut quartz by Engan et al[185] in 1967. They are also called pseudo-surface waves, or PSAW's. These waves have velocities higher than that of the slowest bulk wave. The term 'leaky wave' refers to a theoretical solution in which the wavenumber is allowed to be complex, so that the amplitude decreases as $\exp(-\alpha x)$. For selected orientations the attenuation can be very small, negligible in practical terms. Useful leaky waves were found in rotated Y-cuts of lithium niobate[186] in 1970 and of lithium tantalate[181, 187] in 1977, and these were reviewed by Yamanouchi and Takeuchi.[188] Propagation is in the X-direction, and the mechanical displacement is in a direction approximately normal to the sagittal plane. Compared with SAW, these waves offer higher piezoelectric coupling, higher velocity and higher power handling, and consequently they have been widely used in low-loss RF bandpass filters. The velocity is higher than that of the slow bulk shear wave. For lithium tantalate, the attenuation varies with the cut angle and becomes practically negligible at about 36°, for a free or metalized surface. In lithium niobate, negligible attenuation requires an angle of 41° for the free-surface case, or 64° for the metalized case. Despite this distinction, both cuts have been successfully used in low-loss devices.

Roughly speaking, transducers and gratings behave on these materials almost as if the waves were conventional Rayleigh waves. However, detailed studies in the 1980s revealed a complex picture. In 36°Y-X lithium tantalate there is in fact a SSBW solution in addition to the leaky wave. On a free surface these waves have almost identical velocities and they are difficult to distinguish, but for a metalized surface the leaky wave velocity is reduced and the waves can be distinguished. The SSBW amplitude

Table 2. SAW Materials.

A. Materials in Common Use

Material	v_0 m/s	$\Delta v/v$	TCD ppm/°C	Comments	Ref.
Lithium niobate, LiNbO$_3$				Strong coupling, large TCD – wide band filters, RAC's convolvers	
Ditto, Y-Z	3488	2.4%	94	1969 Minimal diffraction, bad bulk waves – precise filters using MSC's	39
Ditto, 128°Y-X	3992	2.7%	75	1976. Free of bulk waves. TV filters etc.	58
Quartz SiO$_2$, ST cut (47.5°Y-X)	3159	0.06%	0	ST-cut 1970. Weak coupling, low TCD – narrow band filters, resonators, matched filters (chirp, PSK)	43
Lithium tantalate, LiTaO$_3$ X-112°Y	3300	0.35%	18	1978. Intermediate coupling and TCD, free of bulk waves – TV filters etc.	59
PZT ceramic, Z-X	2360	1%	40	Too lossy above 50 MHz. Poor repeatability. – TV filters	
Ditto, X-Y	2400	11%	9	Ditto	
ZnO/glass (data for fundamental mode)	2576	0.7%	11	Fabrication complex, starting in early 1970s. Not very repeatable. TV filters	60
Lithium tetraborate Li$_2$B$_4$O$_7$, 45°X-Z	3350	0.45%	0	1981. Moderate coupling, zero TCD. Slightly water soluble (special processing)	63
LiNbO$_3$, 64°Y-X, LSAW	4742	5.5%	80	1970. Strong coupling, high velocity – high-frequency low-loss filters	186
Ditto, 41°Y-X LSAW	4792	8.5%	80		
LiTaO$_3$, 36°Y-X LSAW	4212	2.4%	32	1977. As above. Often used for high-frequency low-loss filters. Supplanted by 42° cut	181, 187, 193
Quartz, 36°Y-X+90°	5100	n/a	0	STW, 1977. High-frequency temperature-stable devices – resonators	181

B. Other Established Materials

Material	v_0 m/s	$\Delta v/v$	TCD ppm/°C	Comments	Ref.
Bismuth germanium oxide, Bi$_{12}$GeO$_{20}$, (001, 110)	1681	0.7%	120	Large delays. Also low-diffraction cut	47
LiTaO$_3$, 167° rot	3394	0.75%	64	Low diffraction	47
Berlinite, AlPO$_4$, 80.4° rot	2741	0.25%	0	Similar to ST-X quartz	194
Quartz, LST-cut, -75°Y-X, LSAW	3950	0.05%	0	1985. Similar to ST-X quartz but higher velocity	210
ZnO/sapphire	5500	2%	43	High velocity	62
AlN/sapphire	5910	0.5%	0	1973. High velocity	135

LSAW = leaky surface acoustic wave

theoretically decays as $x^{-1/2}$ for small x and $x^{-3/2}$ for large x.[189] The $x^{-1/2}$ variation has been seen experimentally, as has the exponential decay of the leaky wave.[190] It seems that these complications do not affect practical devices much, perhaps because the devices use transducers rather than free or metalized surfaces. However, spurious signals, possibly associated with the SSBW, have been seen in impedance element filters.[191] A similar situation occurs in 41° Y-X lithium niobate.[190, 192] For lithium tantalate, 42° rotation has been found to give lower loss than 36° in high-frequency devices because the optimum cut angle depends on film thickness.[193]

A summary of SAW materials is presented in Table 2, which includes materials discussed previously in Sec. 3.1 and also some cases not mentioned in the text. Recent reviews of the subject are those of Ebata,[61] Fujishima[195] and Shimizu.[196]

4.3 Recent innovations in materials (1990 - present)

In this Section we describe a number of promising materials, under study but not yet established because, for example, growth techniques need more development. Prominent among these is langasite ($La_3Ga_5SiO_{14}$), a quartz-like material for which growth techniques have advanced recently. SAW properties[197] have been investigated since 1995, showing a variety of orientations with properties like ST-X quartz but with higher coupling.[198] For example, with Euler angles (0, 26, 33°) the SAW theoretically has v_0 = 2464 m/s, $\Delta v/v$ = 0.15% and TCD = 0. Similar results have been found for langanite ($La_3Ga_{5.5}Nb_{0.5}O_{14}$) and gallium phosphate ($GaPO_4$).[198]

Another intriguing newcomer is potassium niobate ($KNbO_3$), which has extremely high coupling ($\Delta v/v$ up to 25%) and a zero-TCD orientation,[199] though experimental results are not as good as predictions because of limited crystal quality. The material is also being tested as a film on strontium titanate substrates,[200] and shows strong non-linearities potentially applicable to convolvers.[201]

In the last few years there has been renewed interest in the search for leaky surface wave solutions, analogous to the high-velocity high-coupling leaky waves in lithium niobate and tantalate. A new leaky wave in lithium tetraborate was reported in 1994 by Sato and Abe.[202] This is a *longitudinal* leaky wave, with particle displacement almost parallel to the propagation vector, giving a very high velocity comparable to twice the SAW velocity. It is also known as a high-velocity PSAW, or HVPSAW. The orientation is 47° rotated Y-cut with propagation normal to X, giving a velocity of 7000 m/s and $\Delta v/v$ = 0.7%. A series of detailed studies was done on this, including numerical transducer analysis with finite film thickness, culminating in an experimental IIDT bandpass filter at 1.5 GHz with 2 dB insertion loss[203] (for the IIDT, see Sec.4.5 below). Many other authors have investigated HVPSAW's, notably da Cunha[204] who quotes several high-velocity low-attenuation solutions in both lithium tetraborate and quartz.

Another approach for high velocities, with a view to high-frequency applications, is the use of a high-velocity non-piezoelectric substrate plus a piezoelectric film. We have already noted (Sec. 3.1) the use of zinc oxide films for TV filters.[61] For high velocities, Koike reported deposition of the film on a sapphire substrate,[62] giving a velocity of 5400 m/s, $\Delta v/v$ = 2.1% and TCD = 43 ppm/°C, and a 1.5 GHz IIDT filter gave 1.3 dB insertion loss. Another film from the past is aluminium nitride (AlN), again on a sapphire substrate.[135] Recent advances[205] have enabled the realization of a 2.4 GHz PSK matched filter; the material gives v_0 = 5910 m/s, $\Delta v/v$ = 0.5% and TCD = 0.

Diamond has recently emerged as a high-velocity material which can be grown on a silicon substrate, again with a ZnO film on top for piezoelectricity. Yamanouchi et al[206] presented initial theoretical and experimental studies in 1989. The SAW velocity on diamond is around 11,000 m/s, but the ZnO reduces this. The SAW velocity and coupling depend on the ZnO thickness, but typically v_0 = 7000 to 10,000 m/s and $\Delta v/v$ = 1 to 2%. Extensive studies were made, for example, by Nakahata,[207] and a

2.5 GHz IIDT filter was made with 2.7 dB loss.[208] Also, a HVPSAW with velocity up to 16,000 m/s has been observed.[209]

It will be seen that there is a wide variety of novel SAW materials showing considerable promise, particularly for high-frequency devices. Even so, the above description is selective, omitting many interesting points. As for many technologies, the topic of materials is fundamental to SAW, but new materials take time to establish because substantial work is needed to develop growth techniques and to find and test suitable orientations. Experimental materials sometimes turn out eventually to have unforeseen problems, but even if only a few of the above are successful we can expect them to have a substantial impact on the future performance of SAW devices. More detail on the subject will be found in articles by Adler, Hickernell and Kosinski in this issue.

4.4 Modern low-loss filters (1985 – present)

Most of the novel SAW device advances in this period concern bandpass filters, and little is lost if we confine attention to filters for mobile phones. It is convenient to divide them into the two areas of IF and RF devices, though this distinction is not rigid. A helpful review has been presented by Ruppel et al.[216]

IF filters

Mobile phone applications call for I.F. filters with bandwidths of typically 50 to 500 kHz, and with low loss. For these the design of DART's, mentioned in Sec. 3.7 above, was investigated.[211, 212] It was recognized in particular that both transduction and reflection could be weighted (usually by withdrawal weighting), calling for a complex design process. Filters consisting of two DART's could give excellent performance, with for example 200 MHz center frequency, 1 MHz bandwidth, 9 dB loss, 55 dB triple-transit suppression and 60 dB stop band suppression.[213] However, now there was a new problem in that narrow-band devices were too large for pocket-sized applications such as mobile phones and pagers.

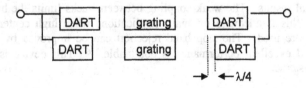

Fig. 17. Two-track DART filter

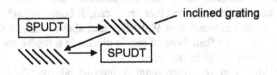

Fig. 18. Z-path filter

A more compact arrangement is obtainable using multi-track DART filters,[212, 214] in which output signals due to direct SAW transits are cancelled; the output arrives only after reflection at the output DART's and, in some cases, at a central grating. Fig.17 illustrates a two-track device with gratings. These arrangements increase the number of frequency-selective stages involved, and therefore reduce the bandwidth for a given device length. Another development[215, 216] was the 'Z-path filter' of Fig.18, in which the wave from an input SPUDT is reflected by an inclined grating into a second track, where it is again reflected by a another grating and finally received by the output SPUDT. In 1994, Ventura et al[217] introduced a novel 'resonant SPUDT', or R-SPUDT. In this subtle DART design, the reflection and transduction weighting functions (applied using withdrawal weighting) are both allowed to change sign when required. In consequence, it is possible to substantially reduce the device length, and this has been widely applied in IF filters.

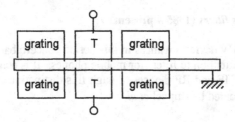

Fig. 19. Transverse-coupled resonator filter (Tiersten and Smythe, 1975). T = single-electrode transducer.

For the smaller fractional bandwidths, a special resonator technology has been used. To obtain adequate stop band rejection in a resonator, it is helpful if the input and output transducers are not in the same track. As shown in Fig.19, the transverse-coupled resonator (TCR) has two parallel tracks, each consisting of grating – transducer – grating. The tracks act as waveguides as well as resonators, and are constructed close enough to give useful acoustic coupling. The principle was originated in 1975 by Tiersten and Smythe,[218] but it was much later, in 1984, that Tanaka[219] published the first experimental results. The weak coupling between tracks limits the bandwidth to about 0.2 %, but this is adequate for many applications when high center frequencies (e.g. 200 MHz) are used. The stop band rejection can be improved by cascading several devices, and excellent performance is obtainable.[216] This device is now common in practical systems.

RF filters

Radio-frequency requirements for mobile phones have also prompted a variety of novel SAW techniques. Because of the relatively wide bandwidth, typically 25-75 MHz, leaky waves on lithium tantalate or niobate are normally used. The low-loss requirement is more urgent here since, for a receiver front end, it bears strongly on the signal-to-noise ratio. Also, center frequencies of 900 MHz and above make designers reluctant to use electrodes any narrower than those of single-electrode transducers, so the complication of electrode reflections has to be accepted.

An early idea was the interdigitated interdigital transducer (IIDT) described by Lewis[162] in 1972 and demonstrated by him[220] in 1982. The device, Fig.20, consists of a regular sequence of transducers connected alternately to the input and output. This scheme received much attention in the 1980s,[216, 221] and Hikita, for example, obtained 4 dB loss at 830 MHz with 30 MHz bandwidth and 50 dB stop band rejection, using a 36°Y-X lithium tantalate substrate.[221] Related devices[216, 222] have used two tracks coupled using self-resonant IDT's, and sometimes reflective gratings at the ends to reduce the loss.

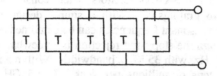

Fig. 20. Interdigitated Interdigital Transducer, IIDT (Lewis, 1972).

Resonator filters, consisting of two or three transducers between two reflective gratings, are also applicable. These are superficially similar to the two-port resonators discussed earlier (Fig.12), but in fact the operation is quite different because they use single-electrode transducers in which electrode reflections are significant. With appropriate design a 2-pole response can be obtained, hence the term 'double-mode SAW' (DMS) filter. Ohmura[223] described lithium tetraborate devices in 1990, and Morita[224] described leaky-wave lithium tantalate and niobate devices in 1992. However, the earliest publication appears to be that of Tanaka[225] in 1986. Using 64° lithium niobate, Morita obtained 2 dB insertion loss at 836 MHz, with 33 MHz bandwidth. The rejection was 40 dB, except for a high-frequency 'shelf' characteristic of these devices. Widespread work on this topic has included, for example, a 1900 MHz filter with 2.4 dB loss.[216]

Fig. 21. A 2.0 GHz bandpass filter on a 1.4 mm square die. From Ref. 22, courtesy F. Hickernell and copyright Academic Press, 1999.

Impedance element filters (IEF's) use SAW components connected electrically, with no acoustic interaction, in a manner reminiscent of ladder filters using bulk wave quartz resonators. This principle was used by Lewis and West[226] in 1985. Hikita et al [227, 229] used banks of single-electrode transducers in which electrode reflections cause the response to be sharply peaked, so that they behave like resonators. They also demonstrated a duplexer, that is, a pair of filters for direct connection to the antenna, with the receiver filter suppressing the transmitter band and vice versa. Starting in 1992, another type of IEF used SAW resonators, each consisting of a single-electrode transducer between two gratings.[230] With a simple design approach, two or more resonators with different resonant frequencies can be combined to give a response with a low-loss passband, though the stop band rejection is somewhat limited. Losses of 2 dB were achieved at 840 MHz, with 35 MHz bandwidth. Within a few years, production of such devices reached tens of millions per year. A 2.5 GHz filter gave 100 MHz bandwidth and 3 dB insertion loss.[231] For improved stop band rejection, a balanced bridge arrangement has been used.[232] These devices use 36°Y-X lithium tantalate or, more recently, 42°Y-X.

For RF applications the power handling capacity has been a concern, since powers in the region of 2 W are sometimes needed. The SAW filters were found to be limited by migration of the aluminium electrodes. In the 1980s there were many studies of this, and substantial improvement was obtained by adding a small proportion of copper to the aluminium, as shown for example by Hikita.[227] Layered materials have also been developed.[233]

Figure 21 shows a practical 2 GHz filter illustrating the very small size, a consequence of the small wavelength (around 2 μm) at these frequencies. Such filters normally use ceramic packages of small size, 3.5×3.5 mm or less, which helps to reduce costs.

4.5 *Recent analysis* (1980 - present)

Various analytic techniques developed up to about 1980 were more or less adequate for the times, as described above (Sec. 3.6). However, later there were more demands on the analysis because of new device developments. Notably, 'conventional' devices such as VSB filters were being applied to more demanding specifications, calling for more accurate analysis; many new devices included internal reflections in the transducers, as seen above; and new modes – leaky waves and STW – were being widely applied. A wide range of new and complex techniques has emerged, many of which are discussed in Ventura's recent survey.[228]

One consequence was the extension of earlier theory to allow for two-dimensional behaviour in SAW transducers, for example, allowing the charge density to vary in two directions. For application to dot arrays, Huang and Paige developed a 2-D electrostatic analysis[234] and a 2-D Green's function[235] in 1982 and 1988, respectively. The Green's function is essentially the potential generated by a point charge on the surface of an anisotropic piezoelectric dielectric. Techniques of this type were also applied to transducers, recognising that the electrostatic charge density on the electrodes is distorted near the ends.[236] Visintini et al[237] applied these concepts to the analysis of sophisticated wide band filters for digital radio systems, incorporating an ASoW synthesis. In some devices diffraction and electrode reflections are present simultaneously, leading to very complex behaviour, and another complication can be the waveguiding behaviour of some transducers, notably in transverse-coupled resonators and some conventional resonators.

Männer[238] discusses these topics, including the very complex behaviour of the Z-path filter of Fig. 18.

Returning to the one-dimensional case, in which all the fields can be considered invariant in the transverse direction, the results of transducer analysis are often expressed in terms of scattering matrices which allow multiple reflections to be analysed conveniently. This was done in 1977 for resonators,[160] but the most popular form now is the slightly different 3×3 P-matrix introduced in 1979 by Tobolka.[239] The P-matrix interrelates the four SAW amplitudes concerned and the voltage and current at the electrical port, thus giving a complete description of the transducer, provided its behaviour is one-dimensional and only one wave type is involved.

For the many transducer types with internal reflections, a convenient and widespread method is the coupling-of-modes (COM) method, complementing the earlier network model. This method, which is approximate, envisages forward- and backward-propagating waves with slowly-varying amplitudes, represented by coupled differential equations. Coupled-mode theory originated with microwave work in the 1950s, but its first use in SAW seems to be that of Suzuki's work in 1976 on gratings.[106] For SAW transducers, the extension to include the necessary transduction terms was done by Koyamada and Yoshikawa[240] and by Hartmann *et al.*[168] Physical justification was given by Akçakaya,[241] with further development by Wright.[242] The COM approach is notable for yielding algebraic expressions for all the required scattering properties (conversion, reflection, admittance) of a unweighted transducer, that is, it gives all the components of the P-matrix.[243] The analysis depends on only a few parameters, governing transduction, reflection, velocity and attenuation. For complex transducers, such as DART's, it is feasible to deduce these from experimental measurements, without considering the complexities of the underlying physics.[244] An alternative method for the P-matrix is the reflective array model (RAM), which relies on cascading techniques but is more closely related to the physics; for some transducers, notably single-electrode transducers and DART's, this can give some dependent parameters theoretically.[245]

In Sec. 3.4 we have already noted theoretical and experimental determinations of the SAW reflection coefficients of metal strips (open- or short-circuit) and grooves. These are of vital interest for reflective gratings and various reflective transducers. These efforts were complemented in the 1980s by rigorous numerical calculations using the finite element method (FEM), particularly by Koshiba,[246] with the ability to include electrical and mechanical loading, the N-SPUDT effect and the presence of leaky and other waves. Calculation of the stop band width of a grating yields the strip reflection coefficient, assuming that the latter does not vary much with frequency. Moreover, the center frequencies of the stop bands for the short-circuit and open-circuit cases give the transduction parameter needed for single-electrode transducer analysis.[247, 248] In this way, COM parameters have been deduced for IDT's using leaky waves on lithium tantalate and niobate.[247] In contrast to the SAW case, the transduction parameter is found to depend on the electrode thickness. Another FEM example was Ventura's analysis, with experimental verification, for the familiar case of reflection of SAW's by aluminium strips on ST-X quartz.[249] This showed that the reflection coefficient was substantially larger than that given by the earlier theory of Datta[107] when the metalization ratio exceeded 0.5, an important factor in the performance of DART's.

An alternative method for grating analysis is based on Floquet expansions, as demonstrated by Xue and Shui[250] for zero-thickness electrodes. More recently, Sato and

Abe[203] applied this to the longitudinal leaky wave on lithium tetraborate, with finite-thickness electrodes.

Rigorous analysis of SAW transducers with irregular electrodes is very complex. Hashimoto[251] used a boundary element method (BEM), incorporating the electrostatic and SAW Green's functions, to analyse a FEUDT with floating electrodes. For general one-dimensional problems with finite electrode thickness, involving mechanical effects, a generalized Green's function can be used. This relates, at the surface, the three components of displacement and the potential to the three components of traction and the charge density. It is therefore a 4×4 matrix of functions, as explained in 1985 by Wang and Chen.[252] Independently, Baghai-Wadji[253] introduced a 2×2 Green's function matrix in which only the transverse component of displacement is allowed for; this is suitable for devices using transverse waves (STW's or leaky waves). Ventura et al[254] applied these ideas, with FEM and BEM, to SAW device analysis. In these papers, the charge density on a electrode is assumed to have the form $p(x)/\sqrt{(a^2 - x^2)}$, where $p(x)$ is a simple polynomial and the edges are taken to be at $x = \pm a$. This form is helpful because it gives the correct behaviour at the edges; it was introduced by Smith and Pedler in the 1970s.[158]

For long regular transducers, special methods can be used to minimize the computation time. As noted above, band-edge frequencies of stop bands can sometimes be used to deduce COM parameters, so the FEM analysis need only be done for the relatively simple cases of infinite gratings. A more general method is the 'harmonic admittance', in which a regular set of electrodes is assumed to have voltages of the form $\exp(jn\gamma)$, where n is the strip number and γ is an independent parameter. This function is applicable to grating analysis, as shown for example by Danicki's early work on transverse surface waves with mechanical loading.[255] The Fourier transform of this is the 'discrete Green's function', which gives the electrode currents when a voltage is applied to one electrode with all other electrodes grounded, a concept introduced in 1993 by Zhang[256] and Hashimoto.[257] Using superposition, this function enables the analysis of a variety of devices with regular electrodes, notably one-port resonators for impedance element filters, though it needs to be calculated separately for all frequencies required. It can be calculated by numerical methods taking account of complications such as mechanical loading, the behaviour of leaky waves and bulk wave generation.[228]

Analysis methods for leaky waves and STW's have been mainly restricted to uniform gratings and single-electrode transducers. These waves behave in a manner more complicated than SAW's, in that the film thickness is more significant, and conversion to bulk waves occurs at and beyond the upper edge of the stop band. Hashimoto[258] has analyzed a metal grating using the effective permittivity, with mechanical loading added using FEM. The numerical results were fitted to an empirical formula deduced by Plessky,[259] from which the COM parameters for transducer analysis can be deduced. This allows the COM parameters to vary with electrode width and thickness, and to model the loss due to bulk wave radiation. This approach has been applied to device analysis for both leaky waves on 36°Y-X lithium tantalate[260, 261] and STW's on quartz.[262]

5. Conclusions

SAW technology, with the advantages noted in Sec. 4.1, is eminently suitable for linear analogue devices. Within this area, the versatility is so great that the devices cover almost all the functions imaginable – delay lines, bandpass filters, pulse compression filters, resonators and so on. This has all been achieved since the IDT emerged in 1965, and the devices have demonstrated steadily-improving reproducible performance with, when needed, cost effectiveness associated with single-stage lithography. The subject is also packed with many fascinating physical topics, involving almost all the phenomena that waves in anisotropic media might exhibit. And new innovations are still occuring today.

Table 3. Frequencies for Mobile and Cordless Phones.[263]

System	Tx band * MHz	Rx band * MHz	RF BW MHz	IF BW kHz
Analogue Cellular Phone (FDMA)				
AMPS (Americas, Australia)	824-849	869-894	25	30
ETACS (UK)	871-904	916-949	33	25
NMT 900 (Europe)	890-915	935-960	25	12.5
Digital Cellular Phone (TDMA)				
GSM (Europe)	890-915	935-960	25	200
IS-95 (CDMA)	824-849	869-894	25	1250
PDC (Japan)	940-956	810-826	16	25
PCN/DCS-1800 (Europe)	1710-1785	1805-1880	75	200
Digital Cordless Phone (TDMA)				
DECT (Europe)	1880-1990	1880-1990	110	1728
PHS (Japan)	1895-1907	1895-1907	12	300
Wireless LAN	2400-2500		100	–

* of handset
AMPS = Advanced mobile phone system
GSM = Global system for mobiles
CDMA = Code division multiple access
PDC = personal digital cellular
DECT = Digital European cordless telephone
PHS = Personal handyphone system
PCN = Personal communication network

Applications for SAW devices have recently expanded enormously with the development of mobile radio systems of various types, particularly mobile phones. The new applications are mostly in the commercial and consumer areas, complementing earlier professional and military applications (Table 1) which are still active. Applications are reviewed by Lam et al,[263] Machui et al[264] and Campbell.[12,20] To bring them into focus, the frequencies of some selected mobile phone systems are given in Table 3. SAW filters are supplied in very large quantities for most, if not all, of these systems. In handsets, the SAW RF filters are usually IEF's or DMS filters, situated at an intermediate point in the RF path. At the front end, the handset has a 'duplexer' consisting of a transmitter filter and a receiver filter, both connected to the antenna. Until recently these were often ceramic resonator filters rather than SAW's because the losses are lower. However, recent SAW duplexers such as those in Refs. 229 and 265 have demonstrated low loss and sharp skirts, and are therefore promising for future applications. In view of the different standards used in different countries (Table 3), some manufacturers are now developing units with several SAW filters in one package, enabling the handset to be switched from one standard to another.

At intermediate frequencies, SAW filters are used for the systems with wider bandwidths, particularly GSM where the bandwidth is around 200 kHz. The TCR filter and the Z-path filter are well established for this. For the wider bandwidth of 1250 kHz in the CDMA system, resonant SPUDT filters are used.

Other application areas, mostly using bandpass filters, include[263] paging, keyless entry systems (for garage doors, cars etc.), fibre optic communications (timing recovery filters), video conferencing equipment, test instrumentation, GPS, LAN's, direct broadcasting satellites (DBS), base station equipment and digital radio.

Quantities involved in SAW production have of course grown remarkably in parallel with the development of wireless applications. Recent informal estimates were that the total production world-wide exceeds 2 billion (2×10^9) devices per year, and that the rate of production is at least doubling every year. The number of manufacturing companies involved is around 60, with about half of these being in Japan.

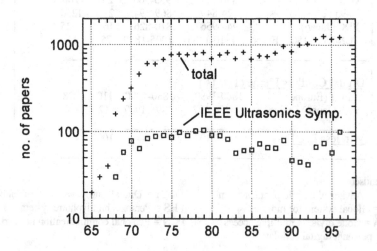

Fig. 22. Number of annual SAW publications, 1965 to 1996. From data in Hickernell, Ref.22.

As an indication of the level of research activity, Fig.22 shows the annual number of publications on SAW worldwide, and the number in the IEEE Ultrasonics Symposium, from data given by Hickernell.[22] The total number has been fairly steady at around 1000 per year for some time. The Symposium number tended to fall off around 1980, perhaps because of the increase in Japanese publications, followed by a rising tendency probably associated with mobile radio and, more recently, sensors.

As stated earlier, key 'background' requirements are piezoelectric materials, lithography and computing. This author would like to add a fourth factor – the hard work and dedication of an international band of very proficient people who had the inspirations, delved into the analysis, conducted experimental tests and finally made the devices perform useful functions. The progress achieved is staggering, and will surely fascinate anyone with technical interests. It has been a privilege to be involved in it.

Acknowledgements

It is a pleasure to acknowledge C.S.Hartmann, F.Hickernell, M.F.Lewis, V.Plessky and C.Ruppel for supplying many helpful comments on the original manuscript and for assitance in obtaining figures.

References

1. I.A.Viktorov. *Rayleigh and Lamb Waves*. Plenum Press, 1967
2. David P. Morgan (ed.). *Surface-Acoustic-Wave Passive Interdigital Devices*. Peter Peregrinus, 1976 (IEE Reprint Series No. 2).
3. H.Matthews (ed.). *Surface Wave Filters*. Wiley, 1977.
4. A.A.Oliner (ed.). *Acoustic Surface Waves*. Springer, 1978.
5. E.A.Ash and E.G.S.Paige (eds). *Rayleigh-Wave Theory and Application*. Springer, 1985.
6. David P.Morgan. *Surface-Wave Devices for Signal Processing*. Elsevier, 1985, 1991.
7. Supriyo Datta. *Surface Acoustic Wave Devices*. Prentice-Hall, 1986.
8. Gordon S. Kino. *Acoustic Waves*. Prentice-Hall, 1987.
9. Colin Campbell. *Surface Acoustic Wave Devices and their Signal Processing Applications*. Boston: Academic Press, 1989.
10. M.Feldmann and J.Henaff. *Surface Acoustic Waves for Signal Processing*. Artech House, 1989 (originally in French, 1986)
11. B.A.Auld. *Acoustic Fields and Waves in Solids* (2 vols.). 2nd ed., Krieger, 1990 (1st ed. Wiley, 1973).
12. Colin K. Campbell. *Surface Acoustic Wave Devices for Mobile and Wireless Applications*. Academic Press, 1998.
13. S.V.Biryukov, Y.V.Gulyaev, V.V.Krylov and V.P.Plessky. *Surface Acoustic Waves in Inhomogeneous Media*. Springer, 1995.
14. K.Hashimoto. *Surface Acoustic Wave Devices in Modern Communication Systems and their Simulation Technologies*. Elsevier (to be published, 2000).
15. A.J. Bahr (ed.) Special issue on 'Microwave acoustics'. *IEEE Trans.*, **MTT-17**, 799-1046 (Nov. 1969).
16. T.M. Reeder (ed.). Special issue on 'Microwave acoustic signal processing'. *IEEE Trans.*, **MTT-21**, 161-306 (Apr. 1973). Also published as *IEEE Trans.*, **SU-20**, 79-230 (Apr. 1973)
17. L.T. Claiborne, G.S. Kino and E.Stern (eds.). Special issue on 'Surface acoustic wave devices and applications'. *Proc. IEEE*, **64**, 579-807 (May 1976)
18. J.H. Collins and L.Masotti (eds.). Special issue on 'Computer-aided design of SAW devices'. *Wave Electronics*, **2**, 1-304 (July 1976)
19. R.C.Williamson and T.W.Bristol (eds). Special issue on 'Surface-acoustic-wave device applications'. *IEEE Trans.*, **SU-28**, 115-234 (1981).
19a. K.Shibayama and K.Yamanouchi (eds.). Proceedings of *International Symposium on Surface Acoustic Wave Devices for Mobile Communication*, Sendai, Dec. 1992. (Pub. by 150th Committee of Japan Society for Promotion of Science).
20. C.K.Campbell. 'Applications of surface acoustic and shallow bulk acoustic wave devices'.

Proc. IEEE, **77**, 1453-1484 (1989)
21. D. P. Morgan. 'History of SAW devices'. *IEEE Intl. Frequency Control Symp.*, Pasadena, 1998, pp. 439-460.
22. Fred S. Hickernell. 'Surface acoustic wave technology – macrosuccess through microseisms'. In *Physical Acoustics*, vol. XXIV, Academic Press, 1999, p.135-207.
23. Lord Rayleigh. 'On waves propagating along the plane surface of an elastic solid'. *Proc. London Math. Soc.*, **17**, 4-11 (1885).
24. A.E.H.Love. *Some Problems of Geodynamics*. Cambridge, 1911; Dover 1967.
25. 25. K.Sezawa. 'Dispersion of elastic waves propagated on the surface of stratified bodies and on curved surfaces'. *Bull. Earthquake Res. Inst. Tokyo*, **3**, 1-18 (1927). **Also**, M.Kanai. 'On the M2 waves (Sezawa waves)'. *ibid.*, **29**, 39-48 (1951) and **33**, 275-281 (1955).
26. L.M.Brekhovskikh. and O.A.Godin. *Waves in Layered Media*. Springer, 1990 (vol. 1), 1992 (vol. 2).
27. W.M.Ewing, W.S.Jardetzky and F.Press. *Elastic Waves in Layered Media*. McGraw-Hill, 1957.
28. R.M.White. 'Surface elastic waves'. *Proc. IEEE*, **58**, 1238-1276 (1970)
29. J.R.Klauder, A.C.Price, S.Darlington and W.J.Albersheim. 'The theory and design of chirp radars'. *Bell Syst.Tech. J.*, **39**, 745-808 (1960).
30. I.N.Court. 'Microwave acoustic devices for pulse compression filters'. *IEEE Trans.*, **MTT-17**, 968-986 (1969).
31. J.E.May. 'Guided wave ultrasonic delay lines'. *in* W.P.Mason (ed.), *Physical Acoustics*, vol. 1A, Academic Press 1964, p.417-483.
32. C.Lardat, C.Maerfeld and P.Tournois. 'Theory and performance of acoustical dispersive surface wave delay lines'. *Proc. IEEE*, **59**, 355-368 (1971).
33. W.S.Mortley. 'Pulse compression by dispersive gratings on crystal quartz'. *Marconi Rev.*, No.159, 273-290 (1965).
34. W.S.Mortley. British patent 988,102 (1963). **Also**, J.H.Rowen, U.S. Patent 3,289,114 (1963).
35. R.M.White and F.W.Voltmer. 'Direct piezoelectric coupling to surface elastic waves'. *Appl. Phys. Lett.*, **7**, 314-316 (1965)
36. R.H.Tancrell, M.B.Schulz, H.H.Barrett, L.Davies and M.G.Holland. 'Dispersive delay lines using ultrasonic surface waves'. *Proc. IEEE*, **57**, 1211-1213 (1969).
37. P. Hartemann and E. Dieulesaint. 'Intrinsic compensation of sidelobes in a dispersive acoustic delay line'. *Electronics Lett.*, **5**, 219-210 (1969).
38. E.A. Ash, R.M. De La Rue, and R.F. Humphryes. 'Microsound surface waveguides'. *IEEE Trans.*, **MTT-17**, 882-892 (1969).
39. J.J.Campbell and W.R.Jones. 'A method for estimating optimal crystal cuts and propagation directions for excitation of piezoelectric surface waves'. *IEEE Trans.*, **SU-15**, 209-217 (1968)
40. M.S.Kharusi and G.W.Farnell. 'Diffraction and beam steering for surface-wave comb structures on anisotropic substrates'. *IEEE Trans.*, **SU-18**, 34-42 (1971).
41. J.C.Crabb, J.D.Maines and N.R.Ogg. 'Surface-wave diffraction on $LiNbO_3$'. *Electronics Lett.*, **7**, 253-255 (1971).
42. F.Seifert, G.Visintini, C.C.W.Ruppel and G.Riha. 'Diffraction compensation in SAW filters'. *IEEE Ultrasonics Symp.*, 1990, p.67-76.
43. M.B.Schulz, B.J.Matsinger and M.G.Holland. 'Temperature dependence of surface acoustic wave velocity on alpha-quartz'. *J. Appl. Phys.*, **41**, 2755-2765 (1970)
44. J.F.Dias, H.E.Karrer, J.A.Kusters, J.H.Matsinger and M.B.Schulz. 'The temperature coefficient of delay time for X-propagating acoustic surface waves on rotated Y-cuts of alpha quartz'. *IEEE Trans.*, **SU-22**, 46-50 (1975).
45. S.J.Kerbel. 'Design of harmonic surface acoustic wave (SAW) oscillators without external filtering and new data on the temperature coefficient of quartz'. *IEEE Ultrasonics Symp.*, 1974, p.276-281.
46. A.J.Slobodnik. 'Surface acoustic waves and SAW materials'. *Proc. IEEE*, **64**, 581-595

(1976).
47. A.J.Slobodnik. 'Materials and their influence on performance'. *in* Oliner, Ref.4, p.226-303 (1978)
48. H.J.Maris. 'Attenuation of ultrasonic surface waves by viscosity and heat conduction'. *Phys. Rev.*, **188**, 1308-1311 (1969).
49. K.Dransfeld and E.Salzmann. 'Excitation, detection and attenuation of high-frequency elastic surface waves'. *in* W.P.Mason and R.N.Thurston (eds.), *Physical Acoustics*, vol. VII, Academic Press, 1970, p.219-272.
50. J.J.Campbell and W.R.Jones. 'Propagation of surface waves at the boundary between a piezolectric crystal and a fluid medium'. *IEEE Trans.*, **SU-17**, 71-76 (1970).
51. E.G.Lean and C.G.Powell. 'Non-destructive testing of thin films by harmonic generation of dispersive Rayleigh waves'. *Appl. Phys. Lett.*, **19**, 356-359 (1971).
52. C.S.Hartmann, private communication.
53. G.W.Farnell. 'Properties of elastic surface waves'. *in* W.P.Mason and R.N.Thurston (eds.), *Physical Acoustics*, vol. 6, Academic Press, 1970, p.109-166. **Also**, G.W.Farnell. 'Elastic surface waves'. *in* Matthews, Ref. 3, p.1-53 (1977).
54. A.J.Slobodnik and E.D.Conway. *Microwave Acoustics Handbook*, vol. 1, *Surface Wave Velocities*. US Air Force Cambridge Research Laboratories, report no. AFCRL-PSRP-414 (1970). Accession No. AD 868360.
55. A.J.Slobodnik, E.D.Conway and R.T.Delmonico. *Microwave Acoustics Handbook*, vol. 1A, *Surface Wave Velocities*. US Air Force Cambridge Research Laboratories, report no. AFCRL-TR-73-0-597 (1973). Accession No. AD 780172.
56. A.J.Slobodnik, R.T.Delmonico and E.Conway. *Microwave Acoustics Handbook*, vol. 2, *Surface Wave Velocities – Numerical Data*. US Air Force Cambridge Research Laboratories, report no. AFCRL-PSRP-609 (1974). Accession No. A006113.
57. G.I.Stegeman. 'Optical probing of surface waves and surface wave devices'. *IEEE Trans.*, **SU-23**, 33-63 (1976).
58. K.Shibayama, K.Yamanouchi, H.Sato and T.Meguro. 'Optimum cut for rotated Y-cut LiNbO$_3$ crystal used as the substrate of acoustic-surface-wave filters'. *Proc. IEEE*, **64**, 595-597 (1976).
59. S.Takahashi, H.Hirano, T.Kodama, F.Mirashiro, B.Suzuki, A.Onoe, T.Adachi and K.Fujinuma. 'SAW IF filter on LiTaO$_3$ for colour TV receivers'. *IEEE Trans. Consum. Electron.*, **CE-24**, 337-348 (1978). **Also**, M.Anhorn and H.E.Engan. 'SAW velocities in X-cut LiTaO$_3$'. *IEEE Ultrasonics Symp.*, 1989, p.363-366.
60. F.S.Hickernell. 'Zinc oxide thin-film surface-wave transducers'. *Proc. IEEE*, **64**, 631-635 (1976)
61. Y.Ebata and H.Satoh. 'Current applications and future trends for SAW in Asia'. *IEEE Ultrasonics Symp.*, 1988, p.195-202.
62. J.Koike, K.Shimoe and H.Ieki. '1.5 GHz low-loss surface acoustic wave filter using ZnO/sapphire substrate'. *Japan J. Appl. Phys.*, part 1, **32**, 2337-2340 (1993). **Also**, H.Ieki and M.Kadota. 'ZnO thin films for high frequency SAW devices'. *IEEE Ultrasonics Symp.*, 1999, paper LL-5.
63. R.W.Whatmore, N.M.Shorrocks, C.O'Hara, F.W.Ainger and I.M.Young. 'Lithium tetraborate: a new temperature-compensated SAW substrate material'. *Electronics Lett.*, **17**, 11-12 (1981). **Also**, B.Lewis, N.M.Shorrocks and R.W.Whatmore. 'An assessment of lithium tetraborate for SAW applications'. *IEEE Ultrasonics Symp.*, 1982, p.389-393.
64. J.L.Bleustein. 'A new surface wave in piezoelectric crystals'. *Appl. Phys. Lett.*, **13**, 412-413 (1968).
65. Y.V.Gulyaev. 'Electroacoustic surface waves in Solids'. *Soviet Phys. JETP Lett.*, **9**, 37-38 (1969).
66. Y.Ohta, K.Nakamura and H.Shimizu. Tech. Rep. IECE Japan, vol. US69-3, 1969 (in Japanese).
67. R.H.Tancrell and M.G.Holland. 'Acoustic surface wave filters'. *Proc. IEEE*, **59**, 393-409 (1971).

68. C.E.Cook and J.Paolillo. 'A pulse compression pre-distortion function for efficient sidelobe reduction in a high power radar'. *Proc. IEEE*, **52**, 377-389 (1964).
69. C.O.Newton. 'Non-linear chirp radar signal waveforms for surface acoustic wave pulse compression filters'. *Wave Electronics*, **1**, 387-401 (1976).
70. M.B.N.Butler. 'Radar applications of SAW dispersive filters'. *Proc. IEE*, **127F**, 118-124 (1980).
71. R.H.Tancrell and P.C.Meyer. 'Operation of long surface wave interdigital transducers'. *IEEE Trans.*, **SU-19**, p.405 (1972). (abstract only)
72. B.R.Potter and C.S.Hartmann. 'Surface acoustic wave slanted device technology'. *IEEE Trans.*, **SU-26**, 411-418 (1979).
73. S.Jen and C.S.Hartmann. 'An improved model for chirped slanted SAW devices'. *IEEE Ultrasonics Symp.*, 1989, p.7-14
74. R.C.Williamson and H.I.Smith. 'Large time-bandwidth product surface-wave pulse compressor employing reflective gratings'. *Electronics Lett.*, **8**, 401-402 (1972). **Also**, R.C.Williamson and H.I.Smith, 'The use of surface elastic wave reflection gratings in large time-bandwidth pulse compression filters'. *IEEE Trans.*, **MTT-21**, 195-205 (1973).
75. T.A.Martin. 'The IMCON pulse compression filter and its applications'. *IEEE Trans.*, **MTT-21**, 186-194 (1973).
76. E.K.Sittig and G.A.Coquin. 'Filters and dispersive delay lines using repetitively mismatched ultrasonic transmission lines'. *IEEE Trans.*, **SU-15**, 111-119 (1968).
77. H.M.Gerard, P.S.Yao and O.W.Otto. 'Performance of a programmanble radar pulse compression filter based on a chirp transformation with RAC filters'. *IEEE Ultrasonics Symp.*, 1977, p.947-951.
78. S.Darlington. 'Demodulation of wide band low-power FM signals'. *Bell Syst. Tech. J.*, **43**, 339-374 (1964).
79. J.M.Alsup, R.W.Means and H.J.Whitehouse. *IEE Intl. Specialist Seminar on Component Performance and System Applications of Surface Acoustic Wave Devices*, Aviemore, Scotland, Sept. 1973.
80. C.Atzeni, G.Manes and L.Masotti. 'Programmable signal processing by analog chirp transformation using SAW devices'. *IEEE Ultrasonics Symp.*, 1975, p.371-376.
81. M.A.Jack, P.M.Grant and J.H.Collins. 'The theory, design and applications of surface acoustic wave Fourier-transform processors'. *Proc. IEEE*, **68**, 450-468 (1980).
82. G.L.Moule, R.A.Bale and T.I.Browning. 'A 1 GHz bandwidth SAW compressive receiver'. *IEEE Ultrasonics Symp.*, 1980, p.216-219.
83. R.C.Williamson, V.S.Dolat, R.R.Rhodes and D.M.Boroson. 'A satellite-borne SAW chirp-transform system for uplink demodulation of FSK communication signals'. *IEEE Ultrasonics Symp.*, 1979, p.741-747.
84. R.Li, G.Stette and P.M.Bakken. 'SAW chirp Fourier transform for multicarrier transmission'. *IEEE Ultrasonics Symp.*, 1993, p.79-84.
85. P.Hartemann and E.Dieulesaint. 'Acoustic surface wave filters'. *Electronics Lett.*, **5**, 657-658 (1969)
86. D.Chauvin, G. Coussot and E. Dieulesaint. 'Acoustic-surface-wave television filters'. *Electronics Lett.*, **7**, 491-492 (1971).
87. R.H.Tancrell and R.C.Williamson. 'Wavefront distortion of acoustic surface waves from apodized interdigital transducers'. *Appl. Phys. Lett.*, **19**, 456-459 (1971).
88. R.H.Tancrell. 'Analytic design of surface wave bandpass filters'. *IEEE Trans.*, **SU-21**, 12-22 (1974).
89. L.R.Rabiner, J.H.McClellan and T.W.Parks. 'FIR digital filter design techniques using weighted Chebyshev approximation'. *Proc. IEEE*, **63**, 595-610 (1975).
90. J.H.McClellan, T.W.Parks and L.R.Rabiner. 'A computer programme for designing optimum FIR linear-phase digital filters'. *IEEE Trans.*, **AU-21**, 506-526 (1973)
91. C.Ruppel, E.Ehrmann-Falkenau, H.R.Stocker and R.Veith. 'Optimum design of SAW filters by linear programming'. *IEEE Ultrasonics Symp.*, 1983, p.23-26.
92. F.G.Marshall and E.G.S.Paige. 'Novel acoustic-surface-wave directional coupler with diverse

applications'. *Electronics Lett.*, **7**, 460-464 (1971)
93. F.G.Marshall, C.O.Newton and E.G.S.Paige. 'Surface acoustic wave multisrip components and their applications'. *IEEE Trans.*, **MTT-21**, 216-225 (1973).
94. C.Maerfeld and G.W.Farnell. 'Nonsymmetrical multistrip coupler as a surface-wave beam compressor'. *Electronics Lett.*, **9**, 432-434 (1973).
95. Y.Ebata and H.Satoh. 'Current applications and future trends for SAW in Asia'. *IEEE Ultrasonics Symp.*, 1988, p.195-202.
96. T.Kodama. 'Broad-band compensation for diffraction in surface acoustic wave filters'. *IEEE Trans.*, **SU-30**, 127-136 (1983).
97. C.Kappacher and G.Visintini. 'Differential delay line on X-cut $LiTaO_3$'. *IEEE Ultrasonics Symp.*, 1990, p.7-10.
98. T.Kodama, K.Sato and Y.Uemura. 'SAW vestigial sideband filter for TV broadcasting transmitter'. *IEEE Trans.*, **SU-28**, 151-155 (1981).
99. R.Ganss-Puchstein, C.Ruppel and H.R.Stocker. 'Spectrum shaping SAW filters for high bit rate digital radio'. *IEEE Trans. UFFC*, **35**, 673-684 (1988).
100. C.S.Hartmann. 'Weighting of interdigital surface wave transducers by selective withdrawal of electrodes'. *IEEE Ultrasonics Symp.*, 1973, p.423-426.
101. A.J.Slobodnik, G.A.Roberts, J.H.Silva, W.J.Kearns, J.Sethares and T.L.Szabo. 'Switchable SAW filter banks at UHF'. *IEEE Trans.*, **SU-26**, 120-126 (1979).
102. L.P.Solie and M.D.Wohlers. 'Use of a SAW multiplexer in FMCW radar systems'. *IEEE Trans.*, **SU-28**, 141-145 (1981)
103. F.Hickernell, private communication.
104. E.A. Ash. 'Surface wave grating reflectors and resonators'. *IEEE Intl. Microwave Symp.*, 1970, p. 385-386.
105. R.C.M.Li and J.Melngailis. 'The influence of stored energy at step discontinuities on the behaviour of surface-wave gratings'. *IEEE Trans.*, **SU-22**, 189-198 (1975)
106. Y.Suzuki, H.Shimizu, M.Takeuchi, K.Nakamura and A.Yamada. 'Some studies on SAW resonators and mulitple-mode filters'. *IEEE Ultrasonics Symp.*, 1976, p.297-302.
107. S.Datta and B.J.Hunsinger. 'First order reflection coefficient of surface acoustic waves from thin-strip overlays'. *J. Appl. Phys.*, **50**, 5661-5665 (1979). **Also**, Datta, Ref. 7, Eqs. (6.10)
108. S.Datta and B.J.Hunsinger. 'An analytical theory for the scattering of surface acoustic waves by a single electrode in a periodic array on a piezoelectric substrate'. *J. Appl. Phys.*, **51**, 4817-4823 (1980).
109. C.Dunnrowicz, F.Sandy and T.Parker. 'Reflection of surface waves from periodic discontinuities'. *IEEE Ultrasonics Symp.*, 1976, p.386-390.
110. P.V.Wright. 'Modeling and experimental measurements of the reflection properties of SAW metallic gratings'. *IEEE Ultrasonics Symp.*, 1984, p.54-63.
111. E.J.Staples, J.S.Schoenwald, R.C.Rosenfeld and C.S.Hartmann. 'UHF surface acoustic wave resonators'. *IEEE Ultrasonics Symp.*, 1974, p.245-252.
112. T.E.Parker and G.K.Montress. 'Precision surface-acoustic-wave (SAW) oscillators'. *IEEE Trans. UFFC*, **35**, 342-364 (1988)
113. J.D.Maines, E.G.S.Paige, A.F.Saunders and A.S.Young. 'Simple technique for the accurate determination of delay-time variations in acoustic surface wave structures'. *Electronics Lett.*, **5**, 678-680 (1969)
114. M.Lewis. 'The surface acoustic wave oscillator – a natural and timely development of the quartz crystal oscillator'. *28th Ann. Freq. Control Symp.*, 1974, p.304-314.
115. L.A.Coldren and R.L. Rosenberg. 'Surface-acoustic-wave resonator filters'. *Proc. IEEE*, **67**, 147-158 (1979)
116. L.O.Svaasand. 'Interaction between elastic surface waves in piezoelectric materials'. *Appl. Phys. Lett.*, **15**, 300-302 (1969).
117. C.F.Quate and R.B.Thomson. 'Convolution and correlation in real time with non-linear acoustics'. *Appl. Phys. Lett.*, **16**, 494-496 (1970).
118. M.Luukkala and G.S.Kino. 'Acoustic convolution and correlation and the associated nonlinearity parameters in lithium niobate'. *Appl. Phys. Lett.*, **18**, 393-394 (1971)

119. Ph.Defranould and C.Maerfeld. 'Acoustic convolver using multistrip beam compressors'. *IEEE Ultrasonics Symp.*, 1974, p. 224-227. **Also,** 'A SAW planar piezoelectric convolver'. *Proc. IEEE*, **64**, 748-751 (1976).
120. N.K.Batani and E.L.Adler. 'Acoustic convolvers using ribbon waveguide beamwidth compressors'. *IEEE Ultrasonics Symp.*, 1974, p.114-116. Also, A.M.Hodge and M.F.Lewis. 'Detailed investigation into SAW convolvers'. *ibid.*, 1982, p.113-118.
121. M.D.Motz, J.Chambers and I.M.Mason. 'Suppression of spurious signals in a degenerate SAW convolver'. *IEEE Ultrasonics Symp.*, 1973, p.152-154. Also, I.Yao. 'High performance elastic convolver with parabolic horns'. *IEEE Ultrasonics Symp.*, 1980, p.37-42.
122. D.P.Morgan, D.H.Warne and D.R.Selviah. 'Narrow aperture chirp transducers for SAW convolvers'. *Electronics Lett.*, **18**, 80-81 (1982).
123. J.B.Green and G.S.Kino. 'SAW convolvers using focused interdigital transducers'. *IEEE Trans.*, **SU-30**, 43-50 (1983).
124. D.P.Morgan. 'General analysis of bilinear SAW convolvers'. *Electronics Lett.*, **17**, 265-267 (1981). **Also,** D.R.Selviah, D.H.Warne and D.P.Morgan. 'Spatial uniformity measurement of SAW convolvers'. *Electronics Lett.*, **18**, 837-839 (1982).
125. J.H.Collins, H.M.Gerard and H.J.Shaw. 'High performance lithium niobate acoustic surface wave transducers and delay lines'. *Appl. Phys. Lett.*, **13**, 312 (1968). **Also,** K.M.Lakin and H.J.Shaw. 'Surface wave delay line amplifiers'. *IEEE Trans.*, **MTT-17**, 912-920 (1969)
126. W.C.Wang and P.Das. 'Surface wave convolver via space charge nonlinearity'. *IEEE Ultrasonics Symp.*, 1972, p.316-321.
127. G.S.Kino. 'Acoustoelectric interactions in acoustic-surface-wave devices'. *Proc. IEEE*, **64**, 724-748 (1976).
128. S.A.Reible. 'Acoustoelectric convolver technology for spread-spectrum communications'. *IEEE Trans.*, **SU-28**, 185-195 (1981).
129. B.T.Khuri-Yakub and G.S.Kino. 'A monolithic zinc oxide on silicon convolver'. *Appl. Phys. Lett.*, **25**, 188-190 (1974).
130. K.Tsubouchi. 'An asynchronous spread spectrum wireless modem using a SAW convolver'. *Proc. Intl. Symp. SAW Devices for Mobile Comm.* (Sendai, 1992), p.215-222.
131. W.D.Squire, H.J.Whitehouse and J.M.Alsup. 'Linear signal processing and ultrasonic transversal filters'. *IEEE Trans.*, **MTT-17**, 1020-1040 (1969). **Also,** S.T.Costanza, P.J.Hagon and L.A.MacNevin. 'Analog matched filter using tapped acoustic surface wave delay line'. *ibid.*, p.1042-1043.
132. E.J.Staples and L.T.Claiborne. 'A review of device technology for programmable surface-wave filters'. *IEEE Trans.*, **MTT-21**, 279-287 (1973).
133. R.D.Lambert, P.M.Grant, D.P.Morgan and J.H.Collins. 'Programmable surface acoustic wave devices using hybrid microelectronic components'. *Radio and Electronic Eng.*, **44**, 343-351 (1974).
134. C.M.Panasik and M.Jurgovan. 'Programmable transversal filter based adaptive line enhancer'. *IEEE Ultrasonics Symp.*, 1992, p.227-230.
135. P.J.Hagon, F.B.Micheletti and R.N.Seymour. 'Integrated programmable analog matched filters for spread spectrum applications'. *IEEE Ultrasonics Symp.*, 1973, p.333-335.
136. F.S.Hickernell, D.E.Olson and M.D.Adamo. 'Monolithic surface wave transversal filter'. *IEEE Ultrasonics Symp.*, 1977, p.615-618.
137. M.J.Hoskins and B.J.Hunsinger. 'Monolithic GaAs acoustic charge transport devices'. *IEEE Ultrasonics Symp.*, 1982, p.456-460.
138. R.L.Miller, C.E.Nothnick and D.S.Bailey. *Acoustic Charge Transport: Device Technology and Applications.* Boston: Artech, 1992.
139. G.A.Coquin and H.F.Tiersten. 'Analysis of the excitation and detection of piezoelectric surface waves in quartz by means of surface electrodes'. *J. Acoust. Soc. Am.*, **41**, 921-939 (1967).
140. S.G.Joshi and R.M.White. 'Excitation and detection of surface elastic waves in piezoelectric crystals'. *J. Acoust. Soc. Am.*, **46**, 17-27 (1969).
141. H.Skeie. 'Electrical and mechanical loading of a piezoelectric surface supporting surface

waves'. *J. Acoust. Soc. Am.*, **48**, 1098-1109 (1970).
142. P.R.Emtage. 'Self-consistent theory of interdigital transducers'. *J. Acoust. Soc. Am.*, **51**, 1142-1155 (1972).
143. B.A.Auld and G.S.Kino. 'Normal mode theory for acoustic waves and its application to the interdigital transducer'. *IEEE Trans.*, **ED-18**, 898-908 (1971)
144. W.S.Jones, C.S.Hartmann and T.D.Sturdivant. 'Second-order effects in surface wave devices'. *IEEE Trans.*, **SU-19**, 368-377 (1972).
145. T.W.Bristol, W.R.Jones, P.B.Snow and W.R.Smith. 'Applications of double electrodes in acoustic surface wave device design'. *IEEE Ultrasonics Symp.*, 1972, p.343-345.
146. W.R.Smith, H.M.Gerard, J.H.Collins, T.M.Reeder and H.J.Shaw. 'Analysis of interdigital surface waves transducers by use of an equivalent circuit model'. *IEEE Trans.*, **MTT-17**, 856-864 (1969).
147. W.R.Smith. 'Experimental dsitinction between crossed-field and in-line three-port circuit models for interdigital transducers'. *IEEE Trans.*, **MTT-22**, 960-964 (1974).
148. C.S.Hartmann, D.T.Bell and R.C.Rosenfeld. 'Impulse model design of acoustic-surface-wave filters'. *IEEE Trans.*, **MTT-21**, 162-175 (1973).
149. A.L.Nalamwar and M.Epstein. 'Immittance characterization of acoustic surface wave transducers'. *Proc. IEEE*, **60**, 336-337 (1972).
150. K.A.Ingebrigtsen. 'Surface waves in piezoelectrics'. *J. Appl. Phys.*, **40**, 2681-2686 (1969)
151. R.F.Milsom, N.H.C.Reilly and M.Redwood. 'Analysis of generation and detection of surface and bulk acoustic waves by interdigital transducers'. *IEEE Trans.*, **SU-24**, 147-166 (1977).
152. K.Blotekjaer, K.A.Ingebrigtsen and H.Skeie. 'A method for analysing waves in structures consisting of metal strips on dipserive media'. *IEEE Trans.*, **ED-20**, 1133-1138 (1973).
153. D.P.Morgan. 'Quasi-static analysis of generalised SAW transducers using the Green's function method'. *IEEE Trans.*, **SU-27**, 111-123 (1980)
154. H.Engan. 'Excitation of elastic surface waves by spatial harmonis of interdigital transducers'. *IEEE Trans.*, **ED-16**, 1014-1017 (1969).
155. B.Lewis, P.M.Jordan, R.F.Milsom and D.P.Morgan. 'Charge and field superpsoition methods for analysis of generalised SAW interdigital transducers'. *IEEE Ultrasonics Symp.*, 1978, p.709-714. **Also**, B.J.Hunsinger and S.Datta. 'A generalised model for periodic transducers with arbitrary voltages'. *ibid.*, 1978, p.705-708.
156. R.C.Peach. 'A general approach to the electrostatic problem of the SAW interdigital transducer'. *IEEE Trans.*, **SU-28**, 96-105 (1981).
157. C.S.Hartmann and B.G.Secrest. 'End effects in interdigital surface wave transducers'. *IEEE Ultrasonics Symp.*, 1972, p.413-416.
158. W.R.Smith and W.F.Pedler. 'Fundamental and harmonic frequency circuit-model analysis of interdigital transducers with arbitrary metalization ratios and polarity sequences'. *IEEE Trans.*, **MTT-23**, 853-864 (1975).
159. S.V.Biryukov and V.G.Polevoi. 'The electrostatic problem for SAW interdigital transducers in an external electric field'. *IEEE Trans. UFFC*, **43**. Part I, pp.1150-1159; Part II, pp.1160-1170 (1996).
160. P.S.Cross and R.V.Schmidt. 'Coupled surface-acoustic-wave resonators'. *Bell Syst. Tech. J.*, **56**, 1447-1482 (1977)
161. W.R.Smith, H.M.Gerard, J.H.Collins, T.M.Reeder and H.J.Shaw. 'Design of surface wave delay lines with interdigital transducers'. *IEEE Trans.*, **MTT-17**, 865-873 (1969).
162. M.F.Lewis. 'Triple-transit suppression in surface-acoustic-wave devices'. *Electronics Lett.*, **8**, 553-554 (1972).
163. M.Feldmann and J.Henaff. 'A new multistrip acoustic surface wave filter'. *IEEE Ultrasonics Symp.*, 1974, p.157-160.
164. R.B.Brown. 'Low-loss device using multistrip coupler ring configuration with ripple cancellation'. *IEEE Ultrasonics Symp.*, 1986, p.71-76.
165. S.A.Dobershtein and V.A.Malyukhov. 'SAW ring filters with insertion loss of 1 dB'. *IEEE Trans. UFFC*, **44**, 590-596 (1997).
166. C.S.Hartmann, W.S.Jones and H.Vollers. 'Wide band unidirectional surface wave

transducers'. *IEEE Trans.*, **SU-19**, 378-381 (1972).
167. K.Yamanouchi, F.M.Nyffeler and K.Shibayama. 'Low insertion loss acoustic surface wave filter using group-type unidirectional interdigital transducer'. *IEEE Ultrasonics Symp.*, 1975, p.317-321.
168. C.S.Hartmann, P.V.Wright, R.J.Kansy and E.M.Garber. 'An analysis of SAW interdigital transducers with internal reflections and the application to the design of single-phase unidirectional transducers'. *IEEE Ultrasonics Symp.*, 1982, p.40-45.
169. M.Lewis. 'Low loss SAW devices employing single stage fabrication'. *IEEE Ultrasonics Symp.*, 1983, p.104-108. Also, 'Group-type unidirectional SAW devices employing intra-transducer reflector banks'. *Electronics Lett.*, **19**, 1085-1087 (1983).
170. T.Kodama, H.Kawabata, Y.Yasuhara and H.Sato. 'Design of low-loss SAW filters employing distributed acoustic reflection transducers'. *IEEE Ultrasonics Symp.*, 1986, p.59-64.
171. K.Yamanouchi and H.Furuyashiki. 'New low-loss SAW filter using internal floating electrode reflection types of single-phase unidirectional transducer'. *Electronics Lett.*, **20**, 989-990 (1984).
172. M.Takeuchi and K.Yamanouchi. 'Coupled mode analysis of SAW floating electrode type unidirectional transducers'. *IEEE Trans. UFFC*, **40**, 648-658 (1993).
173. K.Hanma and B.J.Hunsinger. 'A triple transit suppression technique'. *IEEE Ultrasonics Symp.*, 1976, p.328-331.
174. P.V.Wright. 'The natural single-phase unidirectional transducer: a new low-loss SAW transducer'. *IEEE Ultrasonics Symp.*, 1985, p.58-63
175. T.Thorvaldsson and B.P.Abbott. 'Low loss SAW filters utilizing the natural single phase unidirectional transducer (NSPUDT)'. *IEEE Ultrasonics Symp.*, 1990, p.43-48.
176. E.G.Lean and A.N.Broers. 'S-band surface acoustic delay lines'. *IEEE Intl. Solid State Circuits Conf.*, 1970, p.130-131. **Also**, E.D.Wolf, F.S.Ozdemir and R.D.Weglein. *IEEE Ultrasonics Symp.*, 1973, p.510-516.
177. R.C.Williamson. 'Case studies of successful surface-acoustic-wave devices'. *IEEE Ultrasonics Symp.*, 1977, p.460-468.
178. C.S.Hartmann. 'Systems impact of modern Rayleigh wave technology', *in* E.A.Ash and E.G.S.Paige (eds.), *Rayleigh-Wave Theory and Application*. Springer, 1985, p.238-253.
179. I.Yakovkin. 'Investigation of fundamentals, current status and trends of SAW in Siberia'. *IEEE Ultrasonics Symp.*, 1991, p.1375-1381.
180. W.Buff. 'SAW technology in eastern Europe'. *IEEE Ultrasonics Symp.*, 1991, p.107-110.
181. M.F.Lewis. 'Surface skimming bulk waves, SSBW'. *IEEE Ultrasonics Symp.*, 1977, p.744-752.
182. B.A.Auld, J.J.Gagnepain and M.Tan. 'Horizontal shear surface waves on corrugated surfaces'. *Electronics Lett.*, **12**, 650-651 (1976)
183. T.L.Bagwell and R.C.Bray. 'Novel surface transverse wave resonators with low loss and high Q'. *IEEE Ultrasonics Symp.*, 1987, p.319-324.
184. I.D.Avramov. 'Gigahertz range resonant devices for oscillator applications using shear horizontal acoustic waves'. *IEEE Trans. UFFC*, **40**, 459-468 (1993). **Also**, I.D.Avramov, F.L.Walls, T.E.Parker and G.K.Montress. 'Extremely low thermal noise floor, high power oscillators using surface transverse wave devices'. *ibid.*, **43**, 20-29 (1996).
185. H.Engan, K.A.Ingebrigtsen and A.Tonning. 'Elastic surface waves in alpha quartz: observation of leaky surface waves'. *Appl. Phys. Lett.*, **10**, 311-313 (1967). **Also**, G.W.Farnell. 'Properties of elastic surface waves'. *in* W.P.Mason and R.N.Thurston (eds.), *Physical Acoustics*, vol. 6, Academic Press, 1970, p.109-166.
186. A.Takayanagi, K.Yamanouchi and K.Shibayama. 'Piezoelectric leaky surface waves in $LiNbO_3$'. *Appl. Phys. Lett.*, **5**, 225-227 (1970)
187. K.Nakamura, M.Kazumi and H.Shimizu. 'SH-type and Rayleigh-type surface waves on rotated $LiTaO_3$'. *IEEE Ultrasonics Symp.*, 1977, p.819-822.
188. K.Yamanouchi and M.Takeuchi. 'Applications for piezoelectric leaky surface waves'. *IEEE Ultrasonics Symp.*, 1990, p.11-18.
189. K.Yashiro and N.Goto. 'Analysis of generation of acoustic waves on the surface of a semi-

infinite piezoelectric solid'. *IEEE Trans.*, **SU-25**, 146-153 (1978). **Also**, M.Yamaguchi and K.Hashimoto. 'Simple estimation for SSBW excitation strength'. *J. Acoust. Soc. Jpn., (E)*, **6**, 51-54 (1985).

190. K.Y.Hashimoto, M.Yamaguchi and H.Kogo. 'Experimental verification of SSBW and leaky SAW propagating on rotated Y-cuts of $LiNbO_3$ and $LiTaO_3$'. *IEEE Ultrasonics Symp.*, 1983, p.345-349.

191. M.Ueda, O.Kawachi, K.Hashimoto, O.Ikata and Y.Satoh. 'Low loss ladder type SAW filter in the range of 300 to 400 MHz'. *IEEE Ultrasonics Symp.*, 1994, p.143-146.

192. P.D.Bloch, N.G.Doe, E.G.S.Paige and M.Yamaguchi. 'Observations on surface skimming bulk waves and other waves launched from an IDT on lithium niobate'. *IEEE Ultrasonics Symp.*, 1981, p.268-273.

193. O.Kawachi, G.Endoh, M.Ueda, O.Ikata, K.Hashimoto and M.Yamaguchi. 'Optimum cut of $LiTaO_3$ for high performance leaky surface acoustic wave filters'. *IEEE Ultrasonics Symp.*, 1996, p.71-76.

194. R.M.O'Connell and P.H.Carr. 'High piezoelectric coupling temperature-compensated cuts of Berlinite ($AlPO_4$) for SAW applications'. *IEEE Trans.*, **SU-24**, 376-384 (1977)

195. S.Fujishima. 'Piezoelectric devices for frequency control and selection in Japan'. *IEEE Ultrasonics Symp.*, 1990, p.87-94.

196. Y.Shimizu. 'Current status of piezoelectric substrates and propagation characteristics for SAW devices'. *Japan J. Appl. Phys.*, **32**, 2183-2187 (1993)

197. I.B.Yakovkin, R.M.taziev and A.S.Kozlov. 'Numerical and experimental investigation of SAW in langasite'. *IEEE Ultrasonics Symp.*, 1995, p.389-392. **Also**, K.S.Aleksandrov, B.P.Sorokin, P.P.Turchin, S.I.Burkov, D.A.Glushkov and A.A.Karpovich. 'Effects of static electric field and mechanical pressure on surface acoustic wave propagation in $La_3Ga_5SiO_{14}$ piezoelectric single crystals'. *ibid.*, p.409-412.

198. M.P. da Cunha and S. de A. Fagundes. 'Investigation of recent quartz-like materials for SAW applications'. *IEEE Trans. UFFC*, **46**, 1583-1590 (1999)

199. K.Yamanouchi, H.Odagawa, T.Kojima and T.Matsumura. 'Experimental study of super-high electromechanical coupling of surface acoustic wave propagation in $KNbO_3$ single crystal'. *Electronics Lett.*, **33**, 193-194 (1997). **Also**, K.Yamanouchi and H.Odagawa. 'Super high electromechanical coupling and zero temperature coefficient surface acoustic wave substrates in $KNbO_3$ single crystal'. *IEEE Trans. UFFC*, **46**, 700-705 (1999).

200. K.Yamanouchi, H.Odagawa, T.Kojima and Y.Cho. 'New piezoelectric $KNbO_3$ films for SAW device applications'. *IEEE Ultrasonics Symp.*, 1998, p.203-206.

201. Y.Cho, N.Oota, K.Morozumi, H.Odagawa and K.Yamanouchi. 'Quantitative study on the nonlinear piezoelectric effect of $KNbO_3$ single crystal for super highly efficient SAW elastic convolver'. *IEEE Ultrasonics Symp.*, 1998, p.289-292.

202. T.Sato and H.Abe. 'Propagation properties of longitudinal leaky surface waves on lithium tetraborate'. *IEEE Ultrasonics Symp.*, 1994, p.287-292.

203. T.Sato and H.Abe. 'Propagation properties of longitudinal leaky surface waves on lithium tetraborate'. *IEEE Trans. UFFC*, **45**, 136-151 (1998). **Also**, 'Propagation of longitudinal leaky surface waves under periodic metal grating structure on lithium tetraborate'. *ibid.*, **45**, 394-408 (1998). **Also**, 'SAW device applications of longitudinal leaky surface waves on lithium tetraborate'. *ibid.*, **45**, 1506-1516 (1998).

204. M.P. da Cunha. 'Extended investigation on high velocity pseudo surface waves'. *IEEE Trans. UFFC*, **45**, 604-613 (1998).

205. S.Tomabechi, S.Kameda, K.Masu and K.Tsubouchi. '2.4 GHz front-end multi-track $AlN/\alpha-Al_2O_3$ SAW matched filter'. *IEEE Ultrasonics Symp.*, 1998, p.73-76.

206. K.Yamanouchi, N.Sakurai and T.Satoh. 'SAW propagation characteristics and fabrication technology of piezoelectric thin film/diamond structure'. *IEEE Ultrasonics Symp.*, 1989, p.351-354.

207. H.Nakahata, A.Hachigo, K.Higaki, S.Fujii, S.Shikata and N.Fujimori. 'Theoretical study on

SAW charactersitics of layered structures including a diamond layer'. *IEEE Trans. UFFC*, **42**, 362-375 (1995)
208. A.Hachigo, D.C.Malocha and S.M.Richie. 'Characteristics of IIDT on ZnO/diamond/Si structures'. *IEEE Ultrasonics Symp.*, 1996, p.313-316.
209. D.L.Dreifus, R.J.Higgins, R.B.Henard, R.Almar and L.P.Solie. 'Experimental observation of high velocity pseudo-SAW's in ZnO/diamond/Si multilayers'. *IEEE Ultrasonics Symp.*, 1997, p.191-194.
210. Y.Shimizu, M.Tanaka and T.Watanabe. 'A new cut of quartz with extremely small temperature coefficeint of leaky surface wave'. *IEEE Ultrasonics Symp.*, 1985, p.233-236.
211. C.S.Hartmann and B.P.Abbott. 'Overview of design challenges for single phase unidirectional transducers'. *IEEE Ultrasonics Symp.*, 1989, p.79-89
212. J.M.Hodé, J.Desbois, P.Dufilié, M.Solal and P.Ventura. 'SPUDT-based filters: design principles and optimizaion'. *IEEE Ultrasonics Symp.*, 1995, p.39-50.
213. B.P.Abbott, C.S.Hartmann and D.C.Malocha. 'Matching of single-phase unidirectional SAW transducers and a demonstration using a low-loss EWC/SPUDT filter'. *IEEE Ultrasonics Symp.*, 1990, p.49-54.
214. M.Solal and J.M.Hodé. 'A new compact SAW low loss filter for mobile radio'. *IEEE Ultrasonics Symp.*, 1993, p.105-109.
215. J.Machui and W.Ruile. 'Z-path IF filters for mobile telephones'. *IEEE Ultrasonics Symp.*, 1992, p.147-150.
216. C.Ruppel, R.Dill, A.Fischerauer, G.Fischerauer, W.Gawlik, J.Machui, F Müller, L.Reindl, W.Ruile, G.Scholl, I.Schropp and K.Ch.Wagner. 'SAW devices for consumer applications'. *IEEE Trans. UFFC*, **40**, 438-452 (1993).
217. P.Ventura, M.Solal, P.Dufilie, J.M.Hode and F.Roux. 'A new concept in SPUDT design: the RSPUDT (resonant SPUDT)'. *IEEE Ultrasonics Symp.*, 1994, p. 1-6.
218. H.F.Tiersten and R.C.Smythe. 'Guided acoustic surface wave filters'. *IEEE Ultrasonics Symp.*, 1975, p.293-294. Also, 'Guided acoustic surface wave filters', *Appl. Phys. Lett.*, **28**, 111-113 (1976).
219. M.Tanaka, T.Morita, K.Ono, Y.Nakazawa. 'Narrow bandpass filter using double-mode SAW resonators on quartz'. 38th *Ann. Freq. Control Symp.*,1984, p.286-293.
220. M.F.Lewis. 'SAW filters employing interdigitated interdigital transducers, IIDT'. *IEEE Ultrasonics Symp.*, 1982, p.12-17.
221. M.Hikita, H.Kojima, T.Tabuchi and Y.Kinoshita. '800 MHz high-performance SAW filter using new resonant configuration'. *IEEE Trans.*, **MTT-33**, 510-518 (1985).
222. K.Anemogiannis, F.Mueller, W.Ruile and G.Riha. 'High performance low-loss SAW filters for mobile radio with improved stopband rejection'. *IEEE Ultrasonics Symp.*, 1988, p.77-82.
223. M.Ohmura, N.Abe, K.Miwa and H.Saito. 'A 900 MHz SAW resonator filter on $Li_2B_4O_7$ crystal'. *IEEE Ultrasonics Symp.*, 1990, p.135-138.
224. T.Morita, Y.Watanabe, M.Tanaka and Y.Nakazawa. 'Wideband low loss double mode SAW filters'. *IEEE Ultrasonics Symp.*, 1992, p.95-104.
225. M.Tanaka, T.Morita, K.Ono and Y.Nakazawa. 'Narrow bandpass double mode SAW filter'. *Proc. 15th E.M. Symposium*, Japan, 1986, p.5-10.
226. M.F.Lewis and C.L.West. 'Use of acoustic transducers as generalised electrical circuit elements'. *Electronics Lett.*, **12**, 1211-1212 (1985)
227. M.Hikita, T.Tabuchi, Y.Ishida, K.Kurosawa and K.Hamada. 'SAW integrated modules for 800-MHz cellular radio portable telephones with new frequency allocations'. *IEEE Trans. UFFC*, **36**, 531-539 (1989).
228. P.Ventura, J.M.Hodé, M.Solal and J.Ribbe. 'Numerical methods for SAW propagation characterization'. *IEEE Ultrasonics Symp.*, 1998, p.175-186.
229. M.Hikita, T.Tabuchi and N.Shibagaki. 'Investigation of new low-loss and high-power SAW filters for reverse-frequency-allocated cellular radio'. *IEEE Trans. UFFC*, **40**, 224-231 (1993).
230. O.Ikata, T.Miyashita, T.Matsuda, T.Nishihara and Y.Satoh. 'Development of low-loss bandpass filters using SAW resonators for portable telephones'. *IEEE Ultrasonics Symp.*, 1992,

p.111-115.
231. T.Matsuda, H.Uchishiba, O.Ikata, T.Nishihara and Y.Satoh. 'L and S band low-loss filters using SAW resonators'. *IEEE Ultrasonics Symp.*, 1994, p.163-167.
232. J.Heighway, S.N.Kondratiev and V.P.Plessky. 'Balance bridge SAW impedance element filters'. *IEEE Ultrasonics Symp.*, 1994, p.27-30.
233. Y.Satoh, T.Nishihara and O.Ikata. 'SAW duplexer metalizations for high power durability'. *IEEE Ultrasonics Symp.*, 1998, p.17-26.
234. F.Huang and E.G.S.Paige. 'Reflection of surface acoustic waves by thin metal dots'. *IEEE Ultrasonics Symp.*, 1982, p.72-82.
235. F.Huang and E.G.S.Paige. 'The scattering of surface acoustic waves by electrical effects in two-dimensional metal film structures'. *IEEE Trans. UFFC*, **35**, 723-735 (1988).
236. H.Bachl and A.R.Baghai-Wadji. '3D electrostatic field analysis of periodic two-dimensional SAW transducers with closed-form formulae'. *IEEE Ultrasonics Symp.*, 1989, p.359-362.
237. G.Visintini, A.Baghai-Wadji and O.Manner. 'Modular two-dimensional analysis of surface-acoustic-wave filters'. *IEEE Trans. UFFC*, **39**, Part I, p.61-72 (1992). Also, G.Visintini, C.Kappacher and C.C.W.Ruppel, *ibid.*, Part II, p.73-81.
238. O.Manner, K.C.Wagner and C.C.W.Ruppel. 'Advanced numerical methods for the simulation of SAW devices'. *IEEE Ultrasonics Symp.*, 1996, p.123-130.
239. G.Tobolka. 'Mixed matrix representation of SAW transducers'. *IEEE Trans.*, **SU-26**, 426-428 (1979)
240. Y.Koyamada and S.Yoshikawa. 'Coupled mode analysis of a long IDT'. *Rev. Electrical Comm. Labs.*, **27**, 432-444 (May/June1979).
241. E.Akçakaya. 'A new analysis of single-phase unidirectional transducers'. *IEEE Trans. UFFC*, **34**, 45-52 (1987).
242. P.V.Wright. 'A new generalized modeling of SAW transducers and gratings'. *43rd Annual Symp. on Freq. Control*, 1989, p.596-605.
243. B.P.Abbott, C.S.Hartmann and D.C.Malocha. 'A coupling-of-modes analysis of chirped transducers containing reflective electrode geometries'. *IEEE Ultrasonics Symp.*, 1989, p.129-134.
244. C.S.Hartmann and B.P.Abbott. 'Experimentally determining the transduction magnitude and phase and the reflection magnitude and phase of SAW SPUDT structures'. *IEEE Ultrasonics Symp.*, 1990, p.37-42.
245. D.P.Morgan. 'Reflective array modeling for reflective and directional SAW transducers'. *IEEE Trans. UFFC*, **45**, 152-157 (1998)
246. M.Koshiba and S.Mitobe. 'Equivalent networks for SAW gratings'. *IEEE Trans. UFFC*, **35**, 531-535 (1988). Also, K.Hasegawa and M.Koshiba. 'Finite-element solution of Rayleigh-wave scattering from reflective gratings on a piezoelectric substrate'. *ibid.*, **37**, 99-105 (1990).
247. Y.Suzuki, M.Takeuchi, K.Nakamura and K.Hirota. 'Couple-mode theory of SAW periodic structures'. *Electronics and Commun. Jap.*, part 3, **76**, 87-98 (1993).
248. Z.H.Chen, M.Takeuchi and K.Yamanouchi. 'Analysis of the film thickness dependence of a single-phase unidirectional transducer using the coupling-of-modes theory and the finite-element method'. *IEEE Trans. UFFC*, **39**, 82-94 (1992),
249. P.Ventura. 'Full strip reflectivity study on quartz'. *IEEE Ultrasonics Symp.*, 1994, p.245-248.
250. Q.Xue and Y.Shui. 'Analysis of leaky-surface-wave propagation under a periodic metal grating'. *IEEE Trans. UFFC*, **37**, 13-25 (1990).
251. K.Hashimoto and M.Yamaguchi. 'Derivation of coupling-of-modes parameters for SAW device analysis by means of boundary element method'. *IEEE Ultrasonics Symp.*, 1991, p.21-26.
252. C.Wang and D.P.Chen. 'Analysis of surface excitation of elastic wave field in a half space of piezoelectric crystal'. *Chinese J. Acoustics*, **4**, 233-243 (1985). Also, *ibid.*, p.297-313.
253. A.R.Baghai-Wadji. 'Scattering of piezoelectric surface transverse waves from electrodes with arbitrary h/λ'. *IEEE Ultrasonics Symp.*, 1989, p.377-380.
254. P.Ventura, J.M.Hodé and B.Lopes. 'Rigorous analysis of finite SAW devices with arbitrary

electrode geometries'. *IEEE Ultrasonics Symp.*, 1995, p.257-262. **Also**, P.Ventura, J.M.Hodé and M.Solal. 'A new efficient combined FEM and periodic Green's function formalism for the analysis of periodic SAW structures'. *IEEE Ultrasonics Symp.*, 1995, p.263-268.
255. E.Danicki. 'Propagation of transverse surface acoustic waves in rotated Y-cut quartz substrates under heavy periodic metal electrodes'. *IEEE Trans.*, **SU-30**, 304-312 (1983).
256. Y.Zhang, J.Desbois and L.Boyer. 'Characteristic parameters of surface acoustic waves in a periodic metal grating on a piezoelectric substrate'. *IEEE Trans. UFFC*, **40**, 183-192 (1993).
257. K.Hashimoto and M.Yamaguchi. 'Analysis of excitation and propagation of acoustic waves under periodic metallic grating structures for SAW device modeling'. *IEEE Ultrasonics Symp.*, 1993, p.143-148.
258. K.Hashimoto and M.Yamaguchi. 'Precise simulation of surface transverse wave devices by discrete Green function theory'. *IEEE Ultrasonics Symp.*, 1994, p.253-258.
259. V.P.Plessky. 'A two parameter coupling of modes model for shear horizontal type SAW propagation in periodic gratings'. *IEEE Ultrasonics Symp.*, 1993, p.195-200.
260. K.Hashimoto and M.Yamaguchi. 'General purpose simulator for leaky surface acoustic wave devices based on coupling-of-modes theory'. *IEEE Ultrasonics Symp.*, 1996, p.117-122.
261. J.Koskela, V.P.Plessky and M.M.Salomaa. 'Suppression of the leaky SAW attenuation with heavy mechanical loading'. *IEEE Trans. UFFC*, **45**, 439-449 (1998). **Also**, J.Koskela, V.P.Plessky and M.M.Salomaa. 'Analytic model for STW/BGW/LSAW resonators'. *IEEE Ultrasonics Symp.*, 1998, p.135-138.
262. B.P.Abbott and K.Hashimoto. 'A coupling-of-modes formalism for surface transverse wave devices'. *IEEE Ultrasonics Symp.*, 1995, p.239-245. **Also**, V.L. Strashilov, V.D.Djordjev, B.I.Boyanov and I.D.Avramov. 'A coupling-of-modes approach to the analysis of STW devices'. *IEEE Trans. UFFC*, **44**, 652-657 (1997)
263. C.S.Lam, D.P.Chen, B.Potter, V.Narayanan and A.Vishwanathan. 'A review of the applications of SAW filters in wireless communications'. *Intl. Workshop on Ultrasonic Applications*, Nanjing, China, Sept. 1996
264. J.Machui, J.Bauregger, G.Riha and I.Schropp. 'SAW devices in cellular and cordless phones'. *IEEE Ultrasonics Symp.*, 1995, p.121-130.
265. H.Fukushima, N.Hirasawa, M.Ueda, H.Ohmori, O.Ikata and Y.Satoh. 'A study of SAW antenna duplexer for mobile application'. *IEEE Ultrasonics Symp.*, 1998, p.9-12.

THIN-FILMS FOR SAW DEVICES

FRED S. HICKERNELL

Motorola Inc. (ret.) and the University of Central Florida
5012 E. Weldon, Phoenix, Arizona 85018, USA

Thin-films play a significant role in surface acoustic wave (SAW) technology. Thin metal films are essential for the generation and detection of SAWs on piezoelectric substrates, can provide for directional control, influence the velocity and propagation loss, and can be configured to produce a wide variety of time and frequency signal processing functions for electronic systems. Dielectric films have been used to selectively modify the propagation properties of surface waves, protect the surface, and improve the coupling efficiency and temperature characteristics of SAW devices. Piezoelectric films have extended SAW device developments to amorphous and nonpiezoelectric crystals for enhanced frequency and loss performance, lower cost, and microelectronic integration. Semiconductor, optical, and magnetic films have shown promise for active acoustoelectronic, acoustooptic, and magnetoacoustic devices. Chemically sensitive films are the basis for SAW sensor developments. Active research continues today in improving the properties and utilization of thin films on SAW substrates through advances in processing and characterization.

1. Introduction

Surface Acoustic Wave (SAW) technology has moved from its early beginnings where half-wavelength spaced interdigitated thin-film aluminum electrodes on quartz formed a basic nondispersive delay line to where a wide variety of mass producible delay lines, bandpass filters, and matched filters, using a plethora of complex periodic thin-film metal patterns, are being incorporated in consumer, industrial, and government electronic systems. Recent articles on SAW devices, review their history and success in the electronic world.[1-3] Increased performance requirements have placed more stringent demands on the thin-films essential for SAW devices. Metal films are being developed to maintain high conductivity for high frequency devices while resisting electromigration and stress migration under high power application. Dielectric and piezoelectric films have been developed which extended the basic capabilities of SAW devices and are being used in quantity production. Dielectric films continue to be investigated for customizing the coupling factor, k^2, and simultaneously adjusting the temperature coefficient of frequency (TCF) on piezoelectric substrates. Piezoelectric films have shown promise for use with high velocity substrates to achieve high frequency SAW filters while maintaining reasonable electrode linewidth geometries. All three types of films have been shown to be useful in increasing coupling factor and reducing propagation loss for generalized SAW (GSAW) and higher velocity pseudo-SAW (PSAW) modes. We continue to increase our understanding of the properties and advantages of the use of thin films through theoretical and experimental characterization. Product applications demand that we continue to identify applications for thin-film technology to further improve the capabilities of SAW devices for future use in communications and signal processing applications.

This paper discusses the role of thin-films in SAW technology, how films affect SAW excitation and propagation, and where thin-films can be successfully applied and used in SAW devices. We can enhance processing techniques leading to improved acoustic properties of thin films through design of experiment strategies and simple measurement techniques. There will be an emphasis on the understanding of the SAW properties of thin films and their application through the results of experimental characterization which can be carried out using simple interdigital transducer (IDT) electrode structures. Thin films in this paper are defined as those films less than an acoustic wavelength and are often a small fraction of an acoustic wavelength, such as metal films which are normally less than 0.05 wavelength. In Section 2 there is an overview of the applications of thin-film metals, dielectrics, and piezoelectrics. Section 3 discusses a simple measurement technique for characterizing thin films which will be used throughout the paper to illustrate the effects of film layers on substrates. Sections 4, 5, and 6 illustrate the propagation properties of metal, dielectric, and piezoelectric films respectively, indicating SAW device enhancements. Section 7 illustrates how all three types of films can enhance the propagation properties of Peudo-SAW substrates. Section 8 discusses piezoelectric films on high velocity substrates and gives a specific example of zinc oxide on silicon carbide. Section 9 highlights some opportunities for research and development in which films may impact SAW technology in the future. The paper closes with acknowledgments and references.

2. Applications of Thin Films

2.1. Metal films

Table 1 lists several of the basic functions of thin metal films in SAW technology. Metal films are used to generate, detect, and control the propagation of SAWs on piezoelectric substrates. Every basic SAW device used for electronic applications consists of a periodic metal interdigitated electrode in conjunction with a polished piezoelectric substrate or a piezoelectric film on a nonpiezoelectric substrate. The usefulness of a thin periodic metal interdigital electrode on a polished piezoelectric plate for SAW generation/detection was identified in early experiments by Voltmer and White.[4] The thin-film interdigital electrode and its variations in spacings and amplitude are key to SAW devices and represented the concept that propelled SAW technology into the limelight for electronic signal processing. Typically the wavelength period of the electrodes extends from 100 micrometers to 2 micrometers, encompassing a frequency range from around 40 MHz to 2.0 GHz using standard quarter-wavelength width electrodes. The interdigital electrode with its two degrees of freedom in overlap and spacing produced the basic nondispersive and dispersive delay lines, bandpass filters, resonators, and matched filters for communication and signal processing applications. Connected and floating electrodes within the interdigital electrode provided for directional excitation of the surface waves.

In addition there may also be a continuous or patterned metal film in the propagation path between the interdigital patterns that generate and detect SAW waveforms. Continuous metal films can be used to determine basic substrate and film properties, improve SAW device performance characteristics, and provide for film guidance and SAW energy concentration. Patterned films are used to interrupt the linear flow of the surface wave and redirect all or a portion of the energy. SAW resonators rely on the distributed reflections which come from quarter-wavelength spaced stripes normal to the direction of propagation. Angled quarter-wavelength stripes selectively redirect the wave usually in the performance of a signal processing function. Other patterns have been used to selectively redirect the SAWs such as dots and squares. The multistrip coupler displaces the wave through the induced current in the metal film stripes on the piezoelectric substrate. Section 4 discusses the characteristics of metal films and their applications in greater detail.

Table 1. Thin Metal Patterns on a Piezoelectric Plate for SAW Device Development.

PATTERN TYPE	SAW DEVICE SUCCESS
Interdigital Electrode	
Periodic interdigital electrode	Delay lines, filters, oscillators
Amplitude weighting, withdrawn electrodes	Bandpass filter shape
Graded electrode spatial separation	Dispersive delay lines
Aperiodic electrode reversals	Matched filter correlator
Connected - floating internal electrodes	SPUDT, DART, FEUDT - reduced loss
Contiguous interdigital electrodes	IIDT, GUDT for reduced loss
Transverse coupled electrodes	Low loss narrowband resonator filters
Continuous metal film in SAW path	
Ultra-thin conductive metal film	Determine film and substrate properties
Ultra-thin moderate resistivity film	Acoustoelectric damping of SAW
Split metal path	Bulk wave suppression
Metal waveguiding channels	Waveguides, (microstrip components)
Horn and reduced width strip	Nonlinear elastic-wave convolvers
Patterned metal film in SAW path	
In-line quarter wavelength reflector stripes	Resonators and resonator filters
Angled quarter wavelength reflector stripes	Reflective array devices, ring filters
Dots, squares, triangles, etc.	Reflective array frequency discriminator
Multistrip coupler	Coupler, tap, reflector, concentrator

2.2. Dielectric films

Table 2 lists a number of the useful functions attributed to the application of dielectric films on SAW substrates. A thin amorphous dielectric film, such as glass, can be deposited on a bare piezoelectric wafer to provide the functions of surface passivation, reduction of pyroelectric effects, and smoothing to reduce propagation loss. It can also be used to modify the coupling factor, reduce the temperature coefficient of frequency and put the wafer in a condition where higher order SAW and Pseudo-SAW modes can be used. Thin glass films have also been used as a practical matter on electroded SAW components as a protective coating to prevent shorting by metal particles, to reduce metal migration due to electric and acoustic fields, and to protect environmental degradation due to chemicals and corrosive vapors. A thin dielectric film can be used very effectively for frequency trimming. The presence of a uniform dielectric film layer on the surface of a SAW substrate also alters the propagation characteristics of the surface wave which can provide an enhancement of substrate properties. It was determined early in the investigation of SAWs that there were velocity, loss, coupling factor, and temperature coefficient improvements to be gained through the use of thin films on piezoelectric substrates.[5] While this kind of film technology has been used sparingly in SAW device product development, it still represents a potential source for enhancing such products. For example, a glass film which has a strong positive temperature coefficient of frequency

(TCF = 80), when combined with a high coupling factor piezoelectric substrate with a negative TCF, can improve and even reduce to zero the first order temperature coefficient without reducing the overall coupling efficiency. For certain high dielectric constant oxide and nitride films with appropriate elastic constants, the coupling factor can be doubled.

The presence of a dielectric film layer also presents the possibility for Love wave propagation, low loss waveguidance, higher order SAW modes, and the excitation of Pseudo-SAW modes. Love wave propagation and the accompanying velocity dispersion of a glass film on silicon has been used to develop large time bandwidth dispersive delay lines.[6] Because dielectric films will have lower propagation loss than metal films they are candidates for patterning to produce microacoustic strip-line-type components. The higher order SAW modes increase the frequency capability of the SAW propagation for a given transducer electrode spacing and may result in the added benefits of higher coupling, lower loss, and reduced TCF. The presence of a dielectric film can reduce to essentially zero the propagation losses due to leaky waves over selected frequency regions.

Examples of the measured SAW propagation properties of a class of PECVD dielectric films commonly used in semiconductor integrated circuits is described in Section 5. These films can serve to implement the functions shown in Table 2.

Table 2. Dielectric Films on a Piezoelectric Plate for SAW Device Development.

FILM TYPE	FUNCTION	SAW DEVICE SUPPORT
I. Bare wafer coverage		
Thin amorphous film	Surface passivation	Reduce pyroelectric effect, loss
Amorphous film	Surface isolation	Modify coupling factor
II. Pattern coverage		
Amorphous glass	Surface protection	Reduce mech., environ. degrade
Positive TCF film	Reduce TCF	Temperature stable SAW devices
High dielectric film	Increased coupling factor	Low loss and wide band devices
Thin dielectric film	Frequency trimming	Narrow-band frequency accuracy
High hardness film	Suppress metal migration	High power tolerant devices
III. Prop./excitation		
Slow wave film	Love wave propagation	Dispersive delay lines
Patterned film	Waveguides	Microstrip components
Customized films	Improved GSAW/ PSAW	Higher k^2, high freq., low loss

2.3. Piezoelectric films

Piezoelectric thin films have been in use for over three decades as high frequency bulk acoustic wave (BAW) and surface acoustic wave (SAW) transducers and resonators, and as a SAW medium for acoustooptic, acoustoelectronic, and sensor interactions. Table 3 gives some examples of how piezoelectric films, particularly ZnO films, have impacted SAW devices. Sputtered aluminum nitride (AlN) and zinc oxide (ZnO) films have been the dominant films investigated for acoustic wave device applications. Following the introduction of the interdigital transducer in 1965, thin piezofilms were used for the excitation of surface waves on non-piezoelectric substrates such as glass, silicon, and sapphire. This led to the introduction of zinc oxide transducer based surface acoustic wave (SAW) filters for consumer and communication products.[7] Piezoelectric films have been used with piezoelectric substrates to enhance coupling factors, k^2, and temperature coefficients of frequency (TCF). Piezoelectric films have been used in conjunction with nonpiezoelectric and piezoelectric substrates with optical waveguides for device demonstrations which incorporate acoustooptic interactions.

Table 3. Piezoelectric Films on SAW Substrates for Device Development.

FILM/SUBSTRATE COMBINATION	PRODUCTS (DEMONSTRATED)
Piezofilms on nonpiezoelecrtric substrates	(Efficient SAW excitation/detection)
Piezofilms on piezoelectric substrates	(Increased coupling factor)
Piezofilms on high velocity nonpiezo-subs.	(High frequency low-loss devices)
Piezofilms on semiconductor substrates	(Integrated circuits with SAW devices)
Zinc oxide/glass	Television filters
Zinc oxide/sapphire	Cellular phone filters
Zinc oxide/silicon	Acoustoelectronic convolvers
Zinc oxide/quartz	(Temperature stable filter)
ZnO/diamond/silicon	(High frequency, low loss filters)
Glass/ZnO/diamond/silicon	(High freq., low-loss, TCF stable filter)
Aluminum nitride/sapphire	(High freq., low-loss, TCF stable filter)
Aluminum nitride/silicon	(High freq., resonators and convolvers)
Zinc oxide/optical waveguide film/silicon	(Integrated AO spectrum analyzer)
Zinc oxide/gallium arsenide	(Optical waveguide signal processor)

Basic work with piezofilms continues today toward the realization of the integration of SAW devices with microelectronic integrated circuits. Experimental discovery has given way to theoretical prediction with the availability of matrix based layered film programs. The theoretical predictions lead to experimental verification and if necessary the refining of the theory or adjustment of the films parameters. In the development of process parameters for obtaining high quality piezofilms, there is less trial and error and more design of experiments and statistical analysis being used.[8] This has led to a more comprehensive understanding of which deposition process parameters are the most critical. Obtaining a best film result, which indicates capability, is being replaced by achieving reproducible results. Hand in hand with this is the movement from laboratory depositions to manufacturing products with production type deposition equipment common to the semiconductor industry. With the availability of a wide selection of microstructure measurement techniques we have moved from extrinsic measurements to understanding intrinsic film properties and how they affect acoustic properties. Understanding intrinsic acoustic properties is equally important for dielectric and metal films as well. Section 6 gives examples of the characteristics of zinc oxide films on gallium arsenide and aluminum nitride films on silicon substrates substrates. Zinc oxide has also been used extensively with silicon and high velocity substrates taking advantage of higher order SAW modes.

3. SAW Measurements of Film Properties

Because SAW generation and propagation requires the mechanical movement of material, it is natural that a significant correlation is found between film structure and the SAW properties of velocity, loss, and coupling factor. In SAW propagation through a film layer there is a combination of compressional-tensile, and shearing motions. The shearing motions dominate Rayleigh-type SAW propagation and therefore it is the lateral integrity of the film that is the most important. Since sputtered polycrystalline films tend to grow with a columnar structure, they are most susceptible to lateral discontinuities such as voids, grain boundaries, and dislocations, which disrupt the elastic continuity. Thus, the elastic properties of films can be significantly different from the elastic constants of their bulk counterparts. Using bulk constants to characterize film properties can lead to erroneous results in the design of acoustic devices.

All this has placed new demands on film measurement techniques and requires a more careful selection of the measurements to be used and their relationship to the desired application. In optimizing film deposition conditions for process control and reproducibility, a commonly used measure is the fabrication of a simple transducer test structure and a measurement of electrical parameters and response characteristics. Acoustic device design parameters such as velocity, coupling factor, loss, dielectric constant, and temperature coefficient are best obtained using transducer and resonator patterns which approximate the device being developed. To lead to a better understanding of the intrinsic capabilities of the films, the influence of film microstructure on fundamental acoustic properties is essential. There are several measures for characterizing film structure including x-ray diffraction, electron beam diffraction and imaging, atomic force microscopy, Auger, and others which can be correlated with acoustic properties. To determine all of the fundamental elastic, viscosity, and permittivity constants for dielectric films and piezoelectric constants for piezofilms, requires selective film orientations, and independent electric field and acoustic wave propagation mode control. This has been possible where different surface wave modes have been excited on different substrate orientations with different film thicknesses and the resulting characteristics matched by adjusting the basic constants. The original constants for aluminum nitride were found by this technique.[9] Zinc oxide film constants have similarly been determined.[10]

The author has used a simple split-finger interdigital transducer and its harmonic responses to determine the effects of films on SAW substrates. Linear arrays of 100 nm thick thin-film aluminum interdigital electrodes are photolithographically patterned on the SAW substrate. The center-to-center spacing between transducers is 3.81 mm. Each transducer has 10.5 split-electrode finger pairs with a periodicity of 100 micrometers, and a 2.54 mm aperture. Individual split-finger electrodes within a transducer are 15 micrometers wide, and the spaces between adjacent electrodes are 10 micrometers. This transducer pattern facilitates the excitation of a wide range of harmonic waves and permits the delineation of SAW velocity and loss characteristics over an extended frequency region. Frequency and loss measurements between transducer pairs are made using a network analyzer test set with time gating and signal averaging to enhance signal sensitivity at the higher frequencies. In general, measurements are made over the frequency range from 30 MHz to over 1.0 GHz using a HP 8753 network analyzer with a HP 85046 S-parameter test set.

For each film sample the center frequencies of the fundamental and the higher harmonic responses are measured. From the frequency information the SAW phase velocity is calculated and from the insertion loss measurements the transducer conversion loss and propagation loss can be determined. A velocity dispersion characteristic is developed as a function of film thickness-to-wavelength ratio from which elastic constants can be determined. Insertion loss measured between transducer pairs with different separations determines the SAW propagation loss as a function of frequency. Eight pairs of transducers with sequentially increasing separation are normally used. The linear change in loss with increasing transducer spacing is dominated by the propagation loss whether viscous or leaky-wave. A least-squares linear regression of the data points is used to establish a dB/cm value for each frequency of measurement. From the conversion loss the coupling factor can be estimated. The properties can be determined as a function of temperature by use of an oven or precision temperature forcing system.

As an example of a representative network analyzer plot from which SAW velocity and loss parameters are experimentally determined, the frequency vs. amplitude spectrum for ST-X quartz, is shown in Figure 1. The Y-rotated, X-propagating, (ST-X) cut of quartz (Euler angles, 0°, 132.75°, 0°) is used extensively for the development of surface acoustic wave (SAW) devices where temperature stability is of primary importance. Besides the commonly used generalized SAW (GSAW) mode, theoretical considerations have shown that this crystal cut also supports a pseudo-SAW (PSAW) and high velocity pseudo-SAW (HVPSAW) mode.[11,12] The GSAW, PSAW, and HVPSAW modes are clearly evident in Fig. 1 with the strong GSAW and HVPSAW modes observed over a broad frequency range which extended above 1 GHz. The PSAW mode is limited to the low frequency region because of the high value of its leaky-wave loss. The identification of the modes was determined by their frequency, which translates to a particular velocity and harmonic number, and also by their presence or absence under the application of a dampening material on the surface. Dominant displacements of the PSAW and HVPSAW modes are in plane of the substrate and are not as strongly affected by damping of the surface. The respective velocities of the modes calculated from the experimental frequency spectrum of Fig. 1 together with the theoretical values are shown in Table 4. The correlation is very good, within 0.1%.

Table 4. Comparison of Measured and Calculated SAW Velocity Values for ST-Quartz.

SAW Mode	Theoretical Velocity (meters/second)	Experimental Velocity (meters/second)
Generalized SAW (GSAW)	3152	3153
Pseudo-SAW (PSAW)	5078	5072
High Velocity PSAW (HVPSAW)	5745	5746

The GSAW, which has a large component of displacement normal to the substrate, is completely dampened (> 40 dB) by an elastically soft material on the surface. The PSAW and HVPSAW, which have strong shearing and compressional motions respectively in the plane, had reduced average amplitudes of only 8 dB for the PSAW and 4 dB for the HVPSAW when dampened with rubber cement. In Fig. 2 the SAW propagation loss of the three modes in dB/cm is plotted as a function of frequency. The GSAW mode loss increases as the frequency squared and the HVPSAW propagation loss linearly with frequency. The PSAW loss, which is strongly leaky, is above 20 dB/cm at 200 MHz.

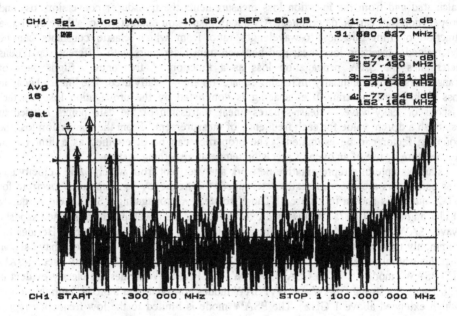

Fig. 1. Network analyzer transmission frequency spectrum of SAW modes generated and detected by paired interdigital transducers on ST-Quartz.

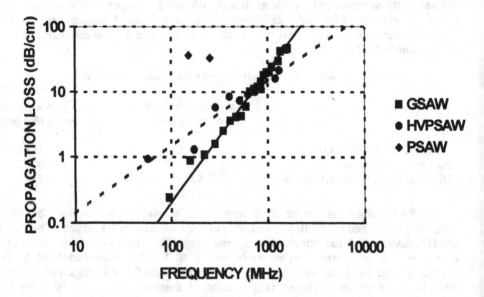

Fig. 2. SAW propagation loss for GSAW, PSAW, and HVPSAW modes on ST-quartz. The continuous line is a frequency squared dependence for the GSAW mode and the dashed line indicates a linear frequency dependence for the HVPSAW mode.

In Sections 5, 6, 7, and 8 the application of this technique to determining the SAW propagation properties of dielectric and piezoelectric films is illustrated. To determine the properties of dielectric films it is convenient to grow the films on a piezoelectric substrate. The piezoelectric films have their own means for SAW generation and detection and can be measured on nonpiezoelectric substrates. It is through this means of experimental measurement that the basic acoustic properties of films can be determined, theoretical expectations can be confirmed, new modes of propagation uncovered for investigation and process techniques refined. First though a discussion of the characteristic properties of metal films with the focus on the properties of vacuum deposited aluminum.

4. Characteristics of Metal Films

4.1. *The interdigital electrode*

Table 5 gives a list of properties that should be considered in achieving an ideal transducer electrode metallization. Aluminum, a low mass metal of good conductivity, chemically stable, easy to deposit and pattern, and a good acoustic impedance match to quartz and lithium niobate, has been the thin-film metal of choice over the years. Other metals such as gold have been used, particularly where the device requires a high reflectivity from mass loading on a low coupling factor piezoelectric substrate. As power levels increased it was essential to address the adverse effects of electromigration (mass transport due to high current densities) and acoustomigration effects (mass transport of metal due to high mechanical stress). It is common knowledge that sputtered polycrystalline aluminum is superior to evaporated polycrystalline aluminum in reducing metal migration. Stress and aging become an important feature in narrowband resonators and delay lines as applied to SAW oscillators. The source of low frequency phase noise in SAW devices can be affected by the type of alloy used in the aluminum, and copper additions have been useful in this regard. The temperature characteristics of the metal must be understood and accounted for, particularly for use with temperature stable substrates. Even the shape of the metal edges of the aluminum stripes and metal propagation loss become important in high frequency devices which are dominated by numerous finger pairs. Theoretical and experimental modeling of the effect of linewidth shape and space to gap ratio on the response of high frequency filters and resonators is an important area of investigation.

Table 5. The Ideal SAW Electrode Metallization

- Easy to deposit and pattern by standard semiconductor processing techniques
- High electrical conductivity for low resistive losses
- Good adhesion and chemically stable
- Good acoustic impedance match to the substrate
- Low propagation loss
- Electro- and acousto-migration resistant
- Good temperature coefficient of frequency
- Stress free and low noise

Metal failure modes are critical for all SAW devices, particularly high frequency filters for hand-held telecommunication applications, where the electrodes must withstand power levels near 1 Watt. When there is a concentration of current flow, stress energy from the acoustic wave, such as occurs in resonators, or where a high amplitude voltage pulse appears across the electrodes, the filter will degrade and fail due to electrode metal opens or shorts[13]. In the case of electro- and acousto-migration, the electrodes may deteriorate over a long period of time, whereas, electrical static or dynamic discharge is instantaneous. These failure modes have been investigated since the early 1970's and recently with telecommunication duplexer requirements for lower loss, higher frequencies, and power, the issues have become more critical.[14] In terms of structuring aluminum metal films to reduce migration effects, good progress has come from impurity introduction, sandwich structures of aluminum with refractory metals, and epitaxial growth. Electromigration and acoustomigration effects are highly dependent on film microstructure and the combination of sputtering and alloying produces a smaller more uniform grain size with tight grain interconnection which reduces migration effects. Epitaxial growth of aluminum films has been highly successful in reducing migration effects where the substrate lattice structure permits such growth.[15] The investigation of metal film structures which resist electromigration and acoustomigration is a very active area of research and development among SAW manufacturers.

4.2. Continuous metal film

A continuous thin-film metal on a substrate will affect the velocity and attenuation of the surface acoustic wave. For example a metal on a piezoelectric substrate shorts the electric field at the surface and lowers the SAW propagation velocity. This gives rise to important effects such as waveguiding, surface energy concentration, bulk-wave suppression and propagation loss decrease or increase. Experimentally certain acoustic and electronic properties of the thin metal film and the substrate can be determined.

Pouliquen and Vaesken investigated the effects on surface waves of continuous metallic thin films of copper, gold, and aluminum as they grew on Y-cut quartz.[16] There were three distinct regions of influence related to the growth, electrical conductivity, and mass of the films. During the initial growth phase (<5 nm), where discontinuous metal islands were formed, there was a small increase attenuation and a slight decrease in phase velocity. As the metal islands connect and the electrical resistance decreases, a region of rapid velocity change and increased attenuation occurs whose magnitudes are dependent upon the coupling factor of the piezoelectric substrate. This occurs for film thicknesses in the range of 5 to 10 nm and is associated with the acoustoelectric effect.[17] Beyond 10 nm the velocity and propagation loss come under the influence of the mass and elastic properties of the film. The initial velocity for the very thin films changes in a linear fashion with thickness and comparison of different films have shown that this change corresponds to metal mass ratios.

Some very practical and basic information can come from a careful study of the velocity and attenuation changes in the region of acoustoelectronic interaction with film thicknesses from 5 to 10 nm. The sharp velocity transition from an unmetallized to metallized surface on a piezosubstrate is the $\Delta v/v$ which defines the coupling factor of the substrate. The attenuation characteristic, which takes the shape of a Gaussian distribution, was used by Bierbaum to obtain the mobility of the films for a comparison with bulk mobilities for seven different metal films.[18] For high conductivity materials the maximum attenuation on quartz was similar but decreased significantly for the low mobility films measured. A practical application of the acoustoelectric attenuation has been the use of an ultra-thin film 6-8 nm for reducing end reflections on high coupling factor materials.

The shorting effect and reduction in SAW velocity has been used for a number of practical device applications. On a strong piezoelectric substrate such as lithium niobate the difference between the velocity of the unmetallized and metallized is 5% and provides a channel for a linear SAW waveguide with width dimensions of only a few wavelengths. Such waveguides, which concentrate the SAW energy from a broader aperture, have been indispensable in the development of higher-efficiency nonlinear elastic convolvers.[19] A metal horn can be used to transition the film from the wider aperture interdigital electrode to the narrower width waveguide. The difference in velocity has also been used very effectively in the suppression of bulk-wave spurious signals in SAW filters.[20] Optic-like metal films such as prisms and lenses have been used to enhance the performance of SAW devices.[21]

One of the original visions for SAW signal processing was the control and confinement of surface waves through waveguides which relied on structures where the propagation velocity in the guiding structure was reduced to a value less than the free surface velocity and any bulk mode velocities. Thin-film guides, where the velocity reduction is effected by the elastic or electric properties of a deposited film were investigated using a heavy mass material such as gold.[22] An early idea was to develop "microsonic" circuits, a highly compact SAW circuit technology which were analogous to microwave microstrip circuits.[23] These circuits were envisioned as having guidance, splitting, coupling, short circuit, and open circuit functions using surface waves on a substrate within distances much less than their microwave counterparts. While considerable speculation and theoretical work was given to this concept, in experimental practice only a few of the device functions were demonstrated. The major barrier was radiation losses in trying to confine the elastic wave energy in curved and terminated (open) structures.

There has been considerable interest in the velocity change and propagation loss of films. Generally the film is thin and the losses are inherent to the substrate. However, in some cases where thicker films are required there is an interest in their effect on velocity and attenuation. Figure 3 shows the velocity and loss characteristic for a gold film on quartz. The data is taken from papers by Cambon and co-workers.[24,25] The velocity and loss are plotted as a function of the film thickness to acoustic wavelength ratio. As film thickness increases the velocity shows a linear decrease in value proportional to the mass loading while the loss in dB per wavelength shows a near linear increase. The loss values were taken at three different frequencies, yet there is good continuity of values when normalizing the propagation loss to film thickness to wavelength ratio.

The relative change in velocity and the loss of other metal films has been measured on different substrates. Table 6 gives some examples of these measurements which are normalized. The normalized velocity values are taken in the linear region of velocity change using a slope parameter measure ($\Delta v/v$) divided by the film thickness to wavelength ratio (t/λ). A comparison is made to the equivalent measure using aggregate elastic constants. As seen, these values are higher than the measured thin film values indicating reduced elastic constant values and possibly reduced density. The column showing the fractional change in the c_{44} constant was determined by assuming small changes in the compressional constant and density. As has been indicated by Jelks and Wagers in their measurements on aluminum and molybdenum, the largest percentage reduction in elastic constants comes from the shear constant c_{44} because of the nature of the thin film growth.[26]

It is interesting to note that in their studies the Poisson ratio remained reasonably constant which can be helpful in defining the two elastic constants if a separate measurement of density is obtained. The propagation losses at a film-thickness to wavelength ratio of 0.02

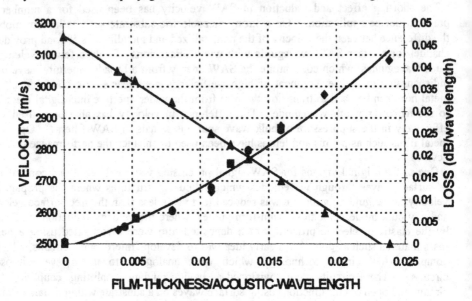

Fig. 3. The SAW velocity and propagation loss change for a deposited gold film on Y-X quartz as a function of film-thickness to acoustic-wavelength ratio.[24,25]

are lowest for the lighter mass materials such as aluminum and increase for copper and gold. An accurate measurement of the elastic properties and attenuation in films is important in the design of SAW devices, particularly where trade-offs must be made between choice of films, thickness for achieving certain velocities, and the resulting propagation losses which will occur. More definitive measurements of this nature need to be made in accurately predicting the performance of SAW devices.

Table 6. SAW velocity change and attenuation for thin-films on piezoelectric substrates The velocity change $\Delta v/v$ and dB/λ loss are taken at the thickness to wavelength ratio 0.02.

Film	Substrate	$(\Delta v/v)/(t/\lambda)$ Experimental	$(\Delta v/v)/(t/\lambda)$ Theoretical	$(\Delta c_{44}/c_{44})$	Loss* (dB/λ)	Reference
Al	Y-X quartz	-0.29	-0.23	-0.13		24,25
Au	Y-X quartz	-9.4, -9.2	-7,4	-0.9	0.035	24,25,16
Ag	Y-X quartz	-3.8	-3.4	-0.15	0.020	24,25
Cu	Y-X quartz	-3.2, -4.4	-2.9	-0.5	0.016	24,25,16
Al	128 YX LNB	-0.34	-0.22	-0.13		26
Mo	128 YX LNB	-2.2	-1.0	-0.66		26
Al	YZ LNB				0.0015	27
Al	37.5 Qtz				0.0032	28

* Propagation loss taken at a film-thickness to acoustic-wavelength ratio of 0.02

4.3. Patterned metal film

Placing a patterned thin film metal in the SAW propagation path will selectively alter the direction and amplitude of the surface wave. The concept of metal gratings as reflectors was set forth by Ash in 1970.[29] Since SAWs do not perfectly reflect at an edge boundary, (there is a decomposition into bulk and surface waves), they cannot be fabricated in the same form as the common bulk wave plate resonator. It is necessary to create a collective reflector made up of quarter-wavelength periodic discontinuities to provide the reflection mechanism. Such reflectors if fabricated carefully can achieve the very high Q-factors which approach the basic Q of the material being used. The easiest fabrication method is to use deposited metal for the gratings and transducers. The key advantage of the SAW resonator is its capability for operation at fundamental frequencies ten to one hundred times that of corresponding fundamental mode bulk acoustic wave (BAW) resonators. Quartz is commonly used for SAW resonators because of its high material Q (low SAW propagation loss) and its temperature stability.

While quarter wavelength metal stripes normal to the propagation will result in a distributed reflector, by angling the stripes and adjusting their number, reflective array devices can be developed. Williamson has reviewed the properties and applications of reflective array devices which have included resonators, oscillators, tapped delay lines, bandpass filters, filter banks, and dispersive delay lines.[30] The reflection gratings which have found the most use for signal processing are those which are angled near 45 degrees to the propagation path and reflect the surface waves 90 degrees to the original propagation path. The reflection mechanisms can be due to topographic features such as ion milled grooves or films with mass loading and piezoelectric shorting features such as metal stripes. Finally, other metal patterns such as dots and squares have been used to redirect the traveling surface waves and produce useful device functions.[31]

A number of useful functions are developed from in-line stripe metal patterns constituting a multistrip coupler (MSC). In this case the propagating wave is displaced by electrical coupling along its propagation path and coupling, focusing, tapping, and unidirectional transduction are a number of the useful functions produced. The SAW MSC was first introduced by Marshall and Paige in 1971.[32,33]

5. Dielectric Films

The past two decades have shown considerable progress in the development of equipment and processes for the fabrication of metal and dielectric thin films used in semiconductor devices. The emphasis has been on producing films whose electrical and dielectric properties meet the electronic circuit requirements. Little attention has been paid to the acoustic properties of such films. With the progress made in the deposition of piezoelectrically active zinc oxide and aluminum nitride films, there has been renewed interest in the possibilities for integrating acoustic wave devices with electronic circuitry on semiconductor substrates such as silicon and gallium arsenide. In order to properly design integrated acoustic devices, such as resonators, filters, and delay lines, it is necessary to characterize the acoustic properties of supporting dielectric film layers and the accompanying substrates which will be used in device development. Basic surface-acoustic-wave (SAW) measurements can be used for the development of an acoustic-material parameter database for dielectric films presently used by the semiconductor industry. In the following two sections examples of SAW propagation measurements on commonly used PECVD dielectric films and sputtered glass on gallium arsenide as measured by the author are described.

5.1. *PECVD dielectric films on gallium arsenide*

The acoustic properties of TEOS glass, silicon oxynitride (SiON), silicon nitride (SiN), and silicon carbide (SiC), typical of films used for passivation and dielectric isolation on semiconductor substrates, have been investigated.[34] The growth conditions of the films are shown in Table 7. Figure 4 shows the velocity dispersion characteristics of these four different PECVD films deposited to a thickness of 500 nm on gallium arsenide, determined from transmission frequency spectra like that shown in Fig. 1 using the multiharmonic transducer electrode deposited on the upper film surface. The SAW waves propagate in the (110) direction of (001) GaAs. The films all stiffen the surface and give an increasing SAW velocity with film-thickness to acoustic-wavelength ratio, with TEOS glass having the smallest velocity increase and SiC the greatest. The experimentally determined velocity dispersion data were fit with a calculated theoretical dispersion curve to determine the two independent elastic moduli for the amorphous films. The theoretical velocity dispersion was calculated using software described and provided by Adler and coworkers from McGill University.[35] The material constants for GaAs used in these calculations were those given by Slobodnik and coworkers.[36] A relative dielectric constant was specified, and the measured density of the films calculated from an independent measurement of the substrate mass difference with and without the film layer. The mass-loading, electrical-shorting, and stored-energy effects of the transducer on SAW velocity dispersion were considered in the calculations. The small SAW coupling factor for GaAs, the use of thin electrodes, and the extended frequencies of measurement minimized these effects.

Table 7. Plasma enhanced chemical vapor deposition (PECVD) conditions for silicon nitride, silicon oxynitride, silicon carbide, and TEOS glass.

Parameter	SiN	SiON	SiC	TEOS
Gases	SiH_4 NH_3 N_2	SiH_4 NH_3 N_2O N_2	SiH_4 CH_4 N_2	TEOS:5l/min O_2
Temperature (^{0}C)	380	250	380	350
Pressure (torr)	0.9	0.9	1.5	2.2
RF Power (watts)	25	40	40	1000
Rate (nm/min)	11	20	10	18

All of the films stiffen the GaAs substrate which has a low SAW velocity. As would be anticipated the carbide and nitride films of silicon have larger elastic constants and show a much more rapid increase in velocity with film-thickness to acoustic-wavelength ratio than the oxides. For zero film thickness the values converge to a velocity of 2860 m/s which is in good agreement with the measured and theoretically predicted SAW velocity of gallium arsenide.

Table 8 shows the range of densities and elastic constants which were determined by measuring two to three film thicknesses for each of the four PECVD films. In general as the film thickness increased the film density and elastic constants decreased. This is a result of the softening of the film structure as the growth continues.

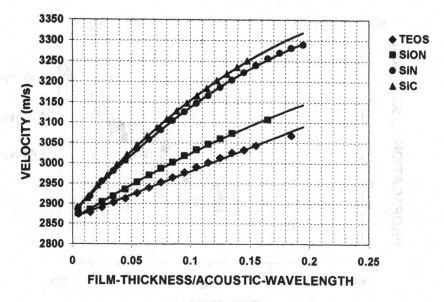

Fig. 4. SAW velocity dispersion characteristic for 500 nm thick PECVD films of TEOS glass, SiON, SiN, and SiC on (001)(110) GaAs.

Table 8. The range of densities and elastic constants for PECVD silicon nitride, silicon oxy-nitride, silicon carbide, and TEOS glass determined from SAW measurements for the films deposited on gallium arsenide.

Film	Density kg/m^3	c_{11} GPa	c_{44} GPa
SiN	2800-2500	190-170	92-55
SiON	2400-2200	106-96	35-30
SiC	2100-2050	180-165	56-50
TEOS	2200	85-76	30-27

Figure 5 shows propagation loss data in units of dB/cm as a function of frequency for silicon nitride films of three different thicknesses, 100, 250, and 500 nm. The high resistivity and low piezo-effect of the GaAs minimizes any contributions to the propagation loss from acoustoelectronic interactions. Above 500 MHz the data shows a fairly uniform power dependent frequency relationship which is close to frequency squared. The losses approach very closely the loss of GaAs alone indicating that the presence of the film has not appreciably changed the overall propagation loss. The highest frequency data points for the films are at film thickness-to-wavelength ratios near 0.2.

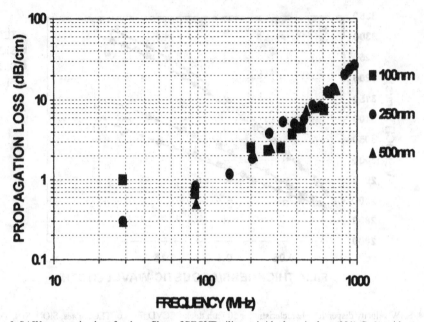

Fig. 5. SAW propagation loss for three films of PECVD silicon nitride deposited on (001) GaAs with propagation along the (110) direction.

Table 9 gives the range of film/substrate propagation loss values at a frequency of 1GHz for the different film/substrate combinations and thickness ranges. The film thickness is given in ascending order and the order of the propagation losses indicates whether the loss is increasing or decreasing with film thickness over this range. The silicon nitride and silicon carbide in combination with the gallium arsenide have the lowest loss values and highest material Q. With the silicon nitride and silicon carbide the loss at a fixed frequency decreased for an increase in film thickness, whereas for the films with oxygen content, SiON and TEOS glass, the propagation loss increases with increasing film thickness. The highest losses are associated with the silicon-oxynitride which had the thickest film that was measured.

Table 9. The range of SAW film/substrate propagation loss values for PECVD films on GaAs at a frequency of 1.0 GHz for the film thickness ranges indicated.

Film	Thickness (nm)	Loss (dB/cm)
SiN	200-1000	25-20
SiON	500-2100	25-40
SiC	200-420	25-15
TEOS	200-1000	20-30

Normalized temperature coefficient of frequency (TCF) values were determined from measurements of the change in synchronous SAW frequency as a function of temperature for the four films on GaAs. In each case the thickest film was used. The frequency change was measured in 25°C increments over the temperature range from -25°C to +100°C. The SAW TCF for (001)(110) GaAs is -45°C. All of the films showed a lowering of the TCF value with film thickness to acoustic wavelength ratio indicating that the films have a positive temperature coefficient of frequency or a negative coefficient which is lower than gallium arsenide. TEOS glass showed the most pronounced change with a TCF value of −15 ppm/°C at a film thickness to wavelength ratio of 0.2. Much slower changes in the TCF were observed for silicon oxynitride, silicon nitride and silicon carbide. Their approximate TCF values were the following at the indicated thickness to wavelength ratios: SiON, -30 ppm/°C at 0.3, SiN, -36 ppm/°C at 0.2, and SiC, -42 ppm/°C at 0.1.

The SAW measurement technique illustrated with the four PECVD films is useful in determining the elastic constants and loss properties of dielectric film layers. Results of these early measurements indicate that dielectric film structures can be developed whose acoustic properties will be useful in the integration of SAW devices directly on semiconductor substrates with electronic circuitry and in the improvement of the SAW properties of piezoelectric substrates. The measurement techniques can also be used for establishing the best deposition conditions for the growth of dielectric films. Using design of experiments and statistical analysis together with SAW measurements becomes a powerful tool for developing the best acoustic-quality dielectric films. This technique is described in the following section.

5.2. *Sputtered glass on 128 Y-X lithium niobate*

It is important to determine the process parameters which produce the best film quality for a particular vacuum system. This can become a time consuming task unless there is a systematic approach to varying the most critical process parameters. Statistically sound design of experiment (DOE) methodologies have been developed by the mathematics community which can effectively be applied.[8] An orthogonal DOE procedure, based upon a set of 9 experiments was used to determine optimum processing parameters for RF diode sputtered glass on 128° Y-X lithium niobate.[37] The glass was RF diode sputtered to a thickness around 600 nm for each experiment. There were four dependent sputter system variables, substrate temperature, background sputter gas pressure, RF power to the target, and gas flow rate. Three independent properties were measured, film stress, SAW velocity, and SAW propagation loss. Velocity and loss were measured with the series of SAW interdigital electrode patterns on the glass surface described in Section 3. The SAW velocities varied from 3880 to 3910 m/s for a film-thickness to acoustic-wavelength ratio of 0.1 and the variation in SAW propagation loss was between 18 dB/cm and 40 dB/cm in the 600 MHz region. A multilinear regression of the data, at a significance level where there is less than a 10% chance that a variable was left in the model when it has no effect, gave three polynomial expressions for stress, velocity, and propagation loss with their dependencies on the significant sputtering parameters. From these expressions the best deposition conditions could be determined for the RF diode vacuum system used.

To obtain a single matrix of values for analysis, a mid-range film-thickness to wavelength-ratio value was chosen to set the velocity value, and a mid-range frequency was chosen to set the representative propagation loss value. The matrix data is shown in

Table 10. The temperature is in degrees Centigrade, the pressure is in milliTorr, the power is in Watts, and the aperture measure of gas flow rate is in diameter inches. The stress is in dynes/cm squared times 10 to the minus ninth power, the SAW velocity, taken at a film thickness to wavelength ratio of 0.1, is in meters per second, and the SAW propagation loss taken at 430 MHz is in decibels per centimeter.

Table 10. Matrix of Values for Analysis

No	AVETEMP $u1$	PRES $u2$	POWER $u3$	APERT $u4$	STRESS $y1$	VELOCITY $y2$	PLOSS $y3$
1.	162.5	7	200	2	1.863	3905	9.8
2.	169	8	250	4	2.900	3922	11.8
3.	157	4	150	1	1.474	3885	16.2
4.	262.5	7	250	1	2.738	3883	14.8
5.	254	8.5	150	2	1.305	3898	12.3
6.	258	4	200	4	2.007	3913	11.8
7.	69.5	7	150	4	2.210	3918	23.7
8.	79	7.5	200	1	1.321	3888	16.4
9.	88	4	250	2	1.557	3906	10.8

The stepwise regressions for a second degree polynomial in the parameters $x1 - x4$ corresponding to normalized $u1-u4$ parameters gave the following:

1. $y1 = 1.9371 + 0.3641\, x3 + 0.2659\, x4 + 0.5131\, x1\, x3$
(Significance: 0.10)

2. $y2 = 3909.3261 - 3.1463\, x1 + 16.1558\, x4 - 7.8315\, x4\, x4$
(Significance: 0.06)

3. $y3 = 14.0043 - 2.0165\, x1 - 2.3595\, x3 + 4.1477\, x1\, x3$
(Significance: 0.10)

The polynomial expressions above indicated the following:

- The stress, $y1$, seems to depend mostly on power, $x3$, aperture, $x4$, and somewhat on temperature as it interacts with power, $x1x3$.

- The velocity depends mostly on aperture, $x4$, and somewhat on temperature, $x1$.

- The propagation loss depends on temperature, $x1$, and power, $x3$.

- Pressure, $x2$, does not seem to be important in this case to affect any of the factors.

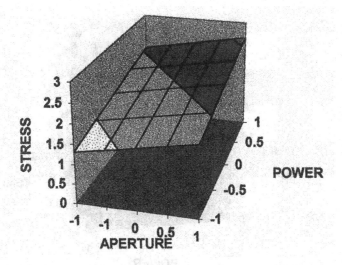

Fig. 6. Relationship of film stress to temperature and power for sputtered glass on 128° Y-X LiNbO$_3$. Aperture and power are normalized, stress is in dynes/cm^2 times 10^{-9}.

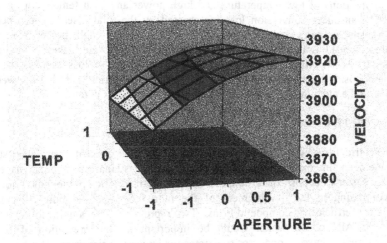

Fig. 7. Relationship of SAW velocity to temperature and aperture for sputtered glass on 128° Y-X LiNbO$_3$. Temperature and aperture are normalized, velocity is in m/s.

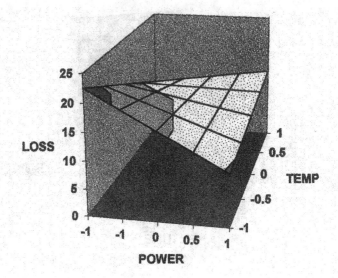

Fig. 8. Relationship of SAW propagation loss to temperature and power for sputtered glass on 128° Y-X LiNbO$_3$. Power and temperature are normalized, loss is in dB/cm.

Figures 6-8 show the stress, velocity, and loss as they relate to the dominant process factors. For low stress films a low flow rate (small aperture) and low power are the best conditions. The velocity characteristic can be tailored by the flow rate with a wide aperture producing a high velocity. The propagation loss is a more complex function with the low loss under conditions of low temperature and high power and high temperature and low power. The transducer conversion loss represented by the total insertion loss between adjacent transducer pairs was lowest under conditions of high power and lower temperature. Depending upon the most critical operational parameters, the best deposition conditions can be specified. Since transducer and propagation loss seemed the most critical, the deposition condition chosen for process standardization were high power, low temperature, and a medium aperture controlling the sputter gas flow rate.

6. Piezoelectric Films

The piezoelectric films which have received the most development over the past four decades have been zinc oxide and aluminum nitride which as binary compounds have been considerably easier to deposit than the more complex ferroelectric compounds. However, work is continuing in the development of the more complex and potentially higher coupling factor ferroelectric compound films. Two representative examples follow of ZnO on GaAs and AlN on Si which would be important in the integration of acoustic components directly on semiconductor substrates for integrated circuit applications. What the examples are intended to show is the high quality of polycrystalline films which can be achieved through sputtering techniques as evidenced by their performance. In the case of AlN on Si some interesting relations to microstructure and composition are also given.

6.1. Zinc oxide on GaAs

Zinc oxide and gallium arsenide have similar SAW velocities. It is well known that the presence of a ZnO film layer on GaAs will increase the effective coupling factor by a factor of ten or more.[38] It has also been shown that an intermediate dielectric layer between the ZnO film and the GaAs substrate leads to a further increase in coupling factor and can enhance the TCF.[39] Zinc oxide can be grown directly on GaAs or on a buffer layer grown on the gallium arsenide. The presence of a buffer layer of silicon nitride or oxynitride promotes better ZnO film adherence, and can be used to modify the velocity characteristic. Two examples are described to illustrate the velocity and loss properties observed.

Figure 9 shows the SAW velocity dispersion for a 1700 nm film of ZnO grown on a 300 nm silicon nitride layer and an 1800 nm silicon-oxynitride layer. The zinc oxide was dc triode sputtered and deposition conditions have been discussed elsewhere.[40] The velocity is plotted as a function of the ratio of the additive thickness of the two film layers divided by the acoustic wavelength. The influence of the thicker dielectric layer is seen in the 0 to 0.5 thickness to wavelength region. Above 0.5 the velocity characteristic is basically the same for both with the velocity being controlled by the zinc oxide film structure. The fact that the velocity values merge is a strong indication that the zinc oxide films areessentially equivalent. The ZnO elastic constants agreed with those given by Carlotti, et. al.[10]

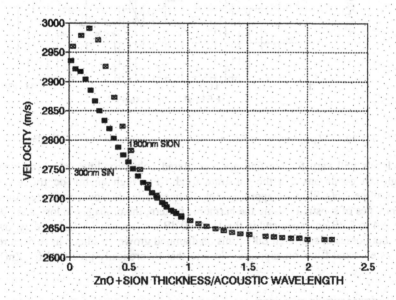

Fig. 9. SAW velocity dispersion for 1700 nm ZnO on GaAs with SiON and SiN buffers.

The SAW propagation loss is quite different for the two composite layers as shown in Figure 10. The loss increases rapidly with increasing frequency for both buffer layers at a rate greater than frequency squared. The thicker dielectric layer yields a higher loss which is in keeping with both the thickness and inherent higher loss of SiON over SiN. In both cases a frequency is reached in which the loss decreases and then starts to rise at a rate close to frequency squared. It is believed that at this point the zinc oxide film is controlling the loss, and the losses associated with the higher loss amorphous dielectric have less influence on the loss characteristic. In the figure is shown a solid line which is a prediction of the SAW propagation loss for single crystal ZnO. Two additional data points are shown based on results reported for epitaxial ZnO on sapphire which are very near theoretical. The loss level of the fine grain polycrystalline film is almost three times that of the predicted single crystal value.

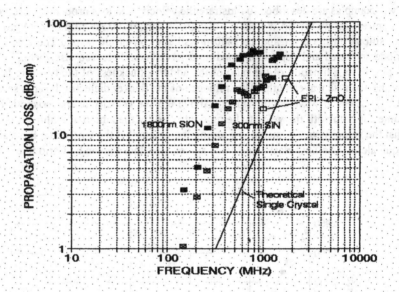

Fig. 10. SAW propagation loss for 1700 nm ZnO on GaAs with SiON and SiN buffers.

Loss levels will vary for the fine grain sputtered polycrystalline ZnO films dependent upon growth conditions and the substrate used. For high quality dc triode sputtered polycrystalline ZnO films of approximately 1 micron in thickness measured on several different types of substrates using the harmonic pattern, the propagation losses at 500 MHz are in the 4 to 8 dB/cm range and in the 15 to 30 dB/cm range at 1.0 GHz. Very low SAW propagation losses of sputtered ZnO films on diamond grown on silicon have been reported as low as epitaxial films on sapphire, close to that theoretically predicted for ZnO single crystal.

6.2. Aluminum nitride on silicon

Aluminum nitride and silicon have similar SAW velocities and, therefore, SAW devices with low velocity dispersion can be fabricated. The SAW coupling factor of AlN on Si is lower than that of zinc oxide, but AlN has some very attractive features such as chemical stability, mechanical strength, high thermal conductivity, and good dielectric properties. It is favored over ZnO by the semiconductor industry because of the stability of its constituent elements.

Velocity dispersion and propagation loss characteristics were measured on magnetron sputtered aluminum nitride on silicon with a thin (300 nm) intermediate silicon nitride layer. The resulting characteristics varied with the process parameters which resulted in changes in the film structure as detected by x-ray, surface micrography, and Auger analysis. A design of experiments (DOE) with multilinear regression was used to confirm the effect of the different sputtering process parameters on the films acoustic properties and establish the best deposition conditions.[8] The best films were fine grain polycrystalline with grain sizes in the 30 to 50 nm range, a strong 002 x-ray peak, less than 0.5 percent detectable oxygen content, and a surface roughness of less than 10 nm for a 1.0 micrometer film. Such films were realized in a production-type single-wafer sputtering system under elevated substrate temperatures in the 300 to 500°C range, low background sputter pressures and flow rates, high nitrogen content of the plasma, and deposition rates of 0.5 nm/s or higher. The films had stress levels less than 1.0 GPa. Electron diffraction of the cross-section of high quality polycrystalline films up to 2 microns in thickness showed a near epitaxial spot pattern indicating good lattice continuity between the grains.

Figure 11 shows the velocity characteristic of the high quality c-axis normal AlN film. The velocity dispersion characteristic follows within less than two percent the theoretically expected characteristic, based on the elastic constants originally determined from epitaxial films grown on sapphire.[9] Also shown is a velocity characteristic based upon a Brillouin scattering analysis of the film.[41] The experimentally observed characteristic follows the Brillouin measurements very closely indicating good agreement between the two methods of measurement. Films of poorer quality had velocity characteristics which deviated from theoretical and had a reduced coupling factor and higher propagation losses.

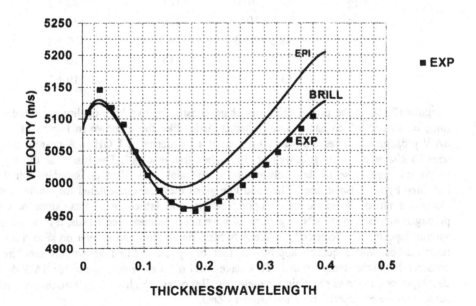

Fig. 11. SAW velocity characteristic for 1000 nm AlN on Si with a 200 nm SiN buffer.

The measured elastic constants for aluminum nitride single crystal, epitaxial film, and sputtered fine grain polycrystal are shown in Table 11. The single crystal constants were determined by standard ultrasonic pulse-echo techniques.[42] The constants for the epitaxial film were determined from SAW measurements made on several films.[9] The agreement is reasonably consistent among the different constants, but differences can be seen due to the nature of the type of material and method of growth. On an average the epitaxial film constants are the closest to the sputtered film constants and the crystalline constants are greater than both as would be expected. The Brillouin measurement of the sputtered constants also gave bulk longitudinal and shear velocity values. For the c-axis normal films, the velocity of the longitudinal mode traveling parallel to the surface was 10500 m/s and normal to the surface was 11550 m/s. The shear horizontal mode traveling parallel to the surface was 6040 m/s. The Rayleigh-wave velocity was 5620 m/s. As the film structure degraded due to the presence of oxygen, all the elastic constants reduced in value with those controlling shear wave motion lowering the fastest. This was consistent with the nature of the film growth and the crystallographic changes due to the introduction of oxygen into the lattice.

Table 11. Measured Elastic Constants of Aluminum Nitride

Const. (GPa)	Crystal [42]	Epitaxial [9]	Sputtered [41]
C_{11}	410.5 ± 10.0	345	360 ± 6
C_{12}	148.5 ± 10.0	125	122 ± 8
C_{13}	98.9 ± 10.0	120	123 ± 5
C_{33}	388.5 ± 10.0	395	410 ± 8
C_{44}	124.6 ± 10.0	118	116 ± 2
C_{66}	131 ± 10.0	110	119 ± 2

Figure 12 shows the propagation loss characteristic for two high quality sputtered AlN films on silicon which had velocity characteristics like those shown in Figure 11. The SAW propagation losses are near 10 dB/cm at a frequency of 1 GHz. A near frequency squared characteristic is followed at the higher frequencies. The losses at the higher frequencies are close to those observed for epitaxial AlN films on sapphire which is indicated by the dashed line. The solid line gives the losses calculated for single crystal aluminum nitride substrate based upon velocity and lattice properties, since acoustic propagation losses in an AlN single crystal have not been measured. Thus, it is possible to sputter deposit piezoelectrically active AlN films on an amorphous layer on silicon at low temperatures whose quality approaches that of epitaxial films grown on sapphire at considerably higher temperatures. This makes AlN films extremely useful for SAW device developments either as the main piezoelectric film or as a high velocity intermediate layer for a second piezoelectric film layer such as ZnO.

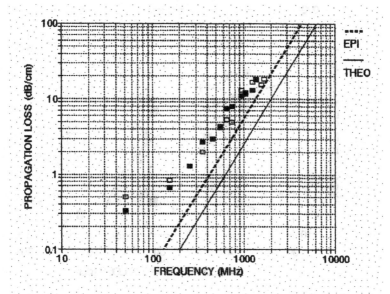

Fig. 12. Propagation loss for 1000 nm sputtered AlN on Si with a 200 nm SiN buffer.

The coupling factor for the sputtered AlN films can be estimated from the insertion loss between adjacent transducers which consists of the transducer conversion loss for two transducers and propagation losses due to absorption, surface roughness, thermoelastic losses, beam diffraction, reflections etc. All of the propagation losses will be minor compared to the transducer conversion losses and may be neglected because of the proximity of the IDTs and the fundamental frequency of measurement which is near 50 MHz. Figure 13 is a plot of the total untuned insertion loss between adjacent interdigital transducers at 150 MHz and its relation to the normalized x-ray intensity. The data is from films with thicknesses from 1000 nm to 3000 nm. For x-ray intensities near 10 cps/nm the total insertion loss is near 60 dB. As the x-ray intensity rises to above 100 cps/nm, the loss is 40 dB or less. This change in loss represents an order of magnitude change in the film coupling factor from approximately 0.01% to 0.1%. The reason for this dramatic change in piezoelectric activity is crystallographic changes due to reduction in oxygen incorporation.

It was determined through Auger analysis that the x-ray intensity was closely linked to the oxygen content of the films (Fig. 14). When the oxygen content of the AlN films was 0.5% or less, the x-ray counts were in the region of 100 to 200 cps per nm of film thickness. With increasing oxygen content above 1% the x-ray intensity decreased rapidly and the SAW transducer conversion loss rapidly increased indicating a strong decrease in piezoelectric activity. The cause appears to be a crystallographic structural change caused by how the oxygen substitutes in the lattice. This change in structure, which changes the piezoelectric nature of the film, is reported to occur near 0.75 atomic percent.[43] This model is consistent with what is experimentally observed with the change in acoustic properties of the films. It also shows the use of x-ray in determining film quality and in understanding the microstructure of oriented polycrystalline films.

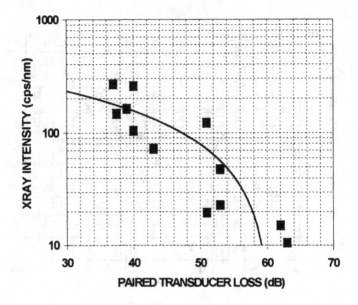

Fig. 13. Relation between normalized x-ray intensity and SAW insertion loss between adjacent IDTs for sputtered AlN on Si.

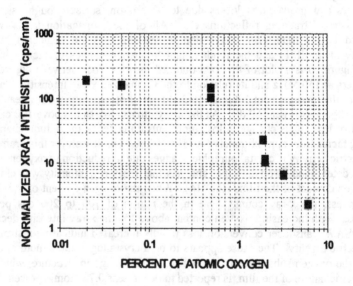

Fig. 14. Normalized x-ray peak intensity as a function of percent of atomic oxygen in sputtered AlN films on Si.

7. Films on Substrates with Pseudo-SAW Propagation

Rotated cuts of lithium niobate and lithium tantalate are used extensively in the production of SAW filters for telecommunication products. Particular cuts have been chosen usually for their higher velocity, higher coupling efficiency and/or lower temperature coefficient of frequency than the cuts along principal crystallographic axes. Thin films do play a role in the propagation properties of such cuts which in general can support three modes of propagation, GSAW, PSAW, and HVPSAW. The presence of films, which soften the surface, can move the PSAW modes into regions where the propagation properties and displacement profiles change. In the following sections the influence of metal, dielectric, and piezoelectric films on the propagation properties of the important cuts, 41° Y-X $LiNbO_3$, 64° Y-X $LiNbO_3$, and 36° Y-X $LiTaO_3$ are characterized.

7.1. *Metal films on substrates supporting PSAW propagation*

The properties of 41° Y-X $LiNbO_3$, 64° Y-X $LiNbO_3$, and 36° Y-X $LiTaO_3$ for PSAW and HVPSAW have been characterized theoretically with and without a shorting surface.[12] The propagation loss for the three substrates behaves in a different fashion depending upon the presence or absence of a shorting surface. This can be seen in Fig. 15 where propagation loss is plotted as a function of the ratio of frequency using the harmonic IDT pattern with (closed data points) and without (open data points) a 100 nm thick Al film in the region between the transducers. As predicted theoretically for the two cuts of lithium niobate, the loss is lower for the free surface of the 41° cut and lower for the metallized surface of the 64° cut. Overall, the propagation loss is considerably lower for the 41° cut as compared to the 64° cut. The change in loss with frequency is close to linear, indicating that the dominant loss mechanisms are associated with the leaky wave loss. There is also a tendency for the loss to show a decrease for the 64° cut as the frequency nears 1.0 GHz.

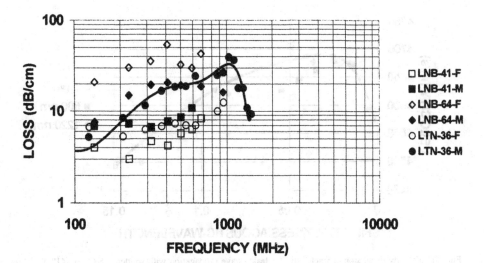

Fig. 15. PSAW propagation loss versus frequency using the harmonic IDT pattern with (closed data points) and without (open data points) a 100 nm thick aluminum film in the region between the transducers.

The propagation loss for the 36° Y-X LiTaO$_3$ lies between that of the two cuts of lithium niobate with the metallized surface (a trendline is shown on the figure) having an average loss that was 3 times higher than the free surface loss in the 0.1 to 1.0 GHz region. There is a plateau region of loss from 400 to 700 MHz. A most remarkable feature is the strong reduction in loss for the metallized surface which occurs above 1.0 GHz. This appears to be characteristic of the change in surface loading. A similar feature was noted recently for a wax layer on 128° Y-X LiNbO$_3$.[45]

7.2. *Dielectric films on substrates supporting PSAW propagation*

Sputtered glass film layers on rotated cuts of lithium niobate and lithium tantalate were experimentally investigated using the harmonic IDT pattern to determine if any improvements in coupling factor and TCF could be realized.[46,47] In these cases where the presence or absence of a film layer can affect the propagation loss of the leaky-wave modes, some interesting facts have been uncovered. For particular film-thickness to wavelength ratios the strong leaky-wave loss of the Pseudo-SAW modes goes to zero and the only loss is that due to the viscous losses of the film and substrate. By operating in these regimes the presence of the film gives coupling factor and TCF advantages.

7.2.1. *Glass film on 41° LiNbO$_3$*

Figure 16 shows the velocity dispersion characteristics as a function of t/λ ratio for leaky-wave propagation for two thicknesses of sputtered pyrex glass (500 and 1220 nm) on high coupling factor, high velocity, 41° Y-X lithium niobate. With the harmonic pattern on the upper film surface, the glass initially isolates the metal from the surface and the leaky-wave velocity shows a rise in value to a maximum at t/λ values between 0.02 and 0.04. Beyond t/λ = 0.05 the leaky-wave velocity decreases as the characteristic elastic properties of the glass dominate the propagation. The experimental values fall below the theoretical.

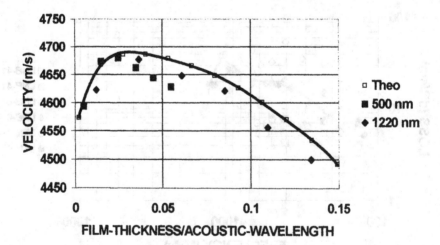

Fig. 16. Velocity dispersion characteristic for leaky-wave propagation with sputtered SiO$_2$ on 41° Y-X lithium niobate as a function of t/λ.

The theoretical leaky-wave propagation loss predicted for a metal at the top surface and no metal at the film-substrate boundary as a function of silicon dioxide film thickness to wavelength ratio (t/λ) is shown in Fig. 17 as a line. The strong minimum occuring near t/λ =0.02 is reflected by the drop in the experimental loss values. Experimentally as a function of t/λ on the log-log scale the propagation loss in dB/λ decreases to a value of 0.01 at t/λ in the range 0.02 to 0.03 and then increases. For the 1220 nm film there was a second minimum at t/λ between 0.08 an 0.09 before increasing loss again.

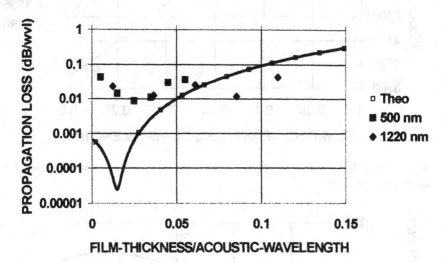

Fig. 17. Propagation loss characteristics in dB/λ for 41° Y-X $LiNbO_3$ leaky waves with sputtered SiO_2 as a function of film-thickness to acoustic-wavelength ratio.

7.2.2 Glass film on 64° $LiNbO_3$

The 64° Y-X lithium niobate substrate has been favored for wide bandwidth SAW devices because of its high coupling factor ($k^2 = .11$) in the leaky-wave SAW mode of propagation. The velocity is near 4600 m/s with a reported 0.036 dB/wavelength propagation loss. The TCF has been reported at -70 ppm/°C. The 64° cut also supports a Rayleigh-mode wave with a velocity near 3700 m/s and a coupling factor of $k^2 = 0.01$. The change in acoustic properties due to the presence of a silicon dioxide film on the 64° Y-X lithium niobate substrate was determined for the leaky-wave propagation mode.

Figure 18 shows the experimental velocity values for leaky-wave propagation for three thicknesses of sputtered silicon dioxide (795, 1105, and 1980 nm) as a function of t/λ ratio. The theoretically calculated velocity dispersion shown in the figure as a line agreed quite well with the experimentally determined data points. The silicon dioxide initially isolates the surface and the leaky-wave velocity shows a small rise in value to a maximum at t/λ values between 0.02 and 0.04. Beyond $t/\lambda = 0.05$ the leaky-wave velocity decreases as the characteristic elastic properties of the silicon dioxide begin to dominate the propagation. The non-leaky Rayleigh-wave velocities had broad peak values between 3660 and 3670 m/s for $0.1 > t/\lambda < 0.2$.

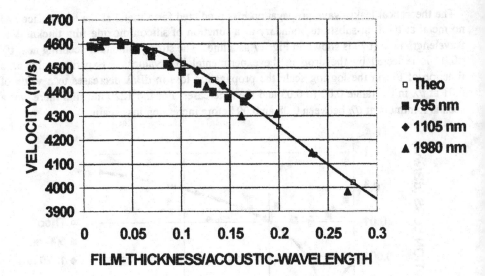

Fig. 18. Velocity dispersion characteristic for leaky-wave propagation with sputtered SiO_2 on 64° Y-X lithium niobate as a function of t/λ.

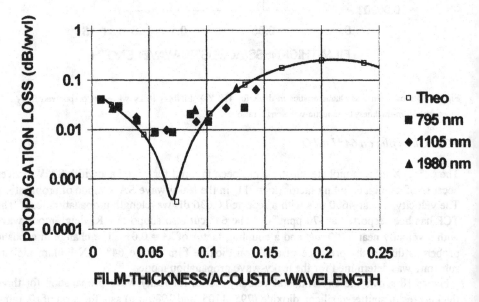

Fig. 19. Propagation loss characteristics in dB/λ for 64° Y-X $LiNbO_3$ leaky waves with sputtered SiO_2 as a function of film-thickness to acoustic-wavelength ratio.

Theoretically the leaky-wave propagation loss as a function of film thickness to wavelength ratio (t/λ) (Fig. 19) has sharp minimum at $t/\lambda=0.07$ influenced by the presence of the silicon dioxide film. Experimentally as a function of t/λ the leaky wave loss in dB/λ

decreases to a value of 0.01 at t/λ near 0.06 and then increases rapidly. The minimum in the leaky-wave loss coincides with film-thickness to acoustic-wavelength ratios just beyond the peak in the velocity characteristic.

7.2.3. Glass film on 36°LiTaO₃

The 36° Y-X cut of lithium tantalate supports leaky-wave propagation with a free surface SAW velocity of 4212 m/s, a metallized velocity of 4112 m/s, a coupling factor of k^2 = 0.05 and a temperature coefficient of frequency in the -30 to -40 ppm/°C range. It has a predicted low leaky-wave propagation loss. The 36° Y-X cut has been used extensively for moderate bandwidth SAW devices requiring good temperature stability.

Silicon dioxide films of 790, 1420, and 2000 nm were deposited on the surface of 36° Y-X LiTaO₃. The leaky-wave velocity characteristic, with the theoretical (line) and the experimental (data points) is shown in Fig. 20 as a function of film-thickness to acoustic wavelength ratio. The SAW velocity is near 4140 m/s at zero film thickness representing the partially metallized surface condition of the IDT and rises to a value near 4200 m/s for film-thickness to acoustic-wavelength ratios between 0.04 and 0.05. There would be little or no velocity dispersion for a SAW device on the 36° tantalate with an SiO₂ film in this region. This response at low t/λ ratios represents an isolation of the piezo-surface from the aluminum IDT electrode. Above t/λ = 0.05 the velocity decreases in response to the lower elastic constants of the SiO₂. The theoretically predicted response and the experimentally determined values are in good agreement.

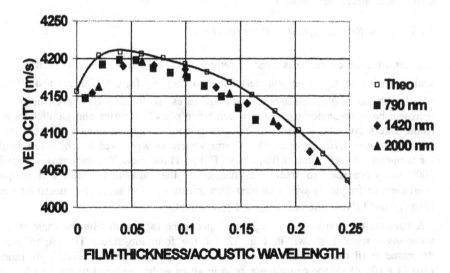

Fig. 20. Leaky-wave velocity characteristic, for 36° Y-X LiTaO₃ with three thicknesses of sputtered SiO₂, as a function of t/λ.

The propagation loss in dB/λ for the 36° Y-X lithium tantalate is shown in Figure 21. With no SiO₂ film present, the propagation loss in dB/λ decreases to a value of 0.004 dB/λ at a t/λ ratio near 0.03 and then starts to increase. With the films, the loss has a similar characteristic with a minimum of approximately 0.01 dB/λ at a t/λ ratio near 0.04 and then increases at a rate to the 1.7 power closely following theoretical. The minimum in the loss occurs near the t/λ ratio where the velocity characteristic is maximum. In this t/λ range the wave propagation is partially controlled by the stiffening effect of the silicon dioxide film.

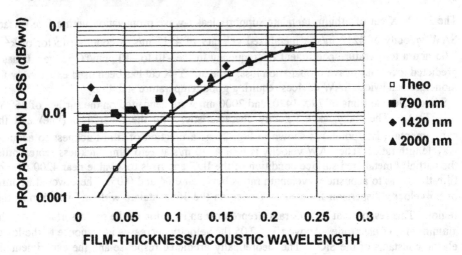

Fig. 21. Leaky-wave propagation loss in dB/λ for 36° Y-X tantalate with a 100 nm aluminum IDT film and three silicon dioxide films as a function of t/λ.

7.2.4. Resonator C_m/C_o and TCF measurements

Resonator transducer patterns, used to determine the ratio of motional capacitance, C_m, to static capacitance, C_o, (approximately equal to coupling factor), had 201 and 301 quarter wavelength interdigital electrodes with apertures of 10 and 20 wavelengths. These represent basic resonator structures which when placed in series and parallel form SAW ladder filters commonly used in hand-held phones. The wavelength of these resonator transducers was 4.76 microns. These same structures were used for the determination of the temperature coefficient of frequency (TCF). Three glass film thicknesses from 800 to 2000 nm were used to track the changes of the capacitance ratio and temperature coefficient of frequency with changing film thickness. The substrates measured were 41° $LiNbO_3$, 64° $LiNbO_3$, 36° $LiTaO_3$, and 128° $LiNbO_3$.

A composite graph of the changes in capacitance ratio with film-thickness to acoustic wavelength ratio is shown in Fig. 22 for the four substrates. The capacitance ratio decreased in all cases but at different rates. For film thickness to wavelength ratios near zero TCF (0.3-0.35) the capacitance ratio in all cases had reduced to the 2 to 3 % region. For the 41° niobate the C_m/C_o values decreased rapidly. With the other three substrates, the capacitance ratio remained reasonably constant, around 6%, before decreasing.

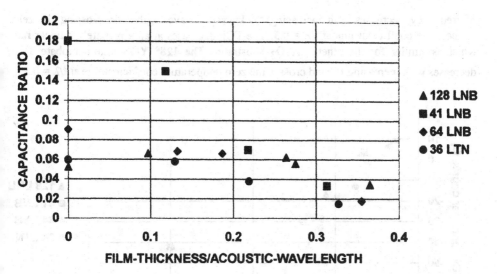

Fig. 22. Capacitance ratio vs. film-thickness to acoustic-wavelength ratio for sputtered SiO_2 on 41° $LiNbO_3$, 64° $LiNbO_3$, 36° $LiTaO_3$, and 128° $LiNbO_3$.

The trend in TCF values for the four substrates as a function of film-thickness to acoustic wavelength ratio is shown in Fig. 23. With the 64° Y-X lithium niobate the temperature coefficient of frequency decreases with increasing t/λ crossing the zero temperature coefficient line at $t/\lambda = 0.33$. For the 36° tantalate the TCF was constant for t/λ ratios up to 0.15 and then decreases as thickness to wavelength increases crossing the zero temperature coefficient line at $t/\lambda = 0.30$. With the 41° Y-X lithium niobate the temperature coefficient

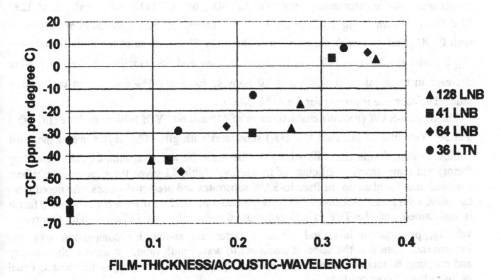

Fig. 23. Temperature coefficient of frequency vs. film-thickness to acoustic-wavelength ratio for sputtered SiO_2 on 41° $LiNbO_3$, 64° $LiNbO_3$, 36° $LiTaO_3$, and 128° $LiNbO_3$.

of frequency decreases with increasing thickness to wavelength ratio crossing the zero temperature coefficient line at t/λ = 0.3. The TCF decreases at approximately a linear rate which is similar for the other LiNbO$_3$ substrates. The 128° Y-X lithium niobate TCF decreases with increasing t/λ and crosses the zero temperature coefficient line at t/λ = 0.35.

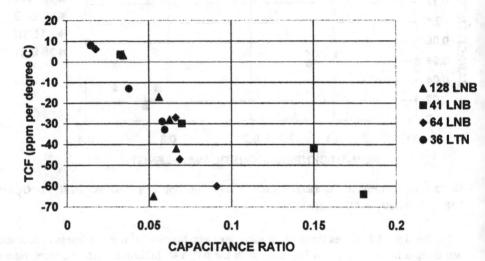

Fig. 24. TCF vs. capacitance ratio for sputtered SiO$_2$ on 41° LiNbO$_3$, 64° LiNbO$_3$, 36° LiTaO$_3$, and 128° LiNbO$_3$.

A composite graph of the results of the temperature coefficient of frequency and capacitance ratio measurements for 41° LiNbO$_3$, 64° LiNbO$_3$, 36° LiTaO$_3$, and 128° LiNbO$_3$ are shown in Fig. 24. For the 128° Y-X lithium niobate advantages are gained in both C_m/C_o and TCF with the addition of the SiO$_2$ film because of the initial increase in C_m/C_o with decrease in TCF. For the 64° niobate and 36° tantalate the corresponding decrease in temperature coefficient is 10 ppm/°C for every 1% decrease in capacitance ratio. The decrease is greater for the 41° LiNbO$_3$.

The pseudo-SAW (PSAW) characteristics of 41° and 64° Y-X lithium niobate (LiNbO$_3$) and 36° Y-X lithium tantalate (LiTaO$_3$) substrates with glass film layers are of the most interest in considering tradeoffs which can be made between capacitance ratio (coupling factor) and temperature coefficient of frequency. Table 12 gives the experimentally best combinations for glass on the Pseudo-SAW substrates and predicted values. As noted from the table, with glass thickness to wavelength ratios of 0.12 to 0.19, a high coupling factor is maintained and the TCF is reduced with the exception of LiTaO$_3$. The theoretical velocity, propagation loss, and coupling factor are shown for comparison with the experimental values at the selected thickness to wavelength ratio. In general the velocity and coupling factors are lower than predicted by theory, and the propagation loss is equal or lower than theory predicts.

Table 12. Predicted and realized properties for glass on pseudo-SAW substrates.

Substrate	t/λ	Loss (dB/λ)	Velocity (m/s)	k^2 (%)	TCF ppm/°C
36° LiTaO$_3$					-32
Theory	0.13	0.02	4100	7.6	
Experiment	0.13	0.023	4070	6.5	-32
64° LiNbO$_3$					-80
Theory	0.19	0.20	4350	10.8	
Experiment	0.19	0.18	4250	7.5	-38
41° LiNbO$_3$					-80
Theory	0.12	0.18	4425	18.7	
Experiment	0.12	0.08	4330	15.0	-42

7.3. Piezoelectric films on substrates supporting PSAW propagation

There has been little work on the use of piezoelectric films for enhancing the properties of rotated cuts of piezoelectric substrates which support PSAW and HVPSAW modes. In general these substrates have been selected for a particular enhanced property and are used without the addition of a dielectric or piezoelectric film layer. As noted in the previous section there are improvements which can be made through the use of dielectric films such as glass, but in general such use has not been incorporated in standard fabrication processes. Zinc oxide films on quartz have been investigated to enhance the coupling factor and retain the temperature stability.[48] In working with piezo-films on piezoelectric substrates it is important to observe the orientation of the substrate with respect to the film since enhancement or degradation of the coupling may occur depending upon orientation.[49] In the following section experimental work of ZnO on ST-quartz is discussed.

7.3.1. Zinc oxide film on ST-quartz

You will recall in Section 3 the characteristics of ST-quartz, which is an example of a material supporting GSAW, PSAW, and HVPSAW modes. How does the presence of a ZnO film affect these modes? The presence of a zinc oxide film on ST-quartz, was investigated and compared to theoretical calculations which included the viscosity of the quartz.[50] The correlation between experimental and theoretical results was good. For the Pseudo-SAW modes the presence or absence of the ZnO film made a significant difference in the observed propagation loss.

The presence of a zinc oxide film on the ST-X quartz alters the velocity, loss and coupling constant of the three modes. A 2.0 micron film was sputter deposited on the ST-X quartz and the array of harmonic interdigital transducers applied to the upper ZnO surface for the measurement of velocity and propagation loss characteristics. For GSAW propagation, the calculated velocity dispersion characteristic from the measured excitation frequencies is shown in Fig. 25 as a function of ZnO film-thickness to acoustic-wavelength ratio. The theoretically predicted velocity dispersion characteristic (not shown in Fig. 25) followed closely with the film elastic constant values up to film-thickness to acoustic-wavelength values of 0.7. Beyond the t/λ = 0.7 value, the velocity dispersion curve was better fit by slightly lower film elastic contants. This is considered to be more representative of the overall structure of the film and surface topography.

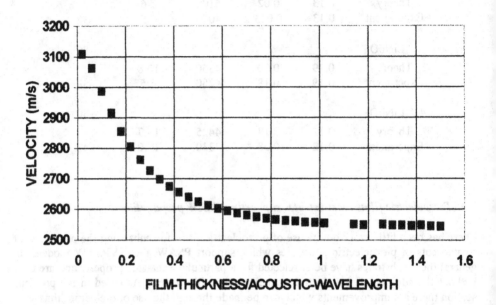

Fig. 25. Experimental GSAW velocity data points for a 2000 nm film of zinc oxide on ST-quartz.

The experimentally measured GSAW propagation loss with frequency is shown as points measured at discrete frequencies in Fig. 26. The theoretical loss characteristics are plotted with the open square points and line representing the loss using the bulk ZnO elastic constants and the dotted points, the ZnO film elastic constants. The measured loss is a factor of 10 times in dB/cm over the predicted loss at frequencies below 800 MHz, which does not include the viscosity of the ZnO. Theoretically and experimentally at the lower frequencies the loss goes as the square of the frequency indicating the dominance of viscosity losses. Above 800 MHz there is a theoretically predicted decrease in propagation loss which is also observed experimentally. This loss is indicative of the acoustic wave energy becoming primarily in the "lossless" zinc oxide. The frequency at which the downturn in the propagation loss occurs, follows more closely the characteristic associated with the use of the film elastic constants than bulk crystal constants. The increase in loss at the highest frequencies is due to the ZnO film viscosity losses. For single crystal ZnO the calculated values of SAW propagation loss would be 8.5 dB/cm at 1.0 GHz and 34 dB/cm at 2.0 GHz. In order to compare exactly the experimental loss pattern with a predicted pattern it will be necessary to specify the viscosity constants of ZnO film layers. This in theory should be possible but was beyond the scope of the present investigation.

There were insufficient measured frequency data points to develop a velocity dispersion characteristic for the PSAW mode. The measured velocity was 4996 m/s at a film-thickness to acoustic-wavelength (t/λ) ratio of 0.02 and 4977 m/s at $t/\lambda = 0.06$. This compares to the predicted values of 4923 m/s and 4383 m/s at the respective t/λ ratios using bulk ZnO elastic contants. The experimental values differ significantly from the predicted values. The reason for this is not yet understood.

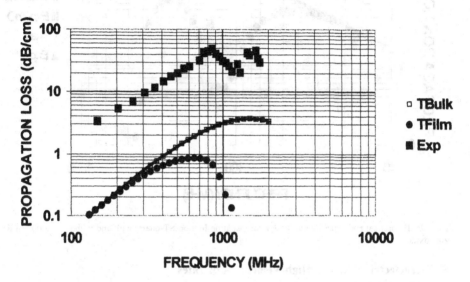

Fig. 26. GSAW propagation loss for a 2000 nm film of zinc oxide on ST-quartz with the theoretically predicted propagation loss using bulk and thin-film ZnO elastic constants.

There was a very strong decrease in the measured propagation loss for the PSAW mode observed at 50 and 150 MHz when the ZnO was deposited on the surface. Figure 27 shows the measured and predicted loss with and without the presence of the 2000 nm ZnO film. Without the ZnO the measured propagation losses are around 30 dB/cm and with the ZnO only 3 dB/cm. This correlates with the predicted loss as seen in Fig. 27 where theoretical characteristics are plotted with and without ZnO and experimental points shown. Without a film layer the calculated propagation loss increases linearly with frequency as shown by the continuous line of data points. With the ZnO film layer, shown as dotted plot points, the predicted loss characteristic has minimum value under 1 dB/cm at approximately 130 MHz. At frequencies above 400 MHz the predicted loss drops substantially coincident with the PSAW mode transitioning into a second order GSAW mode. The measured propagation losses were 3 dB/cm with the ZnO film as noted by the experimental points at 50 and 150 MHz . The single transducer conversion loss for the PSAW mode with the interdigital transducer at the ZnO surface showed a decrease of 3 dB at the first harmonic indicating an increase in the coupling factor at a thickness to wavelength ratio of 0.06. The calculated coupling factor for the PSAW mode increases for small values of thickness to wavelength ratio (t/λ) to 0.05 percent at $t/\lambda = 0.02$. Without the film layer the predicted coupling factor is 0.033 percent.

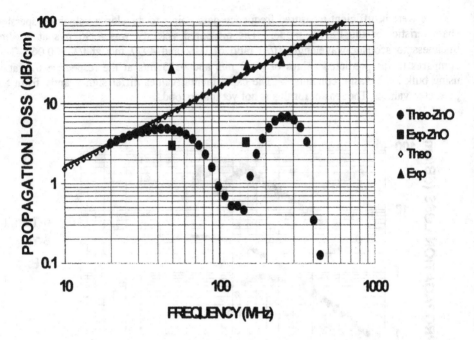

Fig. 27. Theoretical and experimental PSAW propagation loss on ST-quartz with and without a 2000 nm film of zinc oxide.

8. Piezoelectric Films on High Velocity Substrates

As communication frequencies continue to increase substrate properties become very important. Recently there has been considerable activity with ZnO films deposited on thick diamond films grown on silicon to take advantage of the high SAW velocity and low propagation loss of diamond. The advantages which have been found for ZnO on high-velocity low-loss substrates such as diamond are: (a) higher frequency capabilities for a given electrode periodicity, (b) the inherent low propagation loss characteristics of the substrate, (c) the existence of high-coupling-factor, higher-order, generalized SAW modes and high velocity Pseudo-SAW modes, and (d) a higher power handling capability. By using high velocity SAW modes, the capabilities for the excitation frequency are doubled over those of the more standard SAW substrates. Disadvantages of piezo-films on substrates are, their dispersive velocity characteristic, losses which may be dominated by the film properties, and the added film process control of SAW properties and film thickness.

Some recent examples of capabilites for the use of diamond substrates are the following. ZnO on diamond/silicon filters with 3.5 dB insertion loss and a 24 MHz bandwidth at 2.4 GHz have been reported by workers at Kobe Steel.[51] With the addition of a glass layer, a filter at 2.5 GHz has been reported by Sumitomo with a temperature coefficient of frequency of 1 ppm/°C and a loss of less than 6 dB.[52] Such capability demonstrations lead to a strong interest in ZnO or other piezoelectic films on high-velocity low-loss substrates.

A listing of some high-velocity low-loss substrates which might be considered for use with piezoelectric films is shown in Table 13. Average values of the longitudinal and shear bulk acoustic wave velocities and corresponding propagation loss values are shown. Silicon is readily available at low cost, but it has relatively high propagation losses and a lower velocity than the other substrate materials. Rutile (TiO_2) has velocities similar to silicon but substantially less propagation losses. Spinel ($MgAlO_4$) has a very low shear wave propagation loss which would infer very low SAW propagation losses. Sapphire (Al_2O_3) is readily available and c-axis in-plane epitaxial ZnO has been grown on R-plane sapphire with different ZnO deposition techniques to give high coupling factors. Silicon carbide is being developed for high temperature semiconductor applications and substrates are available. Finally there is diamond with the significant advantage in a substantially higher velocity and low propagation losses. There has been considerable work with ZnO on diamond/silicon recorded in the literature. Very thick diamond has been grown on silicon and used as the basis for developing low loss high frequency SAW filters. Both GSAW and higher order modes have been used for propagation. The following subsection discusses experimental and theoretical work with ZnO on silicon carbide as illustrative of the effects of ZnO on a high velocity substrate.[53,54]

Table 13. Average Bulk-wave Longitudinal (L) and Shear (S) Velocity and Loss Properties

Substrate	L-Velocity m/s	S-Velocity m/s	L-Loss* dB/cm	S-Loss* dB/cm
Silicon	9050	5420	8.3	3.0
TiO_2 (Rutile)	9400	5400	0.6	0.3
$MgAlO_4$ (Spinel)	9700	5470	0.4	0.1
Al_2O_3 (Sapphire)	11000	7000	0.2	0.5
SiC	13000	7250	0.4	0.3
Diamond	18000	11000	(0.6)	(0.3)

*Average propagation loss at 1.0 GHz

8.1. Zinc oxide on silicon carbide

There were two types of silicon carbide substrates used in the investigation, low resistivity and high resistivity. The 3.5 cm diameter, 0.4 mm thick silicon carbide substrates were the 6H polytype hexagonal crystal. The c-axis was rotated 3.5 degrees from the substrate normal for the low resistivity substrate, and 8.0 degrees off c-axis normal, for the high resistivity substrate. This will affect the SAW propagation, and needed to be taken into account for analysis purposes. The ZnO was DC triode sputter deposited on the SiC substrates from a 4 inch ZnO ceramic target using a 90:10 argon:oxygen gas mixture at a background pressure of 3 microns. The accumulation rate was 25 nm/min with the substrate heated to approximately 350 °C. The ZnO was clear and smooth with an etch rate in a 25% nitric solution of 3.8 nm/s, which indicates an extremely high quality film. For all the experiments a 400 nm ZnO film was grown.

The measured velocity dispersion characteristics of the GSAW (the first-order mode) for a ZnO film on SiC are shown in Fig. 28. The IDT array is at the upper ZnO film surface. Velocities corresponding to high resistivity and low resistivity SiC, the latter with and without a glass oxide layer, are plotted as functions of ZnO film-thickness to acoustic-wavelength ratio (h/λ). The 60 nm glass oxide film lowers the velocity dispersion characteristic below that without the glass film because of the lower velocity of glass. The difference between the two dispersion curves corresponding to low resistivity SiC, with and without oxide layer, grows with normalized ZnO thickness. For the high resistivity SiC lower GSAW velocities were observed compared to the low resistivity SiC, which can be explained by the different declination angles of the substrate normal with respect to the c-axis of SiC in these two substrates.

To compare the experimental results with simulated GSAW characteristics, calculations were performed using a numerical technique developed for multilayered structures based on the transfer matrix method.[35] The published material constants of SiC and film ZnO were used.[55] Both materials, ZnO and SiC, are in the point symmetry group 6mm. For SiC the available set of material constants were incomplete, piezoelectric constant e_{31} and elastic stiffness module C_{13} being absent. Therefore, a simulation of SAW characteristics was accomplished with various values of C_{13} in order to define the proper value which provides the best agreement between theoretical and experimental data. The effect of the piezoelectric properties of the SiC substrate on velocity dispersion, propagation loss and coupling factors of two SAW modes was estimated with $e_{31}=0$ and found negligible in comparison with much stronger piezoelectric coupling in the ZnO film.

In Fig. 28, two calculated velocity dispersion functions of the GSAW are presented which correspond to $C_{13}=0$ and $C_{13}=150 \times 10^9$ N/m^2. An increase of the unknown elastic modulus C_{13} resulted in a decrease of GSAW velocities. However, the difference between the two dispersion functions was minimal and comparable with the observed difference between the experimental velocity data in the high resistivity and low resistivity SiC.

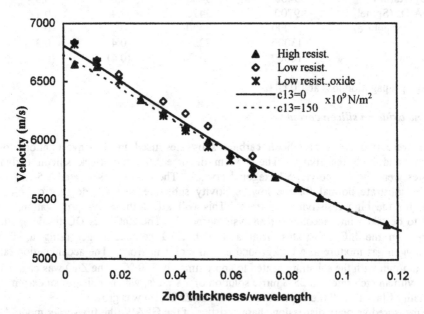

Fig. 28. Measured and calculated velocity dispersion for the 1st GSAW mode. In the calculations 8^0 declination of the substrate normal from c-axis of SiC was taken into account.

The simulation of HVPSAW propagation characteristics was performed using the same numerical technique but with propagation loss as an additional parameter which accounts for a leakage of energy into the bulk due to radiated bulk waves. Since the velocity of the HVPSAW is higher than that of the two shear bulk waves, these two waves must take the energy from the surface into the bulk of SiC substrate and their amplitudes must grow with depth.

The results of velocity and loss simulation obtained for $C_{13}=0$, $C_{13}=50 \times 10^9$ N/m^2 and $C_{13}=150 \times 10^9$ N/m^2 are shown in Fig. 29 together with the experimental data. Velocities (Fig. 29a) and propagation loss in dB per acoustic wavelength (dB/λ) (Fig. 29b) are plotted as functions of normalized ZnO thickness. A declination of the substrate normal by the angle of 8°, which corresponds to the high resistivity SiC substrate, was considered for each C_{13} value analyzed and the experimental propagation loss function was more consistent with a small value for the unknown elastic modulus C_{13} near zero.

The simulated propagation losses for a declination by the angle of 3.5°, which corresponds to the low resistivity SiC substrate, in the case of $C_{13}=0$ was also found to be in a sufficiently good agreement with the measured losses. For small normalized ZnO thicknesses, $h/\lambda<0.02$, the value $C_{13}=0$ adequately described the velocity dispersion of the HVPSAW mode. However, with increasing film thickness neither this C_{13} value, nor any other could provide a good agreement between the calculated and measured velocities, even in conjunction with variation of other elastic constants of SiC. The better agreement, though not perfect, was achieved while the effect of aluminum film was taken into account (see Fig. 29a). The remaining difference between the experimental and simulated HVPSAW velocity dispersion functions can be attributed to the inaccuracy of the ZnO film constants used in calculations, in particular C_{13} and C_{44}. Simulation has shown that the elastic properties of ZnO film analyzed are between that of bulk ZnO and film ZnO.

The decrease of propagation loss at small values of normalized film thickness observed experimentally and confirmed by the calculations can be explained by the following. For any nonzero C_{13}, there are no HVPSAW solutions at the point $h/\lambda=0$, though the non-attenuated Brewster solution can be found to the bulk wave reflection problem. Its velocity V_{BR} is only slightly higher than the limiting value V_1 of the quasilongitudinal bulk mode in the substrate, (e.g. for $C_{13}=150 \times 10^9$ N/m^2, $V_{BR}=12612$ m/s and $V_1=12471$ m/s). With increasing normalized ZnO thickness, the inhomogeneous partial wave is admixed to the Brewster solution resulting in the growth of attenuation coefficient and decrease of velocity. At a certain thickness (for $C_{13}=150 \times 10^9$ N/m^2 it is $h/\lambda=0.003$) it crosses V_1 and the incident quasilongitudinal bulk mode transforms into an inhomogeneous mode with amplitude decreasing with depth. Consequently, the HVPSAW branch appears with further increasing ZnO thickness. If $C_{13}=0$ and the declination angle tends to zero, V_{BR} approaches V_1. Hence, the HVPSAW exists for any ZnO thickness and degenerates into a pure bulk exceptional wave at $h/\lambda=0$.

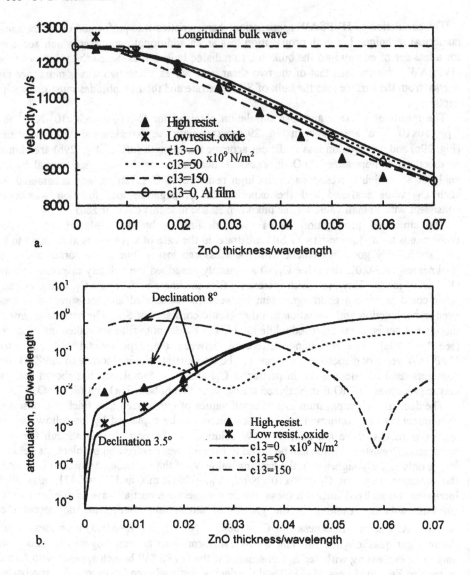

Fig. 29. Measured and calculated velocities (a) and propagation loss (b) of HVPSAW, with declination of the substrate normal from c-axis of SiC taken into account.

Though the comparisons of the experimental and numerical data confirmed that the two observed SAW modes can be identified as the first order GSAW and HVPSAW, the thorough numerical investigation reveals the existence of higher-order surface modes in the ZnO/SiC structure which appear for thicker ZnO films. The results of calculations obtained with zero C_{13} modulus of SiC for normalized ZnO thicknesses up to 0.5 are presented in Fig. 30a, and the corresponding electromechanical coupling factors are

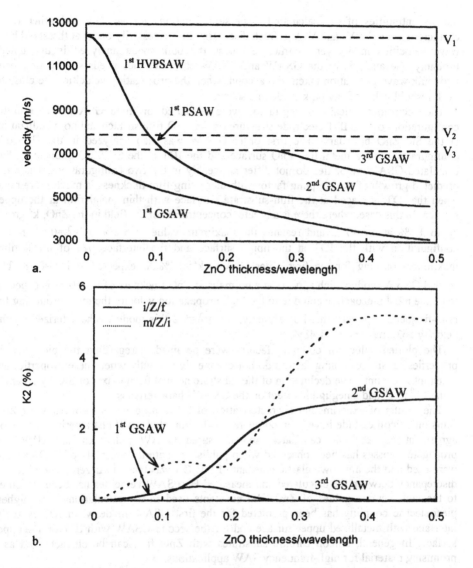

Fig.30. Velocities (a) and electromechanical coupling factors (b) of different surface modes propagating in ZnO/SiC structure, calculated for $C_{13}=0$. V_1, V_2 and V_3 are the threshold bulk wave velocities in SiC substrate. Coupling factors are shown for two different electrode configurations.

plotted in Fig. 30b. In the thickness interval analyzed, two additional GSAW modes (2nd GSAW and 3rd GSAW) and one PSAW mode were found. Love modes with negligible piezoelectric coupling and the higher order HVPSAW modes with strong attenuation found in the same thickness interval are omitted in Fig. 30.

In Fig. 30a, three modes, the 1st HVPSAW, 1st PSAW and 2nd GSAW, appear as a continuous line which crosses the threshold velocities of quasishear bulk waves in the substrate at $V_2=7900$ m/s and $V_3=7228$ m/s. However, a careful numerical analysis reveals the discontinuities of the calculated dispersion functions for velocity and other wave characteristics at $V=V_2$ and $V=V_3$. Such discontinuities necessarily occur at threshold bulk wave velocities in a layered structure unless the bulk mode analyzed is uncoupled. Probably, the analysis of the GSAW and PSAW characteristics needs special treatment with bulk wave generation taken into account when the propagation velocities are close to the threshold value of any bulk mode in a substrate.

The electromechanical coupling factors were calculated for three GSAW modes in the configuration of the IDT electrode structure on the ZnO film surface and no metallization at the SiC/ZnO boundary. The first order GSAW was also analyzed in the electrode configuration with a metal free ZnO surface and the IDT at the SiC/ZnO boundary. The calculated GSAW velocities do not differ noticeably in the two configurations, while the predicted growth of the coupling factor with increasing film thickness is much more rapid when the IDT is placed at the film-substrate interface with thin metal film at the upper surface. In this case, where there is a greater concentration of the field in the ZnO, k^2 grows up to 1.1% at $h/\lambda=0.13$ and reaches the maximum value of 5.2% at $h/\lambda=0.43$. In the configuration with the IDT at the upper surface and the interface free of metal film, maximum coupling for the first order GSAW, $k^2=0.4\%$, is expected at $h/\lambda=0.18$. The second GSAW mode which appears at a normalized ZnO thickness of $h/\lambda>0.11$ is expected to be the most piezoactive and due to the high propagation velocity the most promising for high-frequency applications. For example, for $h/\lambda=0.2$ this mode is characterized by the velocity 5637 m/s and $k^2=1.48\%$.

The plotted values of coupling factors were obtained disregarding the piezoelectric properties of SiC. Coupling factors do not change significantly when these properties are taken into account. The declination of the substrate normal from c-axis of SiC by an angle from 3.5^0 to 8^0 had a negligible effect on the GSAW characteristics.

The results of experimental characterization of SAW propagation properties for ZnO films on silicon carbide have been compared with simulated SAW characteristics. A good agreement between the calculated and measured GSAW velocities and HVPSAW propagation losses has been observed when published material constants of ZnO and SiC were used and the unknown elastic constant of SiC C_{13} was taken near zero. The observed discrepancy between the calculated and measured HVPSAW velocities has been attributed to the non-accurate values of ZnO film constants used in the calculations. The highest piezoelectric coupling has been predicted for the first GSAW mode when IDT is at the interface with metallized upper surface and for the second GSAW with IDT on the upper surface. In general, silicon carbide substrates with ZnO film can be characterized as a promising material for high-frequency SAW applications.

9. Opportunities Awaiting

The opportunities for improvement of films used in SAW devices and the possibilities for extending the capabilities of SAW devices abound. The new millenium will see advancements in the understanding of the acoustic properties of thin films and their application to achieve new and better SAW devices. In the communications area where integrated voice, data, and video systems will prevail, higher frequency, wider bandwidth, higher power, lower loss, and extended signal processing capabilities will be required of SAW devices. The addition of thin films to substrates could give the necessary edge in meeting filter and resonator requirements. Thus it is important that research and development continue in the processing, measurement, and application of thin films to SAW devices.

The requirements for high frequency, low loss, and high power strains the capabilities of the thin metal films used for SAW excitation and detection. There will be a strong requirement for small linewidth metal electrode films of high conductivity and negligible aging and migration effects. The development of alloyed and refractory metal sandwich layers with aluminum need additional development together with epitaxial growth where possible. Are there other high conductivity metal systems which should be investigated?

As frequencies go higher and linewidths become thinner there will be trade-offs made between the use of high velocity substrates such as diamond films with piezofilms versus harmonic GSAW and HVPSAW modes. Even the HVPSAW modes will benefit by the presence of a thin metal or dielectric film layer. There is a myriad of film layers with substrate cuts to be investigated theoretically and explored experimentally. How do we best approach the problem of identifying these film/substrate combinations theoretically and experimentally?

Continued improvements in film processing coupled with definitive measurements will lead to films with improved and reproducible acoustic properties and allowing more ready acceptance for their use with SAW devices. Films must be seen as a part of the regular wafer flow and not some interruption of the process. There is additional work that should be done on clarifying the role of substrate and surface conditions which enhance the early development and preferred growth of specific orientations of piezoelectric films. This will require the measurement of crystallite development at the initial growth stages. How does the presence of an oriented electric field or a thin oriented surface film influence the growth conditions?

There are some areas where better measurement techniques and new measurements will contribute to better films and a better understanding of their capabilities and limitations. Improved techniques for the measurement of film properties in-situ during growth are needed and may be important for process control. Meng et al. used laser technology to do real time stress and elastic constant measurement of aluminum nitride films on silicon.[44] The elastic properties of films are a reflection of film structure and bear a relationship to loss and piezoelectric properties. Are there other acoustically related measurement techniques for in-situ characterization?

It has been possible to isolate piezofilms from their supporting substrates, and this has lead to the development of thin film resonators for filter applications. The acoustic properties of an isolated film such as velocity, material Q, and coupling factor can be measured by resonance techniques. Will process methods for film isolation and control of film orientation lead to a more complete characterization of the fundamental film constants of elasticity, viscosity, and piezoelectricity? Can this technique be applied to metal and dielectric films?

Small levels of impurities 2-4% have been used to enhance the thermal, stress, and resistivity properties of films. Impurity additions have decreased the propagation loss properties at high frequencies. Recent experiments with sputtered aluminum nitride films indicated an increase in electromechanical coupling factor, stress relief, and smoother growth surface with the addition of hydrogen in the lattice.[56] Is it possible through experimental measurements and theory to bring about a better understanding of the role of impurities in acoustic films? How can we better enhance the elastic, piezoelectric, and loss properties of films by adding doping elements?

Higher coupling factor piezofilms of PZT, lithium niobate, lithium tantalate, and potassium niobate are being developed. The direct measurement of acoustic properties needs to be applied to these films. Will they eventually replace ZnO and AlN as the preferred piezoelectric films for the future?

There has been several investigations over the years on the use of piezofilms, particularly ZnO, for the enhancement of the coupling factor on such weak piezoelectric substrates as quartz and gallium arsenide. Nakamura and Hanaoka looked theoretically at the enhancement of a ZnO film on a high coupling factor material, 128° lithium niobate, and showed there was a dependence on the polarity of the substrate.[49] Experimentation on polarity effects needs to be carried out. How do the polarization properties of film and substrate affect the resulting coupling factor?

One of the early dreams was that of acoustoelectronic devices which through interactions of electric and magnetic fields with films produced integrated electronic functions. As the processing and tailoring of thin films becomes better understood, it may be important to revisit this fertile area. Are there still opportunities for microstrip acoustic components?

Finally there is the burgeoning area of SAW sensors. At the very heart of SAW sensors are the thin films which sniff, feel, listen, taste, see, and hear the environment. What are the wonderful opportunities awaiting the thin film specialist who can enhance the senses through the use of specialized thin films on SAW substrates?

10. Conclusions

We have explored the importance of thin films for SAW devices, looked at the properties of metal, dielectric, and piezoelectric films through harmonic transducer measurements, indicated the importance of acoustic parameters for the development of film layer SAW devices, and considered opportunities for film development in the future. Direct and indirect measurement techniques have added to our knowledge of what constitutes an acoustically good film. Guidelines which can be useful in developing processing conditions for the most acoustically sound films and related measurement techniques for solution to a particular SAW device problem have the following characteristics. To determine the best film deposition conditions, (1) select the major process parameters and the most critical SAW operational factor(s) as the measurement criteria, (2) set up a design of experiments routine (i.e. using uniform or orthogonal designs), and (3) follow with a statistical analysis, such as multiple linear regression, to isolate the dominant process parameter(s) for establishing the final deposition conditions. To obtain the most useful SAW device design parameters use direct measurement techniques and choose transducer patterns which most closely approximate the type of device being developed (i.e. if ladder or lattice filters use a resonator pattern to determine velocity, Q, and capacitance ratio). If it is necessary to use indirect measurements, such as x-ray, SEM, Auger, etc., choose a parameter which is most representative of the operational factor(s) and calibrate.

To experimentally explore new acoustic modes and properties for films on substrates, choose transducer test patterns and measurement techniques which maximize the amount of data for analysis. In most cases for films on substrates for SAW device development, a simple interdigital electrode with harmonic generation capabilities can give acoustic data over a wide range of frequencies. Wherever possible use available theoretical models for the understanding and verification of experimental measurements and to direct further experimental measurement work.

The area of film growth, measurements, and SAW device application covers a time span of well over thirty years. With new substrate materials under development and improved cuts of standard materials being identified, the opportunities for film research, development, and application the next 30 years are even more inviting. It will be an exciting time for high-quality acoustic thin films to make an even greater impact on SAW devices.

11. Acknowledgments

The author expresses thanks to Motorola for the support of film development work, and is especially grateful to coworkers at Motorola, Bob Dablemont, Dan Knuth, Virginia Hernandez, Tom Hickernell, Claudia Jensen, and Gloria Judge, in the area of film processing, measurement, and development. Continued discussions with the wider SAW device community and attendance at symposia have stimulated ideas expressed in this paper. Collaborations with Eric Adler, Irina Didenko, Vitali Fedosov, Luke Mang, Ming Liaw, Natalya Naumenko, and Giovanni Socino in collaboration with colleagues at the University of Perugia, have contributed directly to the work reported in this paper. The author has been blessed to be in the exciting technical area of SAW devices over the past 30 years, which has lead to a worldwide network of friendships and cooperative exchanges of technical and scientific information.

12. References

1. F. S. Hickernell, "Surface acoustic wave technology, macrosuccess through microseisms," Physical Acoustics, vol. XXIV, edited by E. P. Papadakis, Academic Press, NY, (1998) 135-207.
2. D. P. Morgan, "History of SAW devices," Proc. International Freq. Control Symposium, (1998) 439-460.
3. D. P. Morgan, "A history of surface acoustic wave devices," Itn'l Journ. of High Speed Electronics and Systems, (2000) (this issue).
4. R. M. White and F. W. Voltmer, "Direct piezoelectric coupling to surface elastic waves." Appl. Phys. Lett., 7, (1965) 314-316.
5. F. S. Hickernell, "The role of layered structures in surface acoustic wave technology," Int'l Specialist Seminar on Component Performance and Systems Applications of Surface Acoustic Wave Devices, (1973) 11-21.
6. C. Lardat, C. Maerfeld, and P. Tournois, "Theory and performance of acoustical dispersive surface wave delay lines," Proc. of the IEEE, 59 (1971) 355-368.
7. S. Fujishima, H. Ishiyama, A. Inoue, and H. Ieki, "Surface acoustic wave VIF filters for TV," Proc. 30th Annual Frequency Control Symposium, (1976) 119-122.
8. F. J. Hickernell, R. X. Yue, and F. S. Hickernell, "Statistical modeling for the optimal deposition conditions of sputtered piezoelectric films," IEEE Trans. Ultrason., Ferroelectrics, and Freq. Ctrl., 44 (1997) 615-623.
9. Kazuo Tsubouchi and Nobuo Mikoshiba, "Zero temperature-coefficient SAW devices on AlN epitaxial films," IEEE Trans. on Sonics and Ultrasonics, 32 (1985) 634-644.
10. G. Carlotti, D. Fioretto, L. Palmieri, G. Socino, L. Verdini and E. Verona, "Brillouin scattering by surface acoustic modes for elastic characterization of zinc oxide films," IEEE Trans. on Ultrasonics, Ferroelectrics, and Frequency Control, 38 (1991) 56-61.

11. M. Pereira da Cunha and E. L. Adler, "High velocity pseudosurface waves (HVPSAW)," IEEE Trans. Ultrasonics, Ferroelectrics, and Frequency Control, **42** (1995) 840 - 844.
12. M. Pereira da Cunha, "High velocity pseudosurface waves (HVPSAW): further insight," Proc. 1996 IEEE International Ultrasonics Symposium, **43** (1996) 97-106.
13. F. S. Hickernell, P. L. Clar, and I. R. Cook, "Pulsed DC breakdown between interdigital electrodes," Proc. 1972 IEEE Ultrason. Symp. (1972) 388-391.
14. Y. Satoh, T. Nishihara, O. Ikata, M. Ueda, and H. Ohomori, "SAW duplexer metallizations for high power durability," Proc. 1998 IEEE Ultrason. Symp., (1998) 17-26.
15. N. Kimura, M. Nakano, and K. Sato, "High power-durable and low loss single-crystalline Al/Ti electrodes for RF SAW devices," Proc. 1998 IEEE Ultrason. Symp., (1998) 315-318.
16. J. Pouliquen and G. Vaesken, "Effect of a metallic film on the propagation of Rayleigh waves," J. Appl. Phys., **44** (1973) 1524-1526.
17. W. P. Mason, "Transmission and amplification of acoustic waves in piezoelectric semiconductors," Physical Acoustics, vol. IV, edited by W. P. Mason, Academic Press, NY, (1966) 1-45.
18. P. Bierbaum, "Interaction of ultrasonic surface waves with conduction electrons in thin metal films," Appl. Phys. Lett., **21** (1972) 595-598.
19. R. Gautier and C. Maerfeld, "Wideband elastic convolvers," Proc. 1980 IEEE Ultrason. Symp., (1980) 30-36.
20. C. T. Vasile and R. LaRosa, "Broadband bulk-wave cancellation in acoustic-surface-wave devices," Electron. Lett., **8** (1972) 478-480.
21. M. T. Wauk, "Suppression of spurious triple-transit signals in acoustic surface-wave delay lines," Appl. Phys. Lett., **20** (1972) 481-483.
22. L. R. Adkins and A. J. Hughes, "Guided acoustic surface waves: principles and devices," Acoustic Surface Wave and Acousto-optic Devices, edited by Thomas Kallard, Optosonic Press, (1971) 17-29.
23. H. Seidel and D. L. White, "Ultrasonic surface waveguides," (1968) U. S. Patent 3,406, 358 and A. H. Meitzler and H. F. Tiersten, "Elastic surface waveguide," (1968) U. S. Patent 3,409,858.
24. G. Cambon and C. F. Quate, "Dispersive Rayleigh waves on quartz," Electron. Lett., **5** (1969) 402-403.
25. G. Cambon and M. Rouzeyre, "Attenuation of dispersive Rayleigh waves on quartz," Electron. Lett., **6** (1970) 539-540.
26. Jelks and R. S. Wagers, "Elastic constants of electron-beam deposited thin films of molybdenum and aluminum on lithium niobate," Proc. 1983 IEEE Ultrason. Symp. (1983) 319-322.
27. K. L. Davis and J. F. Welter, "SAW attenuation in metal film-coated delay lines," Proc. 1979 IEEE Ultrason. Symp., (1979) 659-662.
28. R. Thomas, T. W. Johannes, W. Ruile, and R. Weigel, "Determination of phase velocity and attenuation of surface acoustic waves with improved accuracy," Proc. 1998 IEEE Ultrason. Symp., (1998) 277-282.
29. E. A. Ash, "Surface wave grating reflectors and resonators," Digest of the IEEE Symposium Microwave Theory and Techniques, (1970) 385-386.
30. R. C. Williamson, "Properties and applications of reflective array devices," Proc. IEEE, **64** (1976) 702-710.

31. L. P. Solie. "Reflective dot array devices," Proc. 1977 IEEE Ultrason. Symp. (1977) 579-584.
32. F. G. Marshall, C. O. Newton, and E. G. S. Paige "Theory and design of the surface acoustic wave multistrip coupler," IEEE Trans. Microwave Theory and Techn. 21 (1973) 206-215.
33. F. G. Marshall, C. O. Newton, and E. G. S. Paige, "Surface acoustic wave multistrip components and their applications" IEEE Trans. Microwave Theory and Techn. 21 (1973) 216-225.
34. F. S. Hickernell and T. S. Hickernell, "Surface acoustic wave characterization of PECVD films on gallium arsenide," IEEE Trans. on Ultrasonics, Ferroelectrics, and Frequency Control, 42 (1995) 410-415.
35. E. L. Adler, J. K. Slaboszewicz, G. W. Farnell, and C. K. Jen, "PC software for SAW propagation in anisotropic multilayers," IEEE Trans. on Ultrasonics, Ferroelectrics, and Frequency Control, 38, pp. 215-223, 1990.
36. A. J. Slobodnik, E. D. Conway, and R. T. Delmonico, "Microwave Acoustics Handbook" Air Force Cambridge Research Laboratories, (1973).
37. F. S. Hickernell and H. D. Knuth, "The use of design of experiments for the optimization of deposited glass on SAW filters," Proc. 1999 IEEE International Frequency Control Symposium, (1999) 950-953.
38. T. W. Grudkowski, G. K. Montress, M Gilden and J. F. Black, "GaAs monolithic SAW devices for signal processing and signal control," Proc. IEEE Ultrason. Symp., (1980) 88-97.
39. K. W. Loh, D. K. Schroder and R. C. Clarke, "The GaAs diode storage correlator," Proc. IEEE Ultrason. Symp., (1980) 98-103.
40. F. S. Hickernell, Proc. First Int'l Symp. on Sputter and Plasma Processes - Tokyo, Japan, (1991) 57-66.
41. G. Carlotti, F. S. Hickernell, H. M. Liaw, L. Palmieri, G. Socino, and E. Verona, "The elastic constants of sputtered aluminum nitride films," Proc. 1995 IEEE Ultrason. Symp., (1995) 353-356.
42. L. McNeil, M. Grimsditch, and R. H. French, J. Am. Ceram. Soc., "Elastic constants of single crystal aluminum nitride," 76 (1993) 1132-1133.
43. J. H. Harris, R. A. Youngman, and R. G. Teller, "On the nature of the oxygen-related defect in aluminum nitride," J. Mater. Res., 5 (1990) 1763-1773.
44. W. J. Meng, J. A. Sell, T. A. Perry, and G. L. Eesley, "Real time stress measurements and elastic constant of aluminum nitride films on Si (111)," J. Vac. Sci. Technol. A, 11 (1993) 1377-1382.
45. F. S. Hickernell and H. D. Knuth, "The elastic properties of wax determined by surface acoustic wave measurements," Proc. 1999 IEEE Ultrason. Symp. (1999).
46. F. S. Hickernell, H. D. Knuth, R. C. Dablemont, and T. S. Hickernell, "The Surface Acoustic Wave Propagation Characteristics of 41° Y-X Lithium Niobate with Thin-Film SiO_2", Proc. 1996 IEEE International Frequency Control Symposium, (1996) 216-221.
47. F. S. Hickernell, H. D. Knuth, R. C. Dablemont, and T. S. Hickernell, "The Surface Acoustic Wave Propagation Characteristics of 64° Y-X $LiNbO_3$ and 36° Y-X $LiTaO_3$ Substrates with Thin-Film SiO_2," Proc. 1995 IEEE Ultrasonics Symposium, (1995) 345-348.

48. M. Kadota "Combination of ZnO film and quartz to realize large coupling factor and excellent temperature coefficient for SAW devices," Proc. 1997 IEEE Ultrasonics Symposium, (1997) 261-266.
49. K. Nakamura and T. Hanaoka, "Propagation characteristics of surface acoustic waves in ZnO/LiNbO$_3$ structures," Jpn. J. Appl. Phys., **32** (1993) 2333-2336.
50. F. S. Hickernell and E. L. Adler, "The experimental and theoretical characterization of SAW modes on ST-X quartz with a zinc oxide film layer," Proc. 1997 IEEE International Frequency Control Symposium, (1997) 852-857.
51. D. L. Dreifus, R. J. Higgins, C. Jensen, S. Young, P. J. Ellis, R. A. Fauber, B. A. Fox, R. B. Henard, and B. R. Stoner, "Low-loss high frequency diamond-based SAW devices," 1998 IEEE Frequency Control Symposium, (1998) Pasadena, California. (unpublished)
52. K. Higaki, H. Nakahata, H. Kitabayashi, S. Fujii, K. Tanabe, Y. Seki, and S. Shikata, "High frequency SAW filter on diamond," 1997 IEEE MTT-S Digest, 829-832.
53. F. S. Hickernell, "The characterization of zinc oxide films on silicon carbide," Proc.1998 International Symposium on Acoustoelectronics, Frequency Control, and Signal Generation, (1998) 103-106.
54. I. S. Didenko, F. S. Hickernell, N. F. Naumenko, "The experimental and theoretical characterization of the SAW propagation properties for zinc oxide on silicon carbide," IEEE Trans, Ultrason., Ferroelect., and Freq. Contr. **47** (2000) 179-187.
55. J. G Gualtieri, J. A. Kosinski, and A. Ballato, "Piezoelectric materials for acoustic wave applications." IEEE Trans, Ultrason., Ferroelect., and Freq. Contr., **41** (1994) 53-58.
56. Y-J Yong and J-Y Lee, "Characteristics of hydrogenated aluminum nitride films prepared by radio frequency reactive sputtering and their application to surface acoustic wave devices," J. Vac. Sci. Technol. A **15** (1997) 390-393.

BULK AND SURFACE ACOUSTIC WAVES IN ANISOTROPIC SOLIDS

ERIC L. ADLER
Department of Electrical and Computer Engineering
McGill University, Montreal, QC, H3A 2A7, Canada

ABSTRACT: In this paper methods for analyzing acoustic propagation characteristics for bulk and surface acoustic waves in anisotropic piezoelectric multilayers are described. The method's conceptual usefulness is demonstrated by examples showing how problems of guided wave propagation in complicated layered surface acoustic wave device geometries are simplified. The formulation reduces the acoustoelectric equations to a first order ordinary matrix differential equation in the variables that must be continuous across interfaces. The solution to these equations is a transmission matrix that maps the variables from one layer face to the other. Interface boundary conditions for a planar multilayer are automatically satisfied by multiplying the individual transmission matrices in the appropriate order thus reducing the problem to imposing boundary conditions appropriate to the remaining free surface. The dimensionality of the problem being independent of the number of layers is a significant advantage. A classification scheme for reducing problem dimensionality, based on an understanding of crystal symmetry properties, further simplifies surface acoustic wave problems.

1. Introduction

The propagation of acoustic waves in anisotropic solids has received considerable attention from engineers, geophysicists, and scientists in signal processing, frequency control, and nondestructive evaluation. Ultrasonic devices are being developed using piezoelectric crystalline materials either on their own in surface acoustic wave (SAW) filters and resonators, or in layered geometries in stacked bulk acoustic wave (BAW) crystal filters and composite transducers. Laminated composites are adopted for their structural properties in aerospace, among other applications. Thus, many device studies involve layered structures made up of anisotropic materials, some of which are also piezoelectric. Surface wave filters have had a major impact in IF and RF communication systems and are currently the favored filter type used in cellular wireless systems. Their small size and weight have contributed to a new generation of small and lightweight cellular telephones.

This work provides a unified approach for describing the characteristics and classifying the different types of BAWs and SAWs used in high-speed electronic applications.[1,2] The approach is tutorial and provides many details not usually found in the research literature. Section 1 is a review of the fundamentals of acoustoelectricity and a statement of the defining equations. Section 2 is a formulation of the acoustoelectric propagation problem for bulk waves in a piezoelectric material as a first-order vector-matrix ordinary differential equation (ODE) with solution given by an exponential matrix. There is a development of the eigenvalue problem for bulk waves in piezoelectric materials, a discussion of the existence and nature of the three orthogonal bulk modes, and the defining equations for electromechanical coupling and power flow. Section 3 is a compilation of different wave guiding geometries, their boundary conditions, a matrix description, and a derivation of the ODE for the SAW geometries. Interfacial boundary conditions are automatically satisfied by matrix multiplication that preserves the dimensionality of the problem independent of the number of layers.[3,4] Section 4 is an examination of the way crystal symmetry affects the types of guided wave solutions and contains a scheme for classifying and simplifying surface acoustic waves.[5]

1.1 Fundamentals of acoustoelectricity

All solids consist of molecules or atoms arranged in a regular three dimensional pattern. These molecules can deform from their equilibrium position. Elastic deformations can be static or dynamic and, in the non quantum case are acoustic.

In BAWs and SAWs atomic structure is irrelevant. A solid is an elastic continuum with a certain crystalline symmetry and with physical properties described by a set of thermodynamic constants: piezoelectric, elastic moduli, density, and permittivity; in a set of constitutive equations: Maxwell's Equations and Hooke's Law. In piezoelectric materials these equations are coupled.[6,7] One describes the dynamics of deformation for a macroscopic solid by a translational equation of motion equivalent to Newton's Law.

The field variables describing the dynamic behavior of solids involve tensor relations and vector fields because of the anisotropic nature of acoustoelectric properties of materials. A vector-matrix formulation simplifies descriptions of the deformation dynamics of both bulk and surface acoustic wave problems when studying only sinusoidal or harmonic steady state acoustic waves.[6,7]

1.2. Definitions and terminology

These equations and definitions form the basis for this review. Vectors will be column vectors unless otherwise indicated, and they will usually be written as transposes of row vectors, u^t. The complex traveling wave variation in x and t for all variables at radian frequency, ω, and phase velocity, v_p, assumed throughout, is of the form

$$f(x,t) = \exp\{j\omega(t - x/v_p)\}$$

$$df/dt = j\omega f$$

$$df/dx = (-j\omega/v_p)f . \quad (1)$$

The harmonic time dependence $exp(j\omega t)$ is used throughout, the electrical engineering phasor notation is adopted, and time differentiation du/dt is replaced by $j\omega u$. We assume small signal conditions and consider only linear equations.[7]

1.2.1. Strain and stress: tensor and vector definitions

If the material particle displacements from equilibrium position R, under harmonic time deformation, is represented by the three-component column vector, $u = [u_x \; u_y \; u_z]^t$, the corresponding velocity field vector $v = j\omega u$. Strain is a second rank tensor or a three by three symmetric matrix, S, with elements:

$$S_{ij} = \tfrac{1}{2}(\partial u_i / \partial R_j + \partial u_j / \partial R_i) \qquad i, j = 1, 2, 3 . \tag{2}$$

In (2) only linear terms have been kept. The columns of the strain matrix are the x, y, and z oriented strain vectors. The strain matrix is the symmetric part of the gradient of the displacement vector, u.[7]

$$S = \nabla_s u = \nabla_s v / j\omega. \tag{3}$$

Stress, T, is also a second rank tensor or a three by three symmetric matrix of traction forces per unit area acting on a rectangular volume element. The columns of T are the stress vectors acting along the Cartesian axes,

$$T = [\, T^x \; T^y \; T^z \,] . \tag{4}$$

Since the stress and strain matrices are symmetric, there are only six independent components. One defines six-element stress and strain vectors by contracting the two subscripts into one: 11, 22, 33 or xx, yy, zz are replaced by 1, 2, 3; and 23, 13, 12 or yz, xz, xy are replaced by 4, 5, 6. The six-component stress and strain vectors are

$$T = [\, T_1 \; T_2 \; T_3 \; T_4 \; T_5 \; T_6 \,]^t$$

$$S = [\, S_1 \; S_2 \; S_3 \; S_4 \; S_5 \; S_6 \,]^t . \tag{5}$$

T and S are used for both vectors and tensors.

1.2.2. Translational equation of motion

The vector form of Newton's force equation is

$$\nabla . T = \rho \, \partial^2 u / \partial t^2$$

$$\partial T^x / \partial x + \partial T^y / \partial y + \partial T^z / \partial z = j\omega \rho v \tag{6}$$

with

$$T^x = [\ T_{11}\ T_{12}\ T_{13}\ ]^t = [\ T_1\ T_6\ T_5\ ]^t$$
$$T^y = [\ T_{21}\ T_{22}\ T_{23}\ ]^t = [\ T_6\ T_2\ T_4\ ]^t$$
$$T^z = [\ T_{31}\ T_{32}\ T_{33}\ ]^t = [\ T_5\ T_4\ T_3\ ]^t. \qquad (7)$$

1.2.3. The linear constitutive equations

For a piezoelectric material the electromechanical properties are described by a set of constitutive equations consisting of Hooke's law and the electrical equation of state. These relations are, in general, nonlinear; however for small sinusoidal deformations the linearized equations accounting for piezoelectricity are:[6,7]

$$T = C^E S - e^t E$$
$$D = \varepsilon^S E + eS \qquad (8)$$

where D and E are the electric displacement and electric field vectors, ε^S is the three by three electrical permittivity matrix measured at constant strain, e is the three by six piezoelectric coefficient matrix, and C^E is the elastic stiffness matrix measured at constant electric field. There are other ways of writing constitutive equations, depending on the chosen independent variables.

When viscous and dielectric losses under sinusoidal conditions are considered, the stiffness and permittivity matrices become complex and frequency dependent. The imaginary part of the stiffness matrix is $j\omega\eta$ where η is the material viscosity matrix.

1.2.4. The Divergence Theorem of electromagnetism and the electrostatic approximation

In piezoelectric insulating materials no space charge exists and the divergence of D vanishes.

$$\nabla \cdot D = 0 \rightarrow$$
$$\partial D_x / \partial x + \partial D_y / \partial y + \partial D_z / \partial z = 0. \qquad (9)$$

Since acoustic wave propagation velocities are about five orders of magnitude smaller than the speed of light in piezoelectric materials, the electric field is electrostatic and given by

$$E = -\nabla\varphi. \qquad (10)$$

2. Bulk Acoustic Waves

Here we discuss the bulk wave solution formalism for anisotropic piezoelectrics, derive the system differential equations and the transmission matrix, formulate the BAW characteristic eigenvalue equation, and define piezoelectric stiffening and electromechanical coupling. We examine the Poynting vector, group velocity and power flow angle. Given examples illustrate special features of BAW solutions as they relate to crystal types and piezoelectricity. We describe BAW resonator input impedance calculations.

2.1. Bulk wave solution formalism for anisotropic piezoelectrics

Bulk acoustic waves are uniform plane waves with no spatial variations normal to the propagation direction. Planes of constant phase are normal to the direction of propagation. The wavelength is the phase velocity to frequency ratio. In piezoelectric materials there are electrical field components as part of the wave. To find the velocities of the three acoustic modes that propagate in anisotropic materials, one solves a three dimensional algebraic eigenvalue problem. The characteristic equation of the BAW eigenvalue problem is a cubic polynomial, solved explicitly.[8,9]

2.1.1. Coordinate transformations

In any problem involving acoustic waves the direction of propagation with respect to crystalline (X, Y, Z) axes is usually known or chosen. Here the propagation is always in the x direction of an (x, y, z) set of axes. The Euler angles, $<\varphi, \theta, \psi>$, refer these axes to the crystalline axes. These angles specify three consecutive counterclockwise rotations: φ about the Z axis to give a first intermediate system labeled x', y', z'; θ about the new x' axis to give a second intermediate system labeled x'', y'', z''; and ψ about the newer z'' axis to give the third and final position labeled x, y, z.[10,11] The thermodynamic constants of the material are then transformed to the chosen orientation using standard coordinate transformation routines.[12] Consequently, waves always propagate along x and these formulas or computational procedures work for any material and orientation.[13]

2.1.2. The system differential equations and transmission matrix for bulk waves

In piezoelectric materials the waves are acoustoelectric, the acoustic and electrical field components are coupled, and all the fields propagate at v_p, the velocity of sound. Bulk waves depend only on x, and the partial differential equations (PDEs) given in Section 1.2 become ODEs. From (1), the complex traveling wave x variation for all variables at radian frequency, ω, and phase velocity, v_p, is $\exp(-j\omega x/v_p)$ and the derivative $dv/dx = -j\omega v/v_p$. The electromechanical equation of motion (6) and the strain relations (3) simplify to

$$d\ T^x/dx = -j\omega T^x/v_p = j\omega\rho v$$

$$dv/dx = j\omega\ S^x \ . \tag{11}$$

The equation of motion in (11) becomes

$$T^x = -\rho v_p\ v\ . \tag{12}$$

The stress, T^x, is proportional to the velocity field, v, and the proportionality constant, ρv_p, defines the mechanical impedance. The material constitutive relations are symbolically identical to those in Section 1, only the C, e, and ε matrix elements are rotated to the desired x direction with respect to the crystalline axes.

With only x variations, the electric field (10) has one component

$$E_x = -d\varphi/dx\ . \tag{13}$$

Selecting the components of $T^x = [T_1\ T_6\ T_5]^t$ and D_x from (8), the constitutive equations; with (11) to eliminate S^x, and (13) to eliminate E_x; result in ODEs: for the x component of stress, particle displacement, and electric potential.

$$T^x = C^{11}\ du/dx + [e_{11}\ e_{16}\ e_{15}]^t\ d\varphi/dx$$

$$D_x = [e_{11}\ e_{16}\ e_{15}]\ du/dx - \varepsilon_{11}\ d\varphi/dx \tag{14}$$

where, in (14) the relations $S^x = du/dx$ and $E_x = -d\varphi/dx$ have been used, the "E" superscript is dropped in C, and

$$C^{11} = \begin{bmatrix} C_{11} & C_{16} & C_{15} \\ C_{16} & C_{66} & C_{56} \\ C_{15} & C_{56} & C_{55} \end{bmatrix} . \tag{15}$$

Expressing (14) in terms of the velocity field vector v and $j\omega\varphi$ gives

$$j\omega T^x = C^{11}\ dv/dx + [e_{11}\ e_{16}\ e_{15}]^t\ d(j\omega\varphi)/dx$$

$$j\omega D_x = [e_{11}\ e_{16}\ e_{15}]\ dv/dx - \varepsilon_{11}\ d(j\omega\varphi)/dx\ . \tag{16}$$

Equation (16) can be rewritten in terms of the four component vector $[v^t\ j\omega\varphi]^t$ as the matrix equation

$$\Gamma^{11}\frac{d}{dx}\begin{bmatrix} v \\ j\omega\varphi \end{bmatrix} = j\omega\begin{bmatrix} T^x \\ D_x \end{bmatrix} \tag{17}$$

with

$$\Gamma^{11} = \begin{bmatrix} C_{11} & C_{16} & C_{15} & e_{11} \\ C_{16} & C_{66} & C_{56} & e_{16} \\ C_{15} & C_{56} & C_{55} & e_{15} \\ e_{11} & e_{16} & e_{15} & -\varepsilon_{11} \end{bmatrix}. \tag{18}$$

The mechanical equation of motion in (11) is repeated here with the simplification for the divergence (9) of D

$$dT^x/dx = j\omega\rho v$$

$$dD_x/dx = j\omega D_x / v_p = 0. \tag{19}$$

It follows from (19) that $D_x = 0$ or a constant, meaning there is no traveling wave component to the electric displacement. For unbounded bulk waves, $D_x = 0$, whereas for driven transducers and resonators, D_x must be spatially invariant to satisfy the circuit driving current constraint. The current density in a transducer or resonator is $j\omega D_x$. The input impedance is purely reactive in a resonator when viscosity is neglected. The real part of the input impedance for a transducer bonded to a substrate is the acoustic power per unit current into the substrate.

From (17) and (19) one forms a single eight dimensional matrix ODE for the vector τ of six mechanical and two electrical variables

$$\tau = \begin{bmatrix} T^x \\ D_x \\ v \\ j\omega\varphi \end{bmatrix}. \tag{20}$$

The system ODE is

$$d\tau/dx = j\omega A\tau \tag{21}$$

where A is the eight by eight matrix

$$A = \begin{bmatrix} 0 & G \\ X & 0 \end{bmatrix}. \tag{22}$$

X is the inverse of the non singular four by four Γ^{11}, and the matrix G is four by four diagonal having the material density ρ in the first three entries and zero in the fourth.

The solution to the ODE that gives the x dependency of the vector τ is just the matrix exponential solution to a first order differential equation

$$\tau(x+h) = \exp(j\omega Ah)\tau(x) = \Phi(h)\tau(x). \tag{23}$$

The matrix $\exp(j\omega Ah)$ is the transmission matrix, $\Phi(h)$, which maps the eight field components over a distance h along x.

2.1.3. Multilayer resonator excitation

The external circuit driving current must equal the current density, $j\omega D_x$, times the area of the BAW transducer or resonator. One calculates a transducer or resonator input impedance by finding the voltage to current ratio after satisfying all the mechanical boundary conditions of the geometry. For layered bulk wave devices there is a transmission matrix for each layer. Here we outline the resonator computation procedure. This and other geometries are described elsewhere in more detail.[2,14]

Each layer in a resonator stack has a frequency dependent transmission matrix, (23). The transmission matrix, Φ, for a stack is calculated as the matrix product of the individual material Φs at that frequency. This matrix product automatically satisfies all interfacial boundary conditions since the elements of τ are continuous at an interface. The upper and lower terminal faces are electroded, constant potential, but mechanically free, $T^x = 0$; and the same current flows through each layer. Finite electrode thicknesses are included as additional layers.[1,2,13,14] To characterize the resonator properties the input impedance is calculated. The electrical boundary conditions are: one surface at zero volts, $\varphi = 0$; the other at the applied voltage, $\varphi = V$. Since D_x is constant along x, the current density in the x direction, $J = j\omega D_x$, is constant throughout the stack. Setting $D_x = J/(j\omega)$ and $T^x = 0$ at the upper and lower faces, (23) with Φ the overall transmission matrix is

$$\begin{bmatrix} 0 \\ 0 \\ 0 \\ J/(j\omega) \\ v \\ 0 \end{bmatrix}_u = \Phi \begin{bmatrix} 0 \\ 0 \\ 0 \\ J/(j\omega) \\ v \\ j\omega V \end{bmatrix}_l \tag{24}$$

It is convenient to interchange the order of the fourth and eighth equations of (24) and write

$$\begin{bmatrix} 0 \\ 0 \\ 0 \\ 0 \\ v \\ J/(j\omega) \end{bmatrix}_u = Q \begin{bmatrix} 0 \\ 0 \\ 0 \\ J/(j\omega) \\ v \\ j\omega V \end{bmatrix}_l \tag{25}$$

where Q is obtained from Φ by interchanging the fourth and eighth rows. The upper four equations of (25) are

$$\begin{bmatrix} 0 \\ 0 \\ 0 \\ 0 \end{bmatrix} = Q_{ul} \begin{bmatrix} 0 \\ 0 \\ 0 \\ J/(j\omega) \end{bmatrix}_u + Q_{ur} \begin{bmatrix} v \\ j\omega V \end{bmatrix}_l \tag{26}$$

where Q_{ul} is the four by four upper left block of Q and Q_{ur} is the four by four upper right block of Q. Hence, one can solve for v and V at the lower face in terms of the current density.

$$\begin{bmatrix} v \\ j\omega V \end{bmatrix}_l = -Q_{ur}^{-1} Q_{ul} \begin{bmatrix} 0 \\ J/(j\omega) \end{bmatrix} = -z \begin{bmatrix} 0 \\ J/(j\omega) \end{bmatrix} \tag{27}$$

From (27) the voltage is obtained in terms of the current and the (4,4) matrix element of z as

$$j\omega V = -z_{44} J/(j\omega) \tag{28}$$

and the input impedance Z_i is then

$$Z_i = z_{44}/(\omega^2 a), \tag{29}$$

with a the electrode area. The input impedance, computed using (23) to (29), has frequency as the only independent variable, incorporates the full acoustoelectric dynamics of all the layers, and thus automatically includes all modes that are excited.

To emphasize, for arbitrary material orientations and piezoelectricity these calculations include all the acoustic and electrical information, do not depend on any prior knowledge of material velocities or electromechanical coupling, and require only the layer orientations and material constants.[2,14]

2.2. Three bulk wave modes

We derive the BAW characteristic or eigenvalue equation for the general piezoelectric case. The effect of piezoelectricity is equivalent to having modified stiffness constants, called piezoelectric stiffening. We define the Poynting vector and outline the calculation of power and power flow angle.[7,8]

2.2.1. The BAW eigenvalue equation

Since D_x vanishes for bulk waves, the electric potential is eliminated from (16) using

$$0 = [e_{11}\ e_{16}\ e_{15}]\ dv/dx - \varepsilon_{11} d(j\omega\varphi)/dx \tag{30}$$

to give an equation in only the stress and velocity field vectors

$$j\omega\ T^x = C^{11}\ dv/dx + [e_{11}\ e_{16}\ e_{15}]^t\ [e_{11}\ e_{16}\ e_{15}]\ /\varepsilon_{11}\ dv/dx . \tag{31}$$

Recalling $d/dx = -j\omega/v_p$ from (1) and using (11) to eliminate the stress in (31) gives, after some reordering, the algebraic eigenvalue equation for v called the stiffened Christoffel equation for the BAW problem,[7]

$$[\ C^{11} + [e_{11}\ e_{16}\ e_{15}]^t\ [e_{11}\ e_{16}\ e_{15}]\ /\varepsilon_{11} - \rho\ v_p^2\ I\ ]\ v = 0$$

$$[\ \underline{C}^{11} - \rho\ v_p^2\ I\ ]\ v = 0 \tag{32}$$

The coefficient matrix in (32) is called the stiffened stiffness matrix, $\underline{C}^{11}$. It is a matrix of effective stiffness constants due to piezoelectricity. The eigenvalues and eigenvectors of the stiffened stiffness matrix divided by density yield the phase velocities and polarizations of the three modes. Solving this eigenvalue equation gives the three bulk mode solutions along an arbitrary direction. The eigenvalues of $\underline{C}^{11}/\rho$ are the squares of the three phase velocities, and the eigenvectors are the corresponding eigenvectors of v.

These are the explicit solutions for bulk waves. The modes are mutually orthogonal by virtue of the real and symmetric nature of the stiffened matrix, $\underline{C}^{11}$. In general there are one quasi-longitudinal mode and two quasi-shear modes. Pure longitudinal or pure shear modes occur only along certain directions of propagation. The three velocities are distinct, but degeneracies do occur for some orientations and crystal symmetries. One finds the remaining components of τ from the eigenvectors for v using (30) for the potential and (12) for the stress. All field variables can now be expressed either in terms of the components of τ or the eigenvectors for v.

For non piezoelectric materials (31) simplifies to $j\omega\ T^x = C^{11}\ dv/dx$, the electrical variables do not apply, the τ vector is six dimensional, Λ is six by six, and the eigenvalues and eigenvectors of C^{11}/ρ provide the solution for the BAW.

2.2.2. Piezoelectric stiffening and electromechanical coupling

As seen in (32), piezoelectricity has the effect of modifying the stiffness matrix elements in the BAW eigenvalue equation. The stiffness matrix elements in $\underline{C}^{11}$, are larger than those in C^{11}, no piezoelectricity. Thus, the resulting phase velocities are increased due to piezoelectricity. The change in stiffness, a measure of the strength of the piezoelectricity, is quantified by an electromechanical coupling constant, k_T^2, for each of the modes, [8]

$$k_T^2(i) = ([e_{11}\ e_{16}\ e_{15}].v(i))^2/(\varepsilon_{11}\rho v_p^2(i)) \qquad i=1,2,3 \tag{33}$$

where $v(i)$ is the normalized polarization vector for mode i. The coupling constant is a key parameter in the selection of materials for designing transducers and resonators.[8]

2.2.3. The Poynting vector: power, group velocity and power flow angle

The Poynting vector for a mode, P, for piezoelectrics is the power flow per unit area. It is normal to the slowness surface for the mode and parallel to the direction of the power flow and the group velocity. The slowness surface is the contour generated by calculating the reciprocal of phase velocity for all directions. The angle it makes with the propagation direction is called the power flow angle. The real form of the Poynting Theorem for piezoelectrics is[7]

$$P = Re\{-[v_x\ v_y\ v_z]^*.[T^x\ T^y\ T^z] - (j\omega\varphi)^* D^t\}. \tag{34}$$

The first part is the mechanical contribution, and (34) can be expanded to display the components along the three coordinate directions as a row vector,

$$P = Re\left\{-\begin{bmatrix}v^t & j\omega\varphi\end{bmatrix}^* \begin{bmatrix}T^x & T^y & T^z \\ D_x & D_y & D_z\end{bmatrix}\right\}. \tag{35}$$

The asterisk denotes complex conjugation, and rms values are adopted for the field variables. Recalling that D_x vanishes for bulk waves (19), the x component of P for bulk waves is entirely mechanical and since v is orthonormal

$$P_x = Re\{-(v^t)^*.T^x\} = Re\{-(v^t)^*\rho v_p v\} = \rho v_p. \tag{36}$$

The electric term contribution to power flow in bulk waves is normal to the propagation direction having only y and z components.

To calculate the three components of power in general, all the stress vectors and the electric displacement fields are needed. These are expressible in terms of v, $j\omega\varphi$ and the Γ matrices

$$\begin{bmatrix}T^x \\ D_x\end{bmatrix} = -(\Gamma^{11}/v_p)\begin{bmatrix}v \\ j\omega\varphi\end{bmatrix}$$

$$\begin{bmatrix}T^y \\ D_y\end{bmatrix} = -(\Gamma^{21}/v_p)\begin{bmatrix}v \\ j\omega\varphi\end{bmatrix}$$

$$\begin{bmatrix}T^z \\ D_z\end{bmatrix} = -(\Gamma^{31}/v_p)\begin{bmatrix}v \\ j\omega\varphi\end{bmatrix}. \tag{37}$$

Thus, all components of P can be written in terms of $[v^t \ j\omega\varphi]$ and the Γ matrices

$$P_x = Re\{ [v^t \ j\omega\varphi]^* \ (\Gamma^{11}/v_p)[v^t \ j\omega\varphi]^t \}$$
$$P_y = Re\{ [v^t \ j\omega\varphi]^* \ (\Gamma^{21}/v_p)[v^t \ j\omega\varphi]^t \}$$
$$P_z = Re\{ [v^t \ j\omega\varphi]^* \ (\Gamma^{31}/v_p)[v^t \ j\omega\varphi]^t \} \ . \tag{38}$$

Since the potential term is expressible in terms of v from (30), every component in (38) is expressible in terms of the eigenvectors of v, the phase velocity, and the material constants in the Γ matrices. The components of the group velocity, v_g, are the Poynting vector components divided by the density:[7,8]

$$v_g^x = v_p \ , \quad v_g^y = P_y/\rho \ , \quad v_g^z = P_z/\rho \ . \tag{39}$$

The power flow angle,

$$\theta_p = \cos^{-1}(v_p/|v_g|) \ , \tag{40}$$

is the angle between the direction of propagation and the direction of power flow. There will be a value for each of the three BAW modes.

Software for carrying out computer aided analysis for BAWs in materials of arbitrary anisotropy has been available since 1985.[8]

The examples that follow are typical of the computational capabilities of such software.

2.2.4. Examples

Lithium niobate's XY and YZ planes are chosen to illustrate the BAW propagation properties of piezoelectric materials for the three modes. Note in particular the error in power flow angles if the electrical terms are neglected in a high coupling material.

Figures 1-4 show variation in the XY plane of velocity, polarization, coupling constant, and power flow angles.

Figures 5-8 show variation in the YZ plane of velocity, polarization, coupling constant, and power flow angles.

(a) Phase velocity for XY plane propagation

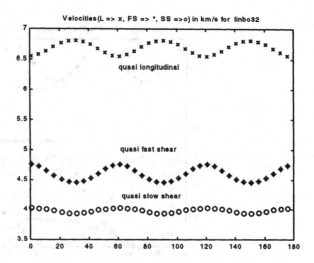

Fig. 1. Velocity (km/s) in XY plane of LiNbO$_3$ (Trigonal 3 m), degrees from X axis. Longitudinal (xxx) , quasi fast shear (***), quasi slow shear (ooo).

(b) Polarization for XY plane propagation

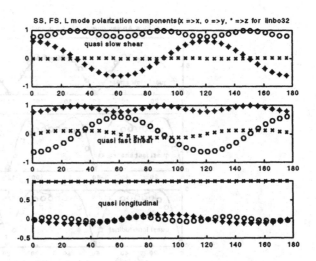

Fig. 2. Polarization of three modes. x, y, z component tags are: x, o, *.

(c) Electromechanical coupling constant. *XY* plane propagation

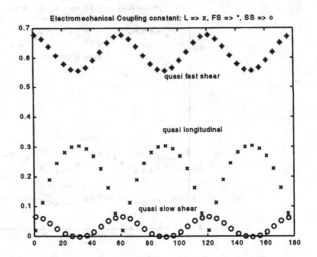

Fig. 3. Coupling constant for *XY* plane propagation. Quasi longitudinal (xxx), quasi fast shear (***), quasi slow shear (ooo).

(d) Power flow angles. *XY* plane propagation

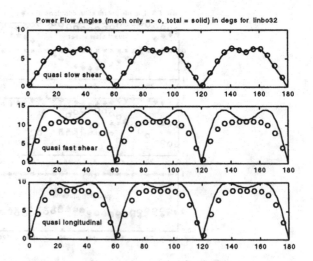

Fig. 4. Power flow angles for the three modes. ooo points neglect electrical contribution. Significant errors arise if electrical contribution is neglected.

Bulk and Surface Acoustic Waves in Anisotropic Solids 667

(e) Phase velocity characteristics for YZ plane propagation

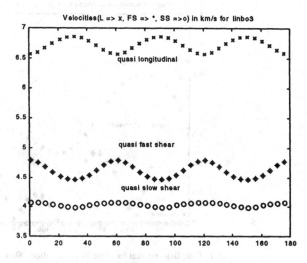

Fig. 5. Velocity (km/s) in YZ plane of LiNbO$_3$ (Trigonal 3 m). Angle in degrees from Y axis.

(f) Polarization for YZ plane propagation

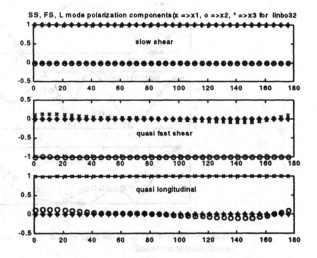

Fig. 6. Polarization of three modes. Slow shear is pure mode.

(g) Piezoelectric coupling. YZ plane propagation

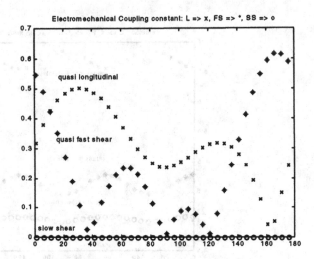

Fig. 7. Coupling constant for same YZ propagation. Slow shear not piezoelectrically stiffened.

(h) Power flow angles for YZ plane propagation

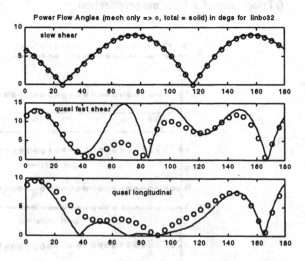

Fig. 8. Power flow angles for the three modes. Note errors which occur if electrical contribution is neglected.

3. Surface Acoustic Waves

In 1885 Lord Rayleigh described the propagation of a sagittally polarized guided wave on a half space.[15] Love in 1911 discussed propagation of shear horizontally polarized waves on a layered geometry and showed that sagittal solutions existed on layered half spaces.[16] In 1927 Sezawa demonstrated the existence of higher order modes.[17] The books by Brekhovskikh and Viktorov contain detailed descriptions of acoustic propagation.[18, 19] Several review articles on SAWs and pseudo SAWs have been published.[3, 20-26] A historical review by David P. Morgan appears as a companion article. As mentioned in the introduction surface wave filters have a vital impact in IF and RF communication systems and are the favorite filters in cellular systems. A new generation of small and lightweight cellular telephones have benefited from SAW filters' miniature size and reproducibility.

3.1. The types of guiding geometries

In half-spaces and geometries with layers, the continuity of particle displacements and normal stresses is the boundary condition that leads to guided wave propagating solutions. For SAW guided modes the energy is contained within a few wavelengths from the half space surface and decays exponentially with depth into the half space. This section is a tutorial review of the strategy for solving these problems. Some structures that act as acoustic waveguides are:

(i) a half-space with one boundary guides a non dispersive surface acoustic wave. There is at least one non leaky SAW mode, and possibly two pseudo SAW modes.[22, 23]

(ii) multilayers on a half space guide modes that are always dispersive and multi modal.

(iii) plates and multilayer plates, guiding structures for plate modes,[27] not SAWs, are not discussed here.

3.2. The matrix method and the ODE

SAW solutions are formulated using linear systems concepts and properties of first-order, linear, ordinary differential equations (ODEs). The steps are: (1) selecting as variables field quantities that must be continuous at interfaces; (2) creating a linear first order ODE in those variables; (3) expressing the solution of the ODE as a transmission matrix; (4) producing a transmission matrix for multilayers as the product of individual layer transmission matrices, automatically satisfying interfacial boundary conditions.

This method was introduced in the section on bulk acoustic waves where the only spatial dependence is in the propagation, x, direction. The complications for SAW problems are: 1) $\partial/\partial z \neq 0$, and the z dependence has to be found; 2) until the guided wave boundary value problem has been solved, the phase velocity v_p is unknown.

3.2.1. Choice of variables

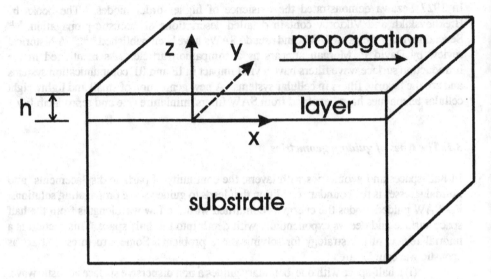

Fig. 9. Coordinate system for multilayer problems.

Figure 9 shows the coordinate system used for guiding structures: the z axis is normal to layer interfaces, the propagation parallel to layer interfaces. The function form for sinusoidal traveling waves along x, at frequency, ω radians per second, having no y, variation, is $f(z)\, exp(-j\omega x/v_p)$. The time dependence $exp(j\omega t)$ assumed throughout is omitted.

For problems unbounded in the xy plane with propagation along x: the problem is two-dimensional; there is no variation in the y direction, $d/dy = 0$; $d/dx = -j\omega/v_p$; and the resulting equations become ODEs. Since the x axis is chosen as the propagation direction, the material constants for the problems are rotated to the Fig. 9 coordinate system. The six mechanical variables continuous across interfaces are the components of the vectors: normal stresses $T^z = [\, T_{31}\ T_{32}\ T_{33}\,]^t = [\, T_5\ T_4\ T_3\,]^t$, and either the particle displacement vector $u = [u_x\ u_y\ u_z]^t$ or velocity $v = [v_x\ v_y\ v_z]^t = j\omega u$. For piezoelectrics the time derivative of the electric potential, $j\omega\varphi$, and the normal electric displacement component, D_z, are chosen as the two continuous electric variables. These eight variables are a sufficient, independent set, and hence form a basis for all other variables in the dynamic and constitutive equations of a material.

3.2.2 Obtaining the ODEs from the PDEs

The equations are similar to those in Section 2, Bulk Acoustic Waves, but since $\partial/\partial z$ is non zero, terms with z variation are included. The PDEs (6) and (9) without y variation become

$$\partial T^x/\partial x + \partial T^z/\partial z = j\omega\rho v$$
$$\partial D_x/\partial x + \partial D_z/\partial z = 0 \qquad (41)$$

and, using (1), $d/dx = -j\omega/v_p$, the corresponding matrix ODE is

$$-j\frac{\omega}{v_p}\begin{bmatrix} T^x \\ D_x \end{bmatrix} + \frac{d}{dz}\begin{bmatrix} T^z \\ D_z \end{bmatrix} = j\omega\rho \begin{bmatrix} v \\ 0 \end{bmatrix}. \qquad (42)$$

The matrix equations corresponding to (17) with the term due to the z variation is

$$\Gamma^{11} \frac{\partial}{\partial x}\begin{bmatrix} v \\ j\omega\varphi \end{bmatrix} + \Gamma^{13} \frac{\partial}{\partial z}\begin{bmatrix} v \\ j\omega\varphi \end{bmatrix} = j\omega \begin{bmatrix} T^x \\ D_x \end{bmatrix} \qquad (43)$$

becoming the ODE

$$-j\frac{\omega}{v_p}\Gamma^{11}\begin{bmatrix} v \\ j\omega\varphi \end{bmatrix} + \Gamma^{13} \frac{d}{dz}\begin{bmatrix} v \\ j\omega\varphi \end{bmatrix} = j\omega \begin{bmatrix} T^x \\ D_x \end{bmatrix}, \qquad (44)$$

and a similar PDE relationship for $[T^z \; D_z]'$ corresponding to (43) is

$$\Gamma^{31} \frac{\partial}{\partial x}\begin{bmatrix} v \\ j\omega\varphi \end{bmatrix} + \Gamma^{33} \frac{\partial}{\partial z}\begin{bmatrix} v \\ j\omega\varphi \end{bmatrix} = j\omega \begin{bmatrix} T^z \\ D_z \end{bmatrix} \qquad (45)$$

becoming the ODE

$$-j\frac{\omega}{v_p}\Gamma^{31}\begin{bmatrix} v \\ j\omega\varphi \end{bmatrix} + \Gamma^{33} \frac{d}{dz}\begin{bmatrix} v \\ j\omega\varphi \end{bmatrix} = j\omega \begin{bmatrix} T^z \\ D_z \end{bmatrix} \qquad (46)$$

where the Γ matrices are obtained from the stiffness, piezoelectric, and permittivity tensors using

$$\Gamma^{ik} = \begin{bmatrix} C_{1i1k} & C_{1i2k} & C_{1i3k} & e_{k1i} \\ C_{2i1k} & C_{2i2k} & C_{2i3k} & e_{k2i} \\ C_{3i1k} & C_{3i2k} & C_{3i3k} & e_{k3i} \\ e_{i1k} & e_{i2k} & e_{i3k} & -\varepsilon_{ik} \end{bmatrix}. \tag{47}$$

Using (42), (44), (46), to eliminate all but T^z, D_z, v, and $j\omega\varphi$, the eight independent components in the vector, τ, for piezoelectrics, or the six independent components T^z, v in the vector τ for non piezoelectrics, results in the differential equation at ω radians/second describing transverse, z, variations for τ in the sinusoidal steady-state,[1]

$$\frac{d\tau}{dz} = j\omega \begin{bmatrix} \Gamma^{13}Z/v_p & \{G-(\Gamma^{11}-\Gamma^{13}Z\Gamma^{31})/v_p^2\} \\ Z & Z\Gamma^{31}/v_p \end{bmatrix} \tau \tag{48}$$

with Z as the inverse of the nonsingular matrix Γ^{33}, or

$$\frac{d\tau}{dz} = j\omega A\tau. \tag{49}$$

The system matrix, A, key to the solution, depends on material constants, rotated to the coordinate system of Fig. 9, and the phase velocity, v_p. It is independent of frequency unless viscosity is considered. In a multilayer (49) applies for each material, and τ must be continuous at each interface. The Z and Γ submatrices are four by four and contain compliance, stiffness, piezoelectric, and permittivity constants. The matrix G is diagonal having the material density, ρ, in the first three places and zero in the fourth. For non piezoelectrics all the matrices are three by three and $G = \rho I_3$.

3.2.3. The solution to the ODE and the transmission matrix

From linear system theory the solution to (49) is the exponential matrix

$$\tau(z+h) = \exp(j\omega Ah)\tau(z) = \Phi(h,\omega,v_p)\tau(z). \qquad (50)$$

The matrix $\Phi(h, \omega, v_p)$ is a transmission or transfer matrix that maps τ across a distance h in the z direction within the same material. It is always nonsingular and $\Phi(0) = I$, the identity matrix. The eigenvalues and eigenvectors of the matrix jA describe the particulars of the z dependence of the acoustoelectric fields.

3.2.4. The eight system eigenmodes

From linear system theory we have:
 (1) The eight eigenvalues, λ_i, of jA define the z dependence of the eight modes.
 (2) The corresponding eigenvectors, p_i, are the eight mode polarizations.
 (3) At any z, τ can be written as the linear combination of modes

$$\tau(z) = \sum_{i=1}^{8} y_i \, p_i \, e^{\lambda_i z}. \qquad (51)$$

With P the eigenvector matrix of jA and Y the mode weighting vector, at the $z = 0$ surface (51) becomes

$$\tau(0) = \sum_{i=1}^{8} y_i(0)\, p_i = P(v_p)Y = PY. \qquad (52)$$

 (4) Since A is real, jA is imaginary and its eight λ_i, j times the eigenvalues of A, occur in four pairs of the form

$$\pm\alpha + j\gamma. \qquad (53)$$

3.3. The electromechanical boundary conditions

3.3.1. Continuity at interfaces for multilayer stacks

For a layered structure, each layer has its own A matrix, satisfies the same type of ODE, and has an associated transmission matrix Φ. Since the elements of the vector τ are chosen to be continuous across interfaces, τ at the upper surface, τ_u, is related to τ at the lowest face, τ_l, by the product of the layer transmission matrices in ascending order. This is valid for arbitrarily oriented material, for plate, and SAW problems where the product of the n layer transmission matrices is

$$\Phi = \Phi_n(h_n) \Phi_m(h_m), \cdots, \Phi_2(h_2) \Phi_1(h_1) . \tag{54}$$

It follows that

$$\tau_u = \Phi \tau_l \tag{55}$$

and in constructing (55) all interfacial boundary conditions are satisfied. Each layer is viewed as an eight-port network with six acoustic and two electric ports, and a multilayer is a cascade of layers equivalent to a single layer with transmission matrix, Φ, which relates the two terminal τ vectors across an entire multilayer. The procedure is modified when shorting planes are inserted between layers.[1,4,13,28]

To obtain the total system solution requires satisfying the appropriate boundary conditions at the remaining two extreme surfaces of the multilayer for plate problems or the remaining one surface for SAWs. These terminal boundary conditions are problem dependent and are discussed in the next section, SAW Propagation. The electrical analog to these two terminal boundary conditions is the specification of a multiport's end terminations.

3.3.2 The boundary conditions at a mechanically free surface, $z = 0$

(1) At a mechanically free surface normal stresses vanish, $T^z(0) = 0$.

(2) For piezoelectric materials an open-circuited surface constrains the normal component of electric displacement, D_z, and the potential to be continuous at the solid air interface. Since Laplace's Equation must be satisfied, the following linear constraint between $D_z(0)$ and $\varphi(0)$ applies at the upper free surface

$$D_z(0) = \frac{\omega}{v_p} \varepsilon_0 \varphi(0) = -j \frac{\varepsilon_0}{v_p} j\omega\varphi(0). \tag{56}$$

(3) At a shorted surface the potential is zero, $\varphi(0) = 0$.

3.4. Solving the SAW propagation problem

Finding the solution to a SAW problem means: satisfying the continuity conditions at every interface and mechanically free surface; and meeting the wave guiding condition, all fields decay exponentially with depth in the half space. The boundary conditions are a set of linear homogeneous algebraic equations having coefficients that are functions of the unknown phase velocity, v_p. A non trivial solution exists if, and only if, the determinant of the coefficients vanishes. An optimization procedure is used to find a value for the velocity that will drive the size of the boundary condition determinant as close to zero as possible. Here, one considers the details of the boundary condition determinant for the half space and the layered half space. Software for solving SAW multilayer problems has been available since 1975.[13,28] Campbell and Jones were the first to publish computed results for the unlayered half space in 1968.[29]

Solutions that decay in the propagation direction, called pseudo SAWs, were reported in the late 60's.[24] Recently a high velocity pseudo SAW has been identified.[22,23] Properties and solution methods for pseudo modes are found in refs. 3 and 22-24. A review of pseudo SAW modes by Mauricio Pereira da Cunha appears as a companion paper; they are not considered here.

3.4.1. A semi infinite substrate

With P the eight by eight matrix having as columns the eigenvectors of jA, τ at the surface, (52), is expressed as the linear combination of normal modes PY where Y is the normal-mode weighting vector. To have a surface guided mode in the half space, $z < 0$, all the fields must vanish as $z \to -\infty$. Only modes that decay with depth satisfy this condition. Since only four of these modes, those with positive real parts to their eigenvalues, decay with depth at the semi infinite substrate surface, the vector $\tau(0)$ is a linear combination of eigenvectors of these four decaying modes only,

$$\tau(0) = P_p Y_p . \tag{57}$$

Here P_P is the eight by four submatrix of P corresponding to decaying modes and Y_p is the corresponding half of the weighting vector Y at the surface.[4] The subscript p identifies the modes that decay with depth; those with positive real parts to their eigenvalues.

The notation $M(2:4,1:3)$ means the matrix containing the second to fourth rows and the first to third columns of M; and $M(2:4,:)$ means the second to fourth rows of M but all its columns. For example from (57) the short-circuited boundary condition obtained by selecting the first three rows and the eighth row is

$$\begin{bmatrix} P_p (1:3,:) \\ P_p (8,:) \end{bmatrix} Y_p = 0 . \tag{58}$$

A non trivial solution requires that the short circuited boundary condition function

$$f_{sc}(v_p) = \det \begin{bmatrix} P_p(1:3,:) \\ P_p(8,:) \end{bmatrix} = 0 .$$ (59)

Using (56) to eliminate D_z from the fourth row of (57), the open circuited boundary condition function is

$$f_{oc}(v_p) = \det \begin{bmatrix} P_p(1:3,:) \\ P_p(4,:) + j\dfrac{\varepsilon_0}{v_p}P_p(8,:) \end{bmatrix} = 0 .$$ (60)

For a non piezoelectric, the boundary condition simplifies to

$$f_{np}(v_p) = \det(P_{pu}) = 0.$$ (61)

Since all the boundary conditions are functions only of the phase velocity, P is a function only of the phase velocity and independent of frequency. A numerical procedure is used to find a phase velocity that satisfies these boundary conditions, (59) for the shorted, and (60) for the open circuited. For the non layered, half space problem there is a unique SAW velocity, independent of frequency. The measure of electromechanical coupling for SAW calculations is given in IEEE SAW standards as: $(2\Delta v/v) = 2\{v_p(oc) - v_p(sc)\}/v_p(oc)$.

3.4.2. A layered semi infinite substrate

The velocity is multi modal and frequency dependent. With the substrate surface at $z = 0$, and the total transmission matrix for a multilayer having all the thicknesses specified, using (55) and (57) $\tau_u = \Phi(\omega, v_p)\tau_l = \Phi\tau(0)$ so that

$$\tau_u = \Phi\tau(0) = \Phi P_p Y_p = BY_p.$$ (62)

The open circuited upper boundary condition function has the same form as (60) but will depend on frequency and velocity since the transmission matrix, $\Phi(\omega, v_p)$, is a function of frequency and phase velocity

$$f_{oc}(\omega,v_p) = \det \begin{bmatrix} \boldsymbol{B}\,(1{:}3,{:}) \\ \boldsymbol{B}\,(4,{:}) + j\dfrac{\varepsilon_0}{v_p}\boldsymbol{B}\,(8,{:}) \end{bmatrix} = 0 \ . \tag{63}$$

At each frequency the boundary condition function is zero at guided mode velocities. Consequently, the full mode velocity versus frequency curves can be calculated.

The boundary condition function for the shorted uppermost surface is (59) with $\boldsymbol{B}$ in place of P_p. The procedure for calculating (54) and (55) is modified when shorting planes are inserted between layers.[1,4,13,28]

The electromechanical coupling constant for layered SAW device calculations, $(2\Delta v/v) = 2\{v_p(\text{oc}) - v_p(\text{sc})\}/v_p(\text{oc})$, has many possible values. For example, for a piezoelectric layer on a half space structure there are two planes that can be shorted resulting in four distinct coupling constants. Since there are many dispersive modes, the coupling will depend on frequency and layer thickness.

Using available software for SAW multilayer structures one may calculate shorted and open circuited velocities and corresponding coupling constants for arbitrarily oriented layer and substrate materials.[28]

This section provides a compilation of different wave guiding geometries, their boundary conditions, a matrix description, and a derivation of the ODE for the SAW geometries given. Interfacial boundary conditions are automatically satisfied by matrix multiplication that preserves the dimensionality of the problem independent of the number of layers.[3,4]

4. Crystal Symmetry, Classification of SAW Solutions, and Examples

The general piezoelectric anisotropic SAW problem in Section 3 can be solved numerically, however, it is advantageous to examine the simplifications offered by special crystal symmetries relative to the propagation geometry.[5,23] These reductions afford a more convenient visualization of the solutions, lead to faster computation, and frequently lead to technologically useful results.

Material crystal symmetry plays a major role in the uncoupling of field components in SAW solutions. Uncoupling means the system solution separates into two independent solutions with orthogonal polarizations and occurs when particular material constants are zero for the chosen orientation of the SAW structure. Rayleigh waves refer to sagittal polarization solutions, only. The common independent solution types, other than the general polarization case, are:

(i) Sagittal polarization: Rayleigh waves for SAW and Lamb waves for plates. These can be piezoelectrically coupled, or stiffened; or purely mechanical.

(ii) Shear Horizontal polarization: Love waves for SAW,[16] or SH waves for plates. These can be purely mechanical or piezoelectrically coupled, such as: the BGS wave for SAW,[30-32] or stiffened SH wave for plates.

4.1. Crystal symmetry and classification of five solutions types

There are five types of possible solutions classified according to the uncoupling of field components in the dynamic equations that arise due to certain thermodynamic coefficients vanishing for some orientation and crystal symmetries.[5]

Type 1. The general anisotropic piezoelectric case of lowest crystal symmetry. The system matrix is eight dimensional, and the mode polarization contains field components in the three coordinate directions. Any arbitrary propagation plane and direction for a piezoelectric material will generally be this type.

Type 2. The general non piezoelectric case of lowest crystal symmetry. The system matrix is six dimensional, no electrical variables are needed, and the mode polarization contains mechanical field components in all three coordinate directions. Any arbitrary propagation plane and direction for an anisotropic non piezoelectric material will generally be this type.

Type 3. Two independent solutions exist. The piezoelectrically stiffened sagittal (x, z) components are uncoupled from the pure mechanical shear horizontal (SH) (y) component. The system matrix for the sagittal, or stiffened Rayleigh wave, is six dimensional. The system matrix for the mechanical SH solution is two dimensional and only involves T_{32} and v_y. This solution is a guided mode only on a layered structure.

Type 4. Two independent solutions exist. The stiffened SH components are uncoupled from the purely mechanical sagittal, or true Rayleigh, components. The system matrix for each solution is four dimensional

Type 5. Non piezoelectric materials. The sagittal (x, z) components are uncoupled from the mechanical shear horizontal (SH) (y) component. The sagittal solution is Rayleigh and has a four dimensional system matrix. The system matrix for the mechanical SH solution is two dimensional and only involves T_{32} and v_y. This solution is a guided mode only on a layered structure.

For symmetry Types 3, 4, and 5 the requirement on the rotated material stiffness matrix elements is:

$$c_{14} = c_{16} = c_{34} = c_{36} = c_{45} = c_{56} = 0 \ . \tag{64}$$

The requirements (64) are satisfied if either the sagittal plane defined by the propagation geometry is a plane of mirror symmetry of the material, or the axis normal to the sagittal plane, y, defined by the propagation geometry, is an axis of twofold rotation symmetry of the material. If both requirements are satisfied the material has a center of inversion, cannot be piezoelectric, and is symmetry Type 5.

For symmetry Type 3, the material is piezoelectric, and in addition to the condition on the stiffness coefficients, (64), the following four piezoelectric matrix elements must be zero:

$$e_{14} = e_{16} = e_{34} = e_{36} = 0 \ . \tag{65}$$

The requirements (64) and (65) are satisfied when the sagittal plane defined by the propagation geometry is a plane of mirror symmetry of the material.

For symmetry Type 4, the material is piezoelectric and, in addition to the condition on the stiffness coefficients, (64), the following six piezoelectric constants must be zero:

$$e_{11} = e_{13} = e_{15} = e_{31} = e_{33} = e_{35} = 0 \ . \tag{66}$$

The requirements (64) and (66) are satisfied when the axis, y, normal to the sagittal plane, defined by the propagation geometry, is an axis of twofold rotation symmetry of the material.

For symmetry Type 5, the material is non piezoelectric and all the piezoelectric elements must be zero. This is satisfied for all non piezoelectric materials if the axis, y, normal to the sagittal plane, defined by the propagation geometry, is either a two fold axis of symmetry or the sagittal plane is a mirror plane.

Table 1 shows: the wave type associated with each of the symmetries, the corresponding eigenvector components and the size of the A matrix. Examples of material orientations for some of the symmetries for a few common materials are given in the fifth column of the table.

Table 1. Symmetry classification for surface waves.

Symmetry Type	Wave Type	Eigenvector components	Size A	Examples for principal directions
1	Generalized Piezoelectric	T^z D_z v $j\omega\varphi$	8	Z cut X propagation on lithium niobate (3 m), X cut Y or Z propagation on quartz (32)
2	Generalized	T^z v	6	Y cut Z propagation on rutile (4/m)
non-piezo	Electrical	D_z $j\omega\varphi$	2	no acoustics involved
3	Stiffened sagittal	T_{31} T_{33} D_z v_x v_z $j\omega\varphi$	6	any direction of XY plane of zinc oxide (6 mm), Y cut Z propagation on lithium niobate
	Mechanical SH	T_{32} v_y	2	as above with Types 3 or 5, or isotropic layer
4	Mechanical sagittal	T_{31} T_{33} v_x v_z	4	Z cut Y propagation on quartz, Y cut X propagation on zinc oxide
	Stiffened SH	T_{32} D_z v_y $j\omega\varphi$	4	above two examples
5	Mechanical sagittal	T_{31} T_{33} v_x v_z	4	Z cut X propagation on diamond (m 3m), any principal direction on a basal plane of silicon
non-piezo	Mechanical SH	T_{32} v_y	2	as above with Type 5 or isotropic layer
	Electrical	D_z $j\omega\varphi$	2	no acoustics involved

Table 2 gives a summary of the symmetry type for SAW propagation in the principal directions on the principal planes for the 32 crystal classes.

Table 2. Symmetry classification for principal directions.

System	Class	XY Plane		XZ Plane		YZ Plane	
		X Prop.	Y Prop.	X Prop.	Z Prop.	Y Prop.	Z Prop.
	1	1	1	1	1	1	1
	$\bar{1}$	2	2	2	2	2	2
Monoclinic	2	4	1	1	1	1	4
	m	3	1	1	1	1	3
	2/m	5	2	2	2	2	5
Orthorhombic	222	4	4	4	4	4	4
	mm2	3	3	4	5	4	3
	mmm	5	5	5	5	5	5
Tetragonal	4 & $\bar{4}$	1	1	1	1	4	1
	4/m	2	2	5	2	5	2
	422, $\bar{4}$ 2m	4	4	4	4	4	4
	4 mm	3	3	4	3	4	3
	4/mmm	5	5	5	5	5	5
Trigonal	3	1	1	1	1	1	1
	$\bar{3}$	2	2	2	2	2	2
	32	1	4	1	4	1	1
	3 m	1	3	1	3	1	1
	$\bar{3}$ m	2	5	2	5	2	2
Hexagonal	6	1	1	4	1	4	1
	$\bar{6}$	1	1	3	1	3	1
	6/m, 6/mmm	Any direction 5		5	5	5	5
	622	4		4	4	4	4
	6 mm	3		4	3	4	3
	$\bar{6}$ m2	4	3	3	3	3	4
		[110]					
Cubic	23, 3 m	4	3	4	4	4	4
	m3, 432, m3m	5	5	5	5	5	5
		(111) Plane				(110) Plane	
		[110]	30° from [110]	[110]		Z Prop.	
Cubic	23, 3 m	1	3		4	3	
	m3, 432, m3m	2	5		5	5	
Isotropic				Type 5			

Using Table 2 and looking at propagation in the crystalline XZ plane:

for quartz (trigonal class 32) the geometry for Z axis propagation is symmetry Type 4 (stiffened SH) whereas for X axis propagation it is Type 1;

for lithium niobate (trigonal class 3 m) the geometry for Z axis propagation is symmetry Type 3 (stiffened sagittal) whereas for X axis propagation it is Type 1;

for zinc oxide (hexagonal class 6 mm) the geometry for Z axis propagation is symmetry Type 3 whereas for X axis propagation it is Type 4;

for diamond (cubic class m 3 m) the geometry for Z or X axis propagation is symmetry Type 5;

for bismuth germanium oxide (cubic class 23) the geometry for X or Z axis propagation is symmetry Type 4 whereas for propagation along a basal plane diagonal it is Type 3.

Some examples by type of symmetry:

Type 1
    generalized, lithium niobate (trigonal 3 m) X propagation in XY plane
        OC velocity = 3.7871 km/s.
        SC velocity = 3.7791 km/s
        Coupling ($2\Delta v/v$) => 0.42 %

Type 2
    non piezoelectric general silicon (cubic m 3 m) (111) plane [110] prop
        generalized velocity = 4.5461km/s

Type 3
    stiffened Rayleigh lithium niobate Z prop in XZ plane Euler <0, 90, 90>
        OC stiffened Rayleigh velocity = 3.4873 km/s
        SC stiffened Rayleigh velocity = 3.4110 km/s
        Coupling ($2\Delta v/v$) => 4.37 %

Type 4
    stiffened SH zinc oxide (hexagonal 6mm) X propagation in XZ plane
        SC SH velocity = 2.8228 km/s
        OC SH velocity = 2.8291 km/s
        Coupling ($2\Delta v/v$) => 0.45 %

    mechanical Rayleigh velocity = 2.6281 km/s

Type 5
    sagittal diamond non piezo (cubic m 3 m) for sagittal X prop in XY plane
        Rayleigh velocity = 10.971 km/s

This section summarizes the impact of crystal symmetry on the uncoupling of the system equations. The two tables provide a quick reference to the uncouplings that occur in the principal planes of the 32 crystal classes.

5. Summary

This presentation provides a unified approach to the analysis of acoustic propagation characteristics for bulk and surface waves in anisotropic piezoelectric materials. It contains details not usually found in the research literature. Following a summary of the defining equations of acoustoelectricity, a vector, τ, of eight field components that must be continuous at material interfaces is chosen. The eigenvalue problem for bulk waves in piezoelectric materials is derived and the equations for electromechanical coupling and power flow are developed. A transmission matrix is the solution to a first order matrix linear ordinary differential equation that describes the z dependence of the continuous vector τ. Interfacial boundary conditions are automatically satisfied by an ordered matrix multiplication of transmission matrices that preserves the dimensionality of the problem independent of the number of layers. The purpose of this work is to solve this problem for multi layers without increasing its dimension. The practicality is that the problem of guided wave propagation in multi layered SAW device geometries is simplified. The final section is a classification of the way crystal symmetry effects the types of guided wave solutions, adding another path to the simplification of surface acoustic waves device analysis.

Acknowledgments

The support of the Natural Sciences and Engineering Research Council of Canada is gratefully acknowledged. The author is indebted to Professor Gerald W. Farnell of McGill University, Montreal and Professor Mauricio Pereira da Cunha of the University of Sao Paulo and acknowledges fruitful discussions with Professor Donald C. Malocha of the University of Central Florida, Orlando and Dr. Frederick S. Hickernell of Motorola, Scottsdale.

References

1. E. L. Adler, "Matrix methods applied to acoustic waves in multilayers", IEEE Trans. Ultrason. Ferroelec. Freq. Contr., vol. 37, No. 6, November 1990, 485–490.

2. E. L. Adler, "Analysis of anisotropic multilayer bulk-acoustic-wave transducers", Electron. Lett., vol. 25, Jan. 1989, 57–58.

3. E. L. Adler, "SAW and pseudo-SAW properties using matrix methods", IEEE Trans. Ultrason. Ferroelec. Freq. Contr., vol. 41, No. 6, November 1994, 876–882.

4. A. H. Fahmy and E. L. Adler, "Propagation of acoustic surface waves in multilayers: a matrix description", Appl. Phys. Lett., Vol. 22, No. 10, May 1973, 495–497.

5. G. W. Farnell and E. L. Adler, "Elastic wave propagation in thin layers", in W. P. Mason and R. N. Thurston (eds.), Physical Acoustics, Academic Press, New York and London, Vol. IX, 1972, 35–127.

6. J. F. Nye, Physical Properties of Crystals, Oxford Univ. Press (Clarendon), London and New York, 1957.

7. B. A. Auld, Acoustic Fields and Waves in Solids, Second Edition, Vol. I, Robert E. Krieger Publishing Company, Malabar, Florida, 1990.

8. C-K. Jen, G. W. Farnell, E. L. Adler and J. E. B. Oliveira, "Interactive computer-aided analysis for bulk acoustic waves in materials of arbitrary anisotropy and piezoelectricity", IEEE Trans. Son. and Ultrason., Vol. SU-32, No. 1, Jan. 1985, 56–60.

9. J. F. Rosenbaum, Bulk Acoustic Waves Theory and Devices, Artech House, Boston, 1988.

10. J. B. Marion, Classical Dynamics of Particles and Systems, Second Edition, Academic Press, New York and London, 1970, 384-386.

11. H. Goldstein, Classical Mechanics, Second Edition, Addison-Wesley, Reading, Mass., 1980

12. A. H. Fahmy and E. L. Adler, "Transformation of tensor constants of anisotropic materials due to rotations of the co-ordinate axes", Proc. IEE, 122, No. 5, May 1975, 591–592.

13. A. H. Fahmy and E. L. Adler, "Multilayer acoustic-surface-wave program", Proc. IEE, 122, No. 5, May 1975, 470–472.

14. E. L. Adler, "Calculating multimode generation in BAW transducers and resonators", to appear in Proceedings of the Ultrasonics Symposium, 1999.

15. Lord Rayleigh, "On waves propagating along the plane surface of an elastic solid", Proc. London Math. Soc., 17, 1885, 4–11.

16. A. E. H. Love, Some Problems of Geodynamics, Cambridge, 1911; Dover, 1967.

17. K. Sezawa, "Dispersion of elastic waves propagated on the surface of stratified bodies and on curved surfaces", Bull. Earthquake Res. Inst. Tokyo, 3, 1927, 1–18.

18. L. M. Brekhovskikh, Waves in Layered Media, Academic Press, 1960, 1980.

19. I. A. Viktorov, Rayleigh and Lamb Waves, Plenum, New York, 1967.

20. G. W. Farnell, "Properties of elastic surface waves", in W. P. Mason and R. N. Thurston (eds.), Physical Acoustics, Vol. VI, Academic Press, New York and London, 1970, 35–127.

21. G. W. Farnell, "Properties of elastic surface waves", in H. Matthews (ed.), Surface Wave Filters, Wiley, 1977.

22. M. Pereira da Cunha and E. L. Adler, "High velocity pseudosurface waves (HVPSAW)", IEEE Trans. Ultrason. Ferroelec. Freq. Contr., vol. 42, No. 5, September 1995, 840–844.

23. M. Pereira da Cunha, "Extended investigation on high velocity pseudosurface waves", IEEE Trans. Ultrason. Ferroelec. Freq. Contr., vol. 45, No. 3, May 1998, 604–613.

24. T. C. Lim and G. W. Farnell, "Character of pseudo surface-waves on anisotropic crystals", J. Acoust. Soc. Amer., 45, November 1969, 845–851.

25. D. P. Morgan, Surface Wave Devices for Signal Processing, Elsevier, 1985, 1991.

26. R. M. White, "Surface elastic waves", Proc. IEEE, vol. 58, 1970, 1238–1276.

27. E. L. Adler, "Electromechanical coupling to Lamb and shear-horizontal modes in piezoelectric plates", IEEE Trans. Ultrason. Ferroelec. Freq. Contr., vol. 36, No. 2, March 1989, 223–230.

28. E. L. Adler, G. W. Farnell, J. K. Slaboszewicz, and C-K. Jen, "PC software for SAW propagation in anisotropic multilayers", IEEE Trans. Ultrason. Ferroelec. Freq. Contr., vol. 37, No. 3, May 1990, 215–223.

29. J. J. Campbell and W.R. Jones, "A method for estimating optimal crystal cuts and propagation directions for excitation of piezoelectric surface waves", IEEE Trans. Son. and Ultrason., Vol. SU-15, 1968, 209–217.

30. J. L. Bleustein, Appl. Phys. Lett., Vol. 13, 1968, 412–413.

31. Y. V. Gulyaev, Soviet Phys. JETP Lett., Vol. 9, 1969, 37–38.

32. Y. Ohta, K. Nakamura, and H. Shimizu, Tech. Rep. IECE Japan, vol. US69-3, (1969).

ANALYSIS OF SAW EXCITATION AND PROPAGATION UNDER PERIODIC METALLIC GRATING STRUCTURES

KEN-YA HASHIMOTO

Dept. Elec. & Mech. Eng., Chiba University, 1-33 Yayoi-cho, Inage-ku
Chiba-shi, 263-8522 Japan

TATSUYA OMORI

Dept. Elec. & Mech. Eng., Chiba University, 1-33 Yayoi-cho, Inage-ku
Chiba-shi, 263-8522 Japan

and

MASATSUNE YAMAGUCHI

Dept. Elec. & Mech. Eng., Chiba University, 1-33 Yayoi-cho, Inage-ku
Chiba-shi, 263-8522 Japan

This paper reviews numerical techniques used for the analysis of excitation and propagation properties of surface acoustic waves (SAWs) under periodic metallic grating structures. First, the finite element method (FEM), the boundary element method (BEM) and the spectral domain analysis (SDA) are compared for the SAW field analysis. Then it is shown how skillfully excitation and propagation properties are characterized by using the FEM/SDA technique. Extended FEM/SDA theories are also detailed for the analysis of multi-finger grating structures.

1. Introduction

Surface acoustic wave (SAW) devices are composed of periodic metallic grating structures such as interdigital transducers (IDTs) and reflectors, and the behavior of SAWs under these structures is substantially responsible to total device performances. Aiming at realizing the simulation and design tools of sophisticated SAW devices, a lot of SAW researchers have made much effort toward developing numerical techniques for the analysis of SAW propagation, reflection, excitation, etc. under the grating structures.

Several techniques were first developed to characterize SAW properties under infinitesimally thin metallic grating structures. Bløtekjær, et al., for example, proposed a theory of analyzing SAW propagation under an infinitely long (acoustically) metallic grating structure,[1,2] in which the charge concentration at the edge of grating fingers was skillfully taken into account. Owing to this technique, various SAW properties got able to be characterized rapidly and accurately. Aoki and Ingebrigtsen extended their theory and applied it to an infinitely long double-finger grating structure (two fingers per period).[3,4]

Effects of the grating finger thickness have also been studied extensively. Because of limited computer resources in the early stages of SAW research, practical approaches were only relatively simple techniques such as the perturbation theory,[5] variational method,[6] etc. This means that their applications are limited to the case where the grating finger is extremely thin; SAW properties change very rapidly with an increase in the grating finger thickness.

In accordance with the vast growth of mobile phone markets, the development of fast and precise simulation tools has become very crucial to respond to urgent necessity of realizing high performance SAW devices. Fortunately, recent rapid progress of computer technologies has made it possible to deal with very large-size problems using personal computers. In fact, a number of research groups have already started projects to develop numerical techniques for rigorous analysis of SAW properties under metallic grating structures.

The first attempt of full wave analysis was carried out by using the finite element method (FEM),[7] which is universal and directly applicable to any structures of arbitrary shape. In addition, its software is commercially available or freely distributed.

Reichinger and Baghai-Wadji proposed a method combining the FEM with the boundary element method (BEM)[8]; the FEM is applied only to the grating finger region, while the BEM is applied to the substrate region with a flat surface. That is, the FEM is suitable for the analysis of the grating fingers having arbitrary shape, and the BEM, compared with the FEM though, markedly reduces the computational time.

Taking the charge distribution on grating fingers into consideration,[9] Ventura, et al. extended the FEM/BEM analysis; the distribution is expanded by the Chebyshev polynomials with a weighting function expressing the charge concentration at the edge of grating fingers.[10] Since this technique significantly reduces the problem sizes, it has been used for the direct analysis of a whole device structure.[11,12]

The authors proposed another method[13] called FEM/SDA method, where the spectral domain analysis (SDA), instead of the BEM, is adopted for the substrate region to analyze infinitely long grating structures. Since field components in the spectral domain are discrete due to the Floquet theorem in this case, the application of the SDA significantly simplifies the formalism. Based on Bl∅tekjær's theory, the FEM/SDA approach also takes account of the charge concentration at the edge of grating fingers and enables to analyze various SAW properties quickly.

The authors also extended the FEM/SDA method to apply to the analysis of SAW excitation and propagation under grating structures consisting of multiple fingers having unequal width, pitch and/or thickness.[14] In the extended method, Aoki's technique and its extension[15] were employed to evaluate the charge distribution on multiple finger grating structures.

It should be recognized that the free softwares based on these methods have been distributed by the authors' group and that they are now widely used amongst SAW researchers.[16,17]

Zhang, et al. pointed out that Bløtekjær's theory can be used to characterize not only acoustic wave propagation but also excitation properties under grating structures.[18] The work stimulated new directions such as the discrete Green function theory,[19] evaluation of the input admittance for the infinitely long IDTs,[20] etc.

This paper reviews the techniques developed for the analysis of SAW excitation and propagation under metallic grating structures.

As a background of the following discussion, the behavior of SAW propagation under periodic grating structures is first explained qualitatively in Section 2.

In Section 3, the stationary property of functionals and the Rayleigh-Ritz method are discussed as fundamentals of the FEM analysis. The application of FEM to the SAW field analysis is also described.

Section 4 deals with the Green function method as a fundamental of the BEM analysis. First, the point matching and moment methods are explained and compared. Then it is shown how the charge concentration at the edge of grating fingers should be taken skillfully into account for quick convergence.

Section 5 describes the original Bløtekjær's theory and the FEM/SDA method which was developed so that effects of the finite thickness of grating fingers may be included. It is shown how efficiently SAW properties can be evaluated by the FEM/SDA method.

In Section 6, it is shown that acoustic wave excitation as well as SAW propagation properties can be characterized by the FEM/SDA method. First, the discrete Green function theory is introduced, and the calculation of the input admittance for the infinitely long IDTs is discussed in detail. Then the parameters for the COM analysis is determined by the FEM/SDA analysis, and it is shown how accurately the properties calculated by the FEM/SDA analysis are modeled by the COM theory.

Section 7 discusses the extended FEM/SDA technique for the application to multi-finger grating structures based on Aoki's theory. It is also shown that the techniques described in Sections 5 and 6 are also directly and effectively applicable to the multi-finger grating structures.

Section 8 describes the extended Bløtekjær's theory for the analysis of obliquely propagating SAWs under grating structures.

It should be noted that all of the analyses discussed here are given only for the Rayleigh-type SAWs. Section 9 deals with the behavior of the shear-horizontal (SH) type SAWs under grating structures, and discusses how the behavior is similar to or different from that of the Rayleigh-type SAWs.

2. Wave Propagation under Periodic Grating Structures

Consider acoustic waves propagating under a periodic structure with the periodicity p as shown in Fig. 1.

If the structure is infinitely long in the propagation direction, the basic periodic

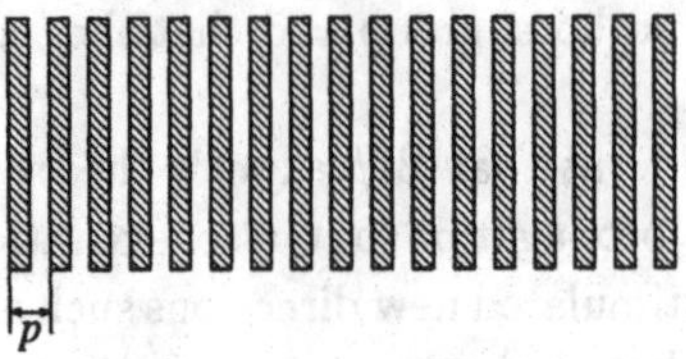

Fig. 1. Periodic structure.

sections are equivalent to each other. So if an eigen mode (grating mode) exists, its field distribution $u(X_1)$ under each period is also similar to each other. Namely, Floquet theorem[21] suggests that $u(X_1)$ should satisfy the following,

$$u(X_1 + p) = u(X_1)\exp(-j\beta p), \quad (1)$$

where β is the wavenumber of the grating mode.

If we define a function $U(X_1) = u(X_1)\exp(+j\beta X_1)$, Eq. (1) suggests that $U(X_1)$ should be a periodic function whose periodicity is p, and $U(X_1)$ is expressed in the Fourier expansion form of

$$U(X_1) = \sum_{n=-\infty}^{+\infty} A_n \exp(-2\pi j n X_1/p). \quad (2)$$

Then we get

$$u(X_1) = \sum_{n=-\infty}^{+\infty} A_n \exp(-j\beta_n X_1), \quad (3)$$

where $\beta_n = 2\pi n/p + \beta$. The equation suggests that the acoustic fields under the periodic structure could be expressed as a sum of sinusoidal waves with discrete wavenumbers β_n. In other words, the wave components with the spatial frequency β_n are generated by spatial modulation of an incident wave with the wavenumber β.

The acoustic waves scattered at each periodic section may interfere with each other. Over wide ranges of frequencies they usually cancel out each other, and the overall reflected field is negligible. At some frequencies, however, the scattered waves may be in phase with one another, and the overall reflection becomes strong. This is called Bragg reflection.

In the structure shown in Fig. 2(a), the phase matching condition (Bragg condition) for reflected waves is given by

$$2p = n\lambda_S \quad \text{or} \quad 4\pi p/\lambda_S = 2n\pi. \quad (4)$$

where λ_S is the SAW wavelength under the grating structure. Using the wavenumber $\beta_S = 2\pi/\lambda_S$ for the SAW, one may rewrite Eq. (4) as

$$\beta_S = n\pi/p \quad \text{or} \quad |\beta_S + 2n\pi/p| = |\beta_S|. \quad (5)$$

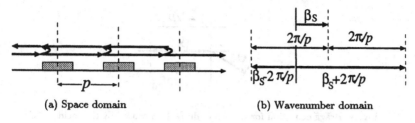

(a) Space domain (b) Wavenumber domain

Fig. 2. Bragg reflection.

This relation can be regarded as follows (see Fig. 2(b)); various wave components with the wavenumbers $\beta_S + 2n\pi/p$ are generated by the spatial modulation of the incident SAW with the wavenumber β_S. Then the Bragg reflection occurs, and a component with the wavenumber $\beta_S - 2\ell\pi/p$ grows up only when it has the phase velocity identical with that for the incident SAW.

The Bragg reflection also occurs between different types of waves. Figure 3 shows the back-scattering of an incident SAW to bulk acoustic waves (BAWs).[22] For this case, the phase matching condition is given by

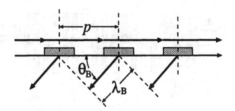

Fig. 3. Back-scattering of incident SAW to BAW.

$$2\pi p/\lambda_S + 2\pi p \cos\theta_B/\lambda_B = 2n\pi \qquad (6)$$

where λ_B is the BAW wavelength, and θ_B is the BAW radiation angle (see Fig. 3). Using the BAW wavenumber $\beta_B = 2\pi/\lambda_B$, one can write Eq. (6) in a form of

$$\beta_S + \beta_B \cos\theta_B = 2n\pi/p. \qquad (7)$$

The relation of Eq. (7) can be interpreted visually by the slowness curve for the BAW in the sagittal plane as shown in Fig. 4.

Using the wavenumbers $\beta_B = 2\pi f/V_B$ and $\beta_S = 2\pi f/V_S$ where V_B and V_S are the phase velocities of BAW and SAW, respectively, one can rewrite Eq. (7) as

$$f = \frac{nV_S/p}{1 + (V_S/V_B)\cos\theta_B}. \qquad (8)$$

The equation shows that the backscattering occurs at frequencies higher than the cut-off frequency f_C given by

$$f_C = \frac{nV_S/p}{1 + (V_S/V_B)}. \qquad (9)$$

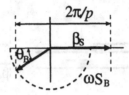

Fig. 4. Bragg condition for the BAW diffraction caused by the incident SAW.

Since $V_S < V_B$, f_C is higher than the SAW resonance frequency given by $nV_S/2p$.

It is of interest to note that although the BAW radiation does not occur at $f < f_C$, this is due to the mutual interference among scattered BAWs, and the BAW scattering itself is not suppressed. Then the energy of the scattered BAW is stored near grating finger edges, and causes a phase delay corresponding to the SAW velocity reduction. This phenomenon is called energy storing effect.[23]

Next, let us discuss the grating structure with finite (acoustic) length. If the Bragg condition is satisfied, the SAW field is evanescent and the fingers far from the incident edge scarcely affect the overall reflection (see Fig. 5(a)). Namely, if

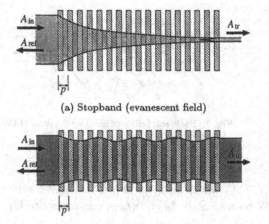

(a) Stopband (evanescent field)

(b) Passband (standing wave field due to reflection at edges)

Fig. 5. Field distribution in grating.

the reflectivity per grating finger is large, the interference among reflected SAWs is localized near the incident edge. Then the overall reflection coefficient Γ is insensitive to the phase matching condition. In other words, the Bragg condition is almost satisfied and the strong reflection occurs at some ranges of frequencies as shown in Fig. 6. This frequency range is called stopband. It is clear from the explanation that the stopband width is proportional to the reflection coefficient per grating finger.

On the other hand, periodic ripples appear in Γ (see Fig. 6) outside the stopband (passband). This is due to the interference between the reflected waves at the left and right edges of the grating as shown in Fig. 5(b). The periodicity is determined by the grating length and effective SAW velocity V_S. In the passband, the frequency response of Γ is governed by frequency dependences of V_S and reflectivity Γ_∞ at each edge. Here Γ_∞ corresponds to the overall reflection coefficient for the semi-infinite grating shown in Fig. 6 because only the incident edge affects the reflection characteristics.

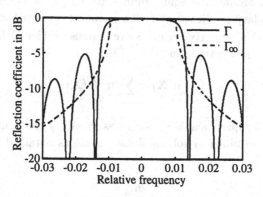

Fig. 6. Typical reflection characteristics of finite (solid-line) and semi-infinite (broken-line) gratings.

Since SAW devices employ piezoelectric substrates, metallic gratings cause electrical reflection in addition to the mechanical one. Namely, since the electric field associated with the SAW is short-circuited by the grating strips, the SAW propagation is perturbed. In addition, the charge induced on the metal strips regenerates the SAWs, and this regeneration is equivalent to a reflection. Then the reflection characteristics are complicated, depending upon the electrical connection among the strips. Detailed discussion will be given in Section 5.1.

3. FEM Analysis

3.1. Stationary property and Rayleigh-Ritz method

Let us define a functional (function of functions) L:

$$L = -\int_V \mathbf{u}^* \bullet (\nabla \bullet \mathbf{T} + \rho\omega^2 \mathbf{u}) dV = \int_V \left\{ \mathbf{S}^* : \mathbf{T} - \rho\omega^2 |\mathbf{u}|^2 \right\} dV - \int_S (\mathbf{u}^* \bullet \mathbf{T}) \bullet d\mathbf{S}, \quad (10)$$

where V and S are the volume and surface area of the structure under consideration, $\mathbf{u}$ is the displacement vector, and $\mathbf{S} = (\nabla \mathbf{u})$ and $\mathbf{T}$ are the strain and stress tensors, respectively. It should be noted that $j\omega L/2$ corresponds to externally supplied power, and L is zero at equilibrium.

When $\mathbf{u} = \mathbf{u}_0 + \delta\mathbf{u}$ and $\mathbf{u}_0$ satisfies the wave equation $(\rho\omega^2 \mathbf{u}_0 + \nabla \bullet \mathbf{T}_0 = 0)$, δL corresponding to variation in L is given by

$$\delta L = -\int_V \mathbf{u}_0^* \bullet (\nabla \bullet \delta\mathbf{T} + \rho\omega^2 \delta\mathbf{u}) dV - \int_V \delta\mathbf{u}^* \bullet (\nabla \bullet \delta\mathbf{T} + \rho\omega^2 \delta\mathbf{u}) dV$$

$$= -\int_S (\mathbf{u}^* \bullet \delta \mathbf{T} - \delta \mathbf{u} \bullet \mathbf{T}^*) \bullet d\mathbf{S}$$
$$+ \int_V (\delta \mathbf{S}^* : \delta \mathbf{T} - \rho\omega^2 |\delta \mathbf{u}|^2) dV - \int_S (\delta \mathbf{u}^* \bullet \delta \mathbf{T}) \bullet d\mathbf{S}. \tag{11}$$

This equation indicates that δL changes parabolically with respect to any kind of perturbation $\delta \mathbf{u}$ around the equilibrium state $\mathbf{u}_0$ provided that $\delta \mathbf{u}$ is chosen to satisfy the boundary conditions at S. This property is called stationary.

Based upon this property, we can solve various problems. Let us express $\mathbf{u}(\mathbf{X})$ as a sum of appropriate trial function $f_n(\mathbf{X})$:

$$\mathbf{u}(\mathbf{X}) = \sum_n \mathbf{u}_n f_n(\mathbf{X}). \tag{12}$$

When $f_n(\mathbf{X})$ is complete and is chosen so as to satisfy the boundary conditions at S, $\mathbf{u}_n$ can be determined by solving linear equations obtained from the stationary condition:

$$\frac{\partial L}{\partial \mathbf{u}_n^*} = 0. \tag{13}$$

This technique is called Rayleigh-Ritz method.[24]

3.2. *FEM*

The FEM is a quite useful formalism of the Rayleigh-Ritz method where the field quantities at selected nodes are used as the unknowns $\mathbf{u}_n$.

Consider the triangles shown in Fig. 7(a). Following the labels indicated in the figure, let us express values of a field quantity $u(x, y)$ at the node $(x, y) = (x_n, y_n)$ as u_n. Then the field distribution $u(x, y)$ within the hatched triangle is approximated as a flat plane given by

$$u(x, y) = a_1 x + a_2 y + a_3, \tag{14}$$

where the coefficient a_i is given by

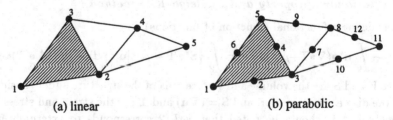

(a) linear   (b) parabolic

Fig. 7. Discretization for two-dimensional analysis.

$$\begin{pmatrix} a_1 \\ a_2 \\ a_3 \end{pmatrix} = \begin{pmatrix} x_1 & y_1 & 1 \\ x_2 & y_2 & 1 \\ x_3 & y_3 & 1 \end{pmatrix}^{-1} \begin{pmatrix} u_1 \\ u_2 \\ u_3 \end{pmatrix}. \tag{15}$$

Instead of flat planes, curved surfaces are also applicable. For example, a parabolic surface is given by

$$u(x,y) = a_1 x^2 + a_2 xy + a_3 y^2 + a_4 x + a_5 y + a_6, \qquad (16)$$

where coefficients a_i are expressed by six $u(x_n, y_n)$ at six nodes on each triangle edge as shown in Fig. 7(b).

If the whole structure under consideration is subdivided into triangles, $u(x, y)$ in the structure can be expressed in terms of u_n. Then by substituting this expression into Eq. (10) and applying the result to the stationary condition of Eq. (13), one may obtain linear equations of the following form:

$$\begin{pmatrix} 0 \\ \hat{\mathbf{T}} \end{pmatrix} = [\mathbf{K} - \omega^2 \mathbf{M}] \begin{pmatrix} \tilde{\mathbf{u}} \\ \hat{\mathbf{u}} \end{pmatrix} \qquad (17)$$

where $\mathbf{K}$ and $\mathbf{M}$ are coefficient matrices, $\hat{\mathbf{u}}$ and $\tilde{\mathbf{u}}$ are vectors composed of $\mathbf{u}$ at the nodal points on S and the others, respectively, and $\hat{\mathbf{T}}$ is a vector composed of the integral of $\mathbf{T}$ at the nodal points on S.

For the analysis of infinitely long grating structures, the simplest way is to subdivide one unit period of the structure into triangles as shown in Fig. 8. For side boundaries S_+ and S_-, the Floquet theorem of Eq. (1) is applied. Namely,

$$\hat{\mathbf{u}}|_{S=S_+} = \hat{\mathbf{u}}|_{S=S_-} \exp(-j\beta p). \qquad (18)$$

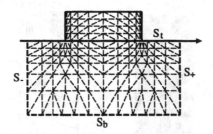

Fig. 8. FEM analysis for one unit period of grating structure.

For the bottom surface S_b, its effect to the SAW propagation may be negligible if the substrate thickness is chosen to be sufficiently large. Thus we can set $\hat{\mathbf{T}} = 0$ not only for the top surface S_t but also S_b. Then by specifying β, Eq. (17) can be solved as an eigenvalue problem to determine ω^2.

This technique has several distinctive features.

1. The FEM can be applied directly to structures with arbitrary shapes. Uniformity of material constants is required only within each cell.
2. Since the final matrices of Eq. (17) are very sparse (most of the matrix elements are zero), the required memory size to store and analyze these data can be reduced significantly.

3. General purpose softwares for the FEM analysis are available as commercial or free products.
4. Various supporting techniques were proposed. They make the FEM analysis user friendly.
5. Fields to be solved should be continuous and bounded.

When piezoelectricity is included, the same procedure is applicable by using the following functional instead of Eq. (10):

$$L = \int_V \{\mathbf{S}^* : \mathbf{T} - \rho\omega^2|\mathbf{u}|^2 - \mathbf{E}^* \bullet \mathbf{D}\} dV - \int_S (\mathbf{u}^* \bullet \mathbf{T} + \phi^*\mathbf{D}) \bullet d\mathbf{S}, \qquad (19)$$

where $\mathbf{D}$ and $\mathbf{E}$ are the electric flux density and field vectors, respectively, and ϕ is the electric potential.

It should be noted that there is a significant problem to apply this technique for the SAW analysis. That is, the FEM implicitly assumes continuity of field variables, but $\mathbf{D}$ changes irregularly at the finger edges. This fact results in ill convergence of the analysis.

As a demonstration, the static capacitance of a solid-type IDT with $w/p = 0.5$ was calculated by using a free software FreeFEM.[25] Figure 9(a) shows the location of the nodes automatically selected by the software for a half section of the unit period. For simplicity, the substrate is assumed to be isotropic with very large permittivity. And the substrate thickness is chosen to be $4p$.

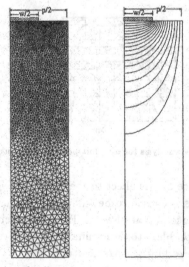

(a) node allocation (b) potential distribution

Fig. 9. Node allocation and electric potential distribution generated by FreeFEM.

Relatively dense discretization was given, and the calculated potential distribution is smooth as shown in Fig. 9(b). However, the calculated capacitance is 1.45%

smaller than the analytical value. The same calculation was performed by increasing the number of elements N_x along the surface. The error reduced to 0.68% for doubled N_x. It is known that the error is proportional to N_x^{-1}.[26]

4. BEM Analysis

4.1. Green function analysis

Let us consider acoustic waves propagating in a periodic metallic grating shown in Fig. 10, whose finger periodicity and width are p and w. An acoustic plane wave propagates toward the X_1 direction. The electrode length is assumed to be much larger than the wavelength, and the effects of finger thickness h and resistivity are neglected.

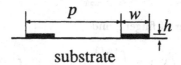

Fig. 10. Periodic metallic-grating.

Since the system is linear, the relationship between charge $q(X_1)$ and electric potential $\phi(X_1)$ on the fingers is given by the following convolution form:

$$\phi(X_1) = \int_{-\infty}^{+\infty} G(X_1 - X_1')q(X_1')dX_1', \qquad (20)$$

where $G(X_1)$ is the Green function, namely the spatial impulse response.

For the system under consideration, $G(X_1)$ is given by

$$G(X_1) = \frac{1}{2\pi} \int_{-\infty}^{\infty} \frac{\exp(-j\beta X_1)}{|\beta|\epsilon(\beta/\omega)} d\beta, \qquad (21)$$

where $\epsilon(S)$ is the effective permittivity[27] and $S = \beta/\omega$ is the slowness.

In the case where the radiation of bulk acoustic waves (BAWs) is negligible, $G(X_1)$ is given analytically by[27,28]

$$G(X_1) = -\frac{1}{\pi\epsilon(0)} \log|X_1| + \frac{jK^2}{2\epsilon(\infty)} \exp(-j\omega S_S|X_1|), \qquad (22)$$

where S_S is the slowness of the SAW on the metallized surface, and K^2 is the electromechanical coupling factor given by

$$K^2 = -2\epsilon(\infty) \left[S_S \frac{\partial \epsilon(S)}{\partial S} \bigg|_{S=S_S} \right]^{-1}. \qquad (23)$$

Let us express the charge distribution $q_m(X_1)$ on the m-th finger as a sum of trial function $f_n(X_1)$, i.e.,

$$q_m(X_1) = \sum_{n=1}^{N} A_{mn} f_n(X_1 - mp). \tag{24}$$

In the equation, coefficients A_{mn} will be determined by the boundary condition of

$$\phi(X_1) = V_m \quad \text{(for } mp - w/2 < X_1 < mp + w/2\text{)}, \tag{25}$$

where V_m is the voltage of the m-th finger.

Substitution of Eq. (24) into Eq. (20) gives the following equation to be solved:

$$\phi(X_1) = \sum_{m=1}^{M} \sum_{n=1}^{N} H_n(X_1 - mp) A_{mn}, \tag{26}$$

where M is the number of fingers, and

$$H_n(X_1) = \int_{-w/2}^{+w/2} G(X_1 - X_1') f_n(X_1') dX_1'. \tag{27}$$

There are two strategies to solve this type of boundary value problems. One is to determine A_{mn} so that the boundary condition of Eq. (25) is satisfied at selected points $X_{k\ell}$ on the fingers, i.e.,

$$V_k = \sum_{m=1}^{M} \sum_{n=1}^{N} H_n(X_{k\ell} - mp) A_{mn} \quad \text{for } (k = [1, M], \ell = [1, N]). \tag{28}$$

Then A_{mn} is determined by solving Eq. (28) as $(M \times N)$ dimensional linear equations with respect to A_{mn}. This technique is called point matching method.

The other one starts from the following functional P:

$$P = -j\frac{\omega}{2} \sum_{k=1}^{M} \int_{kp-w/2}^{kp+w/2} q(X_1)^* \{\phi(X_1) - V_k\} dX_1, \tag{29}$$

corresponding to difference between supplied and dissipated complex powers.

If $f_n(X_1)$ is complete, $P = 0$ at $N \to \infty$. Then, from the stationary property, the following relation must be hold for error minimization:

$$\frac{\partial P}{\partial A_{k\ell}^*} = 0 \quad \text{for } (k = [1, M], \ell = [1, N]). \tag{30}$$

Substitution of Eqs. (24) and (26) into Eq. (29) and combining with Eq. (30) give

$$\sum_{m=1}^{M} \sum_{n=1}^{N} K_{k\ell mn} A_{mn} = V_k L_\ell \quad \text{for } (k = [1, M], \ell = [1, N]), \tag{31}$$

where

$$K_{k\ell mn} = \int_{-w/2}^{+w/2} f_\ell(X_1)^* H_n[X_1 - (m-k)p]dX_1, \qquad (32)$$

$$L_\ell = \int_{-w/2}^{+w/2} f_\ell(X_1)^* dX_1. \qquad (33)$$

Then A_{mn} are determined by solving Eq. (31) as $(M \times N)$ dimensional simultaneous linear equations with respect to A_{mn}. This technique is called method of moment.

By using thus the determined A_{mn}, the total charge Q_m on the m-th finger is given by

$$Q_m = \sum_{n=1}^{N} A_{mn} L_n^*. \qquad (34)$$

Comparison of these techniques is as follows:

1. As shown in Eq. (28), the calculation for the point matching method is simple. However, convergence with respect to N critically depends on the selection of the sampled points $X_{k\ell}$.

2. As shown in Eq. (31), the calculation for the method of moment is rather complicated. However, monotonical convergence is guaranteed provided that $f_n(X_1)$ is complete. This property is also effective to ensure convergence of the calculation.

The selection of $f_n(X_1)$ is important. One well known technique is to use $q(X_{mn})$ at selected nodes $X_1 = X_{mn}$, as unknowns A_{mn}, and to express $q(X_1)$ as

$$q(X_1) = \frac{q(X_{mn}) - q(X_{mn-1})}{X_{mn} - X_{mn-1}}(X_1 - X_{mn-1}) + q(X_{mn-1}) \qquad (35)$$

for $X_{mn-1} \leq X_1 \leq X_{mn}$. This discretization is equivalent to that employed in the FEM described in Section 3.2, and of course we can use nonlinear trial functions. The Green function analysis using this technique is called boundary element method (BEM).

This technique has following distinctive features.

1. Since nodes are located only on the boundary, the total matrix size is significantly smaller than the equivalent problem based upon the FEM analysis.

2. Each element is required to be homogeneous, but not necessary to be bounded. Thus the BEM fits well for the analysis of uniform open structures.

3. General purpose software for the BEM analysis is available as commercial products.

4. Various supporting techniques developed for the FEM are also applicable. They also make the BEM analysis user friendly.

5. Since both the BEM and the FEM employ field variables at the nodes as the unknowns, their combination is easy.

However, since traditional BEM assumes continuity of the field variables, the problem described in Section 3.2 also appears for our purpose. Then, since $q(X_1)$ has a singularity at the finger edges in the form of $1/\sqrt{X_1}$, we should use $f_n(X_1)$ exhibiting this behavior for good convergence. For this reason, the following $f_n(X_1)$ are widely used[10]:

$$f_n(X_1) = \frac{T_{n-1}(2X_1/w)}{\sqrt{1-(2X_1/w)^2}}, \qquad (36)$$

where $T_n(x)$ is the Chebyshev function of the n-th order, which is complete and orthogonal with the weighting function of $1/\sqrt{1-x^2}$. Use of $T_n(X_1)$ enables to perform some integrals that appeared in above calculation analytically.[10]

Figure 11 shows thus calculated static capacitance per period of the infinitely-long solid-type IDT of $w/p = 0.5$ as a function of $1/(N-1)$. For comparison, calculations using the trial function of Eq. (35) instead of Eq. (36) are also shown in the figure. Here, only A_n with even n is taken into account in the calculation because $A_n = 0$ for odd n in this case.

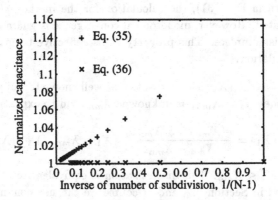

Fig. 11. Calculated static capacitance per period for solid-finger IDT as a function of $1/(N-1)$. ×: Eq. (36) employed, +: Eq. (35) employed.

It is seen that $N = 2$ is sufficient to achieve high accuracy when the weighted polynomial is employed. On the other hand, when Eq. (35) is employed, the calculation converges with $1/(N-1)$ dependence. As mentioned above, this is due to the charge singularity at the finger edges in the form of $1/\sqrt{X_1}$, and many terms are required to express this discontinuity as a sum of continuous functions.

Attention should be paid also to the calculation of $G(X_1)$ and $H_{k\ell mn}$. Even though the calculation procedure for $\epsilon(\beta/\omega)$ is well established, numerical calculation of Eq. (21) is not so simple because $\epsilon(\beta/\omega)$ includes several types of discontinuities.[27] That is, for accurate and rapid evaluation, numerical techniques should be applied after the contributions of discontinuities are estimated separately.

It should be noted that, by combining with the FEM analysis for the electrode region, the above mentioned technique can be applied for the analysis of metallic grating structures with finite film thickness.[11,12] However, this combination also introduces complexity for the evaluation of the Green function and associated integrals.

4.2. Analysis of infinitely long IDT

When M is infinite, the problem to be solved becomes small, and the above calculation can be simplified significantly. For $q(X_1)$ satisfying the Floquet theorem of Eq. (1), Eq. (20) can be rewritten as

$$\phi(X_1) = \int_{-w/2}^{+w/2} G_p(X_1 - X_1')q(X_1')dX_1', \qquad (37)$$

where $G_p(X_1)$ is the periodic Green function[29] given by

$$G_p(X_1) = \sum_{m=-\infty}^{+\infty} G(X_1 - mp)\exp(-jm\beta p). \qquad (38)$$

Substitution of Eq. (21) into Eq. (38) gives

$$G_p(X_1) = \frac{1}{p} \sum_{m=-\infty}^{+\infty} \frac{\exp(-j\beta_m X_1)}{|\beta_m|\epsilon(\beta_m/\omega)}, \qquad (39)$$

where $\beta_m = \beta + 2m\pi/p$. For the derivation, the following identity was used:

$$\frac{p}{2\pi} \sum_{m=-\infty}^{+\infty} \exp[+jm(\xi - \beta)p] = \sum_{m=-\infty}^{+\infty} \delta(\xi - \beta - 2m\pi/p). \qquad (40)$$

Then the analysis can be performed by using $G_p(X_1)$ instead of $G(X_1)$ and applying the procedure described in Section 4.1.[9]

5. FEM/SDA Method

5.1. Bløtekjær's theory for infinitely-long metallic grating with zero thickness

Consider acoustic waves propagating in an infinitely long metallic grating shown in Fig. 10 with $h = 0$. In this case, various properties are skillfully evaluated by using the theory proposed by Bløtekjær, et al.[1,2]

Let us express the charge $q(X_1)$ on the fingers and the electric field $e(X_1)$ within the gaps as follows:

$$q(X_1) = \sum_{m=-\infty}^{+\infty} \frac{A_m \exp(-j\beta_{m-1/2}X_1)}{\sqrt{\cos(\beta_g X_1) - \cos\Delta}}, \qquad (41)$$

$$e(X_1) = \sum_{m=-\infty}^{+\infty} \frac{B_m \mathrm{sgn}(X_1)\exp(-j\beta_{m-1/2}X_1)}{\sqrt{-\cos(\beta_g X_1) + \cos\Delta}}, \qquad (42)$$

where $\Delta=\pi w/p$, $\beta_g=2\pi/p$, $\beta_m=\beta_g(m+s)$ and $s=(\beta_0/\beta_g)$ is the normalized wavenumber of the grating mode. Note that expressions in Eqs. (41) and (42) satisfy the Floquet theorem shown in Eq. (1).

Using the effective permittivity[27] $\epsilon(\beta/\omega)$, one can relate $q(X_1)$ with $e(X_1)$ as,

$$E(\beta_n) = jS_n\epsilon(\beta_n/\omega)^{-1}Q(\beta_n), \qquad (43)$$

In the equation, $S_n=\mathrm{sgn}(\beta_n)$, and $Q(\beta_n)$ and $E(\beta_n)$ are the Fourier expansion coefficients of $q(X_1)$ and $e(X_1)$ given by

$$Q(\beta_n) = \frac{1}{p}\int_{-p/2}^{+p/2} q(X_1)\exp(+j\beta_n X_1)dX_1 = \frac{1}{\sqrt{2}}\sum_{m=-\infty}^{+\infty} A_m P_{n-m}(\cos\Delta), \qquad (44)$$

$$E(\beta_n) = \frac{1}{p}\int_{-p/2}^{+p/2} e(X_1)\exp(+j\beta_n X_1)dX_1 = -\frac{1}{\sqrt{2}}\sum_{m=-\infty}^{+\infty} S_{n-m} B_m P_{n-m}(\cos\Delta), \qquad (45)$$

where $P_m(\theta)$ is the Legendre function of the m-th order.

Substitution of Eqs. (44) and (45) into Eq. (43) gives,

$$\sum_{m=-\infty}^{+\infty}\left(\frac{jS_n\epsilon(\infty)}{\epsilon(\beta_n/\omega)}A_m + S_{n-m}B_m\right)P_{n-m}(\cos\Delta) = 0. \qquad (46)$$

For numerical calculation, assume that the range of summation over m in Eq. (46) is limited to $M_1 \leq m \leq M_2$. When $\epsilon(\beta_n/\omega)$ for a specific substrate material can approximately be described by $\epsilon(\infty)$ for $n \geq N_2 > 0$ or $n \leq N_1 < 0$, the following relation should be satisfied so that Eq. (46) holds for $n \geq N_2 - M_1$ or $n \leq N_1 - M_2$. That is,

$$B_m = -j\epsilon(\infty)^{-1}A_m \qquad (47)$$

for all m. Hence, one obtains

$$\sum_{m=M_1}^{M_2} A_m\left(S_{n-m} - S_n\frac{\epsilon(\infty)}{\epsilon(\beta_n/\omega)}\right)P_{n-m}(\cos\Delta) = 0. \qquad (48)$$

If $M_2=N_2$ and $M_1=N_1+1$, Eq. (48) is automatically satisfied for $n \geq N_2$ and $n \leq N_1$. In this case, since $n = [M_1, M_2-1]$, the total number of the unknowns A_m is greater than that of the equations by one. So relative values of A_m can be determined by solving the simultaneous linear equations with (M_2-M_1) unknowns.

A_m thus determined gives the total charge Q on a finger:

$$Q = \int_{-w/2}^{+w/2} q(X_1)dX_1 = \frac{p}{\sqrt{2}}\sum_{m=M_1}^{M_2} A_m P_{m+s-1}(\cos\Delta), \qquad (49)$$

and the potential Φ:

$$\Phi = -\int_{-\infty}^{-w/2} e(X_1)dX_1 = \frac{2^{-1.5}p}{\epsilon(\infty)\sin(s\pi)}\sum_{m=M_1}^{M_2}(-)^m A_m P_{m+s-1}(-\cos\Delta). \qquad (50)$$

The the strip admittance $Y(s,\omega)$ is defined in terms of the ratio between Q and Φ[1]:

$$Y(s,\omega) = \frac{j\omega Q}{\Phi} \equiv 2j\omega \sin(s\pi)\epsilon_g(s,\omega), \tag{51}$$

where $\epsilon_g(s,\omega)$ is the effective permittivity for the grating structure[19] given by

$$\epsilon_g(s,\omega) = \epsilon(\infty)\frac{\sum_{m=M_1}^{M_2} A_m P_{m+s-1}(\cos\Delta)}{\sum_{m=M_1}^{M_2}(-1)^m A_m P_{m+s-1}(-\cos\Delta)}. \tag{52}$$

It should be noted that $Q = 0$ and $\Phi = 0$ for open-circuited (OC) and short-circuited (SC) gratings, respectively. Hence, the dispersion relations of the grating-modes propagating under the OC and SC gratings are obtained by substituting A_m into Eq. (49) and Eq. (50), respectively.

Figure 12 shows, as a function of M_2, the error in calculated frequencies ω_{o+} and ω_{o-}, respectively, at the upper and lower stopband edges for the OC grating with those ω_{s+} and ω_{s-}, respectively, at the upper and lower stopband edges for the SC grating. As a substrate, $128°$YX-LiNbO$_3$ was chosen. In this case, $\omega_{0-} = \omega_{s+}$ inherently. It is seen that the accuracy increases rapidly with an increase in M_2, and $M_2=1$ is sufficient for most cases.

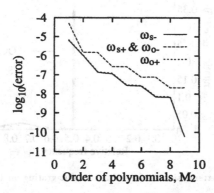

Fig. 12. M_2 dependence of calculation error.

Figure 13 shows the SAW phase velocity V_p calculated for $128°$YX-LiNbO$_3$. The horizontal axis is the normalized frequency fp/V_B where $V_B = 4,025$ m/sec is the slow-shear BAW velocity. The velocity changes irregularly at $fp/V_B \cong 0.5$. This frequency region corresponds to the stopband due to the Bragg reflection. Except for that region, the SAW velocity increases monotonically with an increase in f for the SC grating whereas it decreases for the OC grating.

Figure 14 shows the attenuation α with propagation as a function of the normalized frequency. A steep peak at $f \cong 0.5 V_B/p$ is due to the SAW Bragg reflection, and

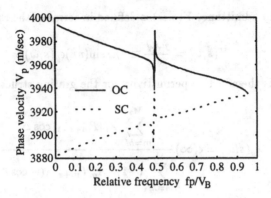

Fig. 13. Phase velocity V_p of SAW in metallic grating on 128°YX-LiNbO$_3$.

the other loss is due to the back-scattering of incident SAW to the BAWs described in Section 2. The scattering into the BAWs does not occur at $f < 0.5V_B/p$ due to its cut-off nature. When $f > 0.62V_B/p$, the attenuation increases very rapidly by the coupling with the longitudinal BAW. The attenuation is relatively small at $0.5 < fp/V_B < 0.62$ because the electromechanical coupling for the slow-shear BAW is very small for 128°YX-LiNbO$_3$.

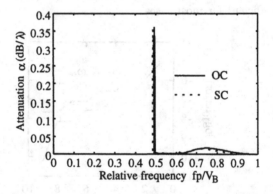

Fig. 14. Attenuation α of SAW in metallic grating on 128°YX-LiNbO$_3$.

Figure 15 shows the change in the frequency dispersion with the metallization ratio Δ. When $f \to 0$, the SAW velocity on the OC grating converges to the SAW velocity on the free surface whereas that on the SC grating converges to the SAW velocity on the metallized surface. This property is independent of Δ. This is originated from negligible phase shift between adjacent periods. That is, as shown in Fig. 16, the effect of the short-circuited electric field is negligible for the OC grating whereas variation of the surface voltage within the gaps is negligible for the SC grating.

On the other hand, at the the stopband edges where $s = 1$, the SAW velocity V_p

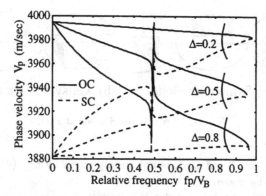

Fig. 15. Frequency dispersion of SAW phase velocity V_p in metallic grating with the metallization ratio Δ as a parameter.

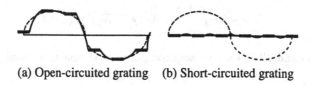

(a) Open-circuited grating   (b) Short-circuited grating

Fig. 16. SAW field distribution at $f \to 0$.

on the OC grating coincides with that on the SC grating. This is because, at $s = 1$, the voltage for each strip is equal even in the OC grating, and thus no current flows among strips independent of electrical mutual connection.

At the stopband edges, the SAW velocities for the OC and SC gratings coincide each other at either one of two frequencies at the stopband edges. This can be explained as follows. At the stopband edges ($s = 0.5$), the standing wave field occurs because of the pure real wavenumber. Since the system considered here possesses symmetrical geometry, allowable resonance patterns can be categorized into the two types shown in Fig. 17; one is symmetric and the other is antisymmetric with respect to the geometrical center for each period. For the antisymmetric resonance shown in (b), the charge distribution induced on the strip is also antisymmetric, and thus the total induced charge is zero. This means that, for the antisymmetrical resonance, no current flows among strips even in the SC grating. On the other hand, the symmetric one shown in (a) has non-zero total charge, and thus the SAW propagation is influenced by the electrical mutual connection.

If none of the stopband edges for the SC grating coincides with those for the OC grating, this indicates the existence of structural and/or crystallographic asymmetries.

5.2. SDA and combination with FEM

Next let us discuss the case where the film thickness h is finite. In that case, the

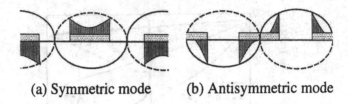

(a) Symmetric mode   (b) Antisymmetric mode

Fig. 17. SAW field distribution at $s = 0.5$.

relation between $\Phi(\beta)$ and $Q(\beta)$ is not simply given by Eq. (43), and appropriate correction must be given.

As shown in Fig. 18, let us define the acoustic complex power $P^\pm$ supplied from the boundary at $X_3 = 0^\pm$. From the Poynting theorem, it is given by

$$P^\pm = \mp j\omega \int_{-p/2}^{+p/2} \mathbf{u}(X_1) \bullet \mathbf{T}(X_1)^* |_{X_3 = 0^\pm} dX_1. \tag{53}$$

where $\mathbf{u}(X_1)$ and $\mathbf{T}(X_1)$ are vectors composed of the particle displacement $u_i(X_1)$

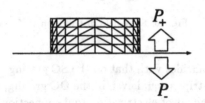

Fig. 18. FEM analysis of finger region.

and the stress $T_{3i}(X_1)$ at the surface ($X_3 = 0$) as

$$\mathbf{u}(X_1) = \{u_1(X_1), u_2(X_1), u_3(X_1)\}, \tag{54}$$
$$\mathbf{T}(X_1) = \{T_{31}(X_1), T_{32}(X_1), T_{33}(X_1)\}. \tag{55}$$

For the strip region, define the vectors $\hat{\mathbf{u}}$ and $\hat{\mathbf{T}}$ composed of the particle displacements and the integration of the stresses at the nodal points of $X_3 = 0^+$. If the driving frequency ω is specified, the application of the FEM to the finger region relates these vectors in the form of

$$\hat{\mathbf{T}} = -[\mathbf{F}]\,\hat{\mathbf{u}}, \tag{56}$$

where $[\mathbf{F}]$ is the matrix derived from the FEM analysis. Then substitution of Eq. (56) into Eq. (53) gives

$$P^+ = j\omega \hat{\mathbf{u}}^* \bullet [\mathbf{F}]^{*t}\,\hat{\mathbf{u}}, \tag{57}$$

where t denotes the transpose of a matrix.

Next, consider the acoustic wave field at $X_3 = 0^-$. Since the field variables satisfy the Floquet theorem of Eq. (1), $\mathbf{T}(X_1)$ of a grating mode with the wavenumber β can be expressed in the form of

$$\mathbf{T}(X_1) = \sum_{n=-\infty}^{+\infty} \mathbf{T}(\beta_n) \exp(-j\beta_n X_1), \tag{58}$$

where $\beta_n = \beta + 2n\pi/p$, and

$$\mathbf{T}(\beta_n) = \frac{1}{p} \int_{-p/2}^{+p/2} \mathbf{T}(X_1) \exp(j\beta_n X_1) dX_1. \tag{59}$$

Carrying out the numerical integration and taking account of Eq. (56), one may rewrite Eq. (59) in the form of

$$\mathbf{T}(\beta_n) = [\mathbf{G}(\beta_n)] \, \hat{\mathbf{T}} = -[\mathbf{G}(\beta_n)] \, [\mathbf{F}] \, \hat{\mathbf{u}}, \tag{60}$$

because $\mathbf{T}(X_1) = 0$ in the un-electroded region. In Eq. (60), $[\mathbf{G}(\beta_n)]$ is a matrix giving the transform in Eq. (59).

Using the definition of the Fourier transform shown in Eq. (59), one may transform $\mathbf{u}(X_1)$, the surface potential $\phi(X_1)$ and the charge $q(X_1)$ into $\mathbf{U}(\beta_n)$, $\Phi(\beta_n)$ and $Q(\beta_n)$, respectively. They are not independent of each other, and are related by

$$\mathbf{U}(\beta_n) = |\beta_n|^{-1} \{ [\mathbf{R}_{11}(\beta_n)] \mathbf{T}(\beta_n) + [\mathbf{R}_{12}(\beta_n)] Q(\beta_n) \}, \tag{61}$$

$$\Phi(\beta_n) = |\beta_n|^{-1} \{ [\mathbf{R}_{21}(\beta_n)] \mathbf{T}(\beta_n) + R_{22}(\beta_n) Q(\beta_n) \}, \tag{62}$$

where $j\omega[\mathbf{R}_{ij}(\beta)]$ is the effective acoustic admittance matrix.[30]

From the Parseval theorem, P^- in Eq. (53) is rewritten as

$$P^- = j\omega p \sum_{n=-\infty}^{+\infty} \mathbf{T}(\beta_n)^* \bullet \mathbf{U}(\beta_n)|_{X_3=0^-}. \tag{63}$$

Since $\mathbf{T}(\beta_n)|_{X_3=0^+} = \mathbf{T}(\beta_n)|_{X_3=0^-}$ from the boundary condition, substitution of Eqs. (60) and (61) into Eq. (63) gives

$$P^- = j\omega p \sum_{n=-\infty}^{+\infty} S_n \beta_n^{-1} \hat{\mathbf{u}}^* \bullet [\mathbf{F}]^{*t} [\mathbf{G}(\beta_n)]^{*t}$$
$$\{ [\mathbf{R}_{11}(\beta_n)] [\mathbf{G}(\beta_n)] [\mathbf{F}] \, \hat{\mathbf{u}} - [\mathbf{R}_{12}(\beta_n)] Q(\beta_n) \}, \tag{64}$$

where $S_n = \mathrm{sgn}(\beta_n)$.

The total power P supplied from the boundary is $P^+ + P^-$. Since $\mathbf{u}(X_1)$ is continuous at the boundary, P should be zero if the solution is rigorous. Although P generally takes a non-zero value because of the numerical evaluation, it must satisfy the following stationary condition:

$$\frac{\partial P}{\partial u(X_\ell)^*} = 0, \tag{65}$$

for each component $u(X_\ell)$ of $\hat{u}$. Substitution of Eqs. (57) and (64) into Eq. (65) gives

$$\hat{u} = \sum_{n=-\infty}^{+\infty} [L(\beta_n)]Q(\beta_n), \quad (66)$$

where

$$L(\beta_n) = [A]^{-1} [G(\beta_n)]^{*t} [R_{12}(\beta_n)],$$

and

$$[A] = \frac{S_n \beta_n}{p}[I] + \sum_{n=-\infty}^{+\infty} [G(\beta_n)]^{*t}[R_{11}(\beta_n)][G(\beta_n)][F].$$

Substituting Eqs. (60) and (66) into Eq. (62), one finally obtains the following relation:

$$\Phi(\beta_n) = S_n \beta_n^{-1} \sum_{\ell=-\infty}^{+\infty} H_{n\ell}Q(\beta_\ell), \quad (67)$$

where

$$H_{n\ell} = -[R_{21}(\beta_n)] \bullet [G(\beta_n)] [F] [L(\beta_\ell)] + R_{22}(\beta_n)\delta_{n\ell}. \quad (68)$$

Equation (67) represents the relationship between the surface potential and the charge, where the mass loading effect has already been taken into account.

In this analysis, field variables are expressed in the spectral β domain. Therefore, this type of analysis is called spectral domain analysis (SDA).

Although the SDA is equivalent to the Green function method mathematically, the SDA has the following features.

1. Since the boundary condition is given in the space coordinate system, boundary conditions must be transformed into the β domain for the application of the SDA. Then the SDA is applicable only to cases with relatively simple surface geometry.

2. The Green function is estimated by the inverse Fourier transform of the system transfer functions, such as $R_{ij}(\beta)$. Thus calculation for the SDA is usually significantly simpler than that for the Green function method.

Since the problem considered here has a simple surface geometry as shown in Fig. 18, use of the SDA is quite efficient.

5.3. *FEM/SDA analysis with Bløtekjær's theory*

By using the relation of Eq. (67) instead of Eq. (43), one may include effects of finite film thickness into Bløtekjær's theory.

Namely, substitution of Eqs. (44) and (45) into Eq. (67) gives the following simultaneous linear equations:

$$\sum_{m=M_1}^{M_2} A_m \left[\epsilon(\infty)S_n \sum_{\ell=-\infty}^{+\infty} H_{n\ell}P_{\ell-m}(r) - S_{n-m}P_{n-m}(r) \right] = 0 \quad (69)$$

for $n = [M_1, M_2 - 1]$. By solving them with respect to A_m and substituting the calculated A_m into Eq. (52), effects of finite film thickness can be included into the effective permittivity for the grating structure.

Figure 19 shows, as an example, the velocity dispersion of the Rayleigh-type SAWs on 128°YX-LiNbO$_3$ calculated by using the FEM/SDA technique, where the shape of the Al fingers was assumed to be rectangular and $w/p = 0.5$. In the figure, $V_B =$ 4,025 m/sec is the slow-shear BAW velocity, and OC and SC indicate the dispersion relations for open- and short-circuited gratings, respectively. With an increase in h/p, the stopband width increases for the OC grating whereas it decreases for the SC grating. This is due to the fact that the electrical reflection by the OC grating has the same polarity as the mechanical reflection whereas that by the SC grating including the electrical regeneration has the opposite polarity. The stopband of the SC grating nearly disappears when $h/p \cong 0.048$.

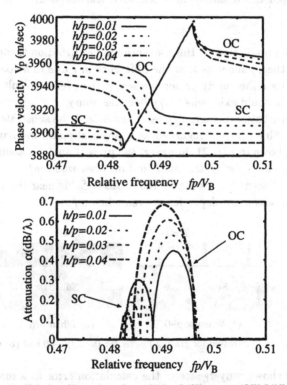

Fig. 19. Velocity dispersion of non-leaky SAWs on 128°YX-LiNbO$_3$.

Figure 20 shows the effective SAW velocity V_e and the reflectivity r per strip as a function of the metallization ratio $\Delta = w/p$, where V_e and r were evaluated from the Bragg frequency and maximum attenuation, respectively. It is seen that the calculation agrees fairly well with the experimental results.[31]

It was shown[32] that, when w and/or h are relatively small, the behavior of r

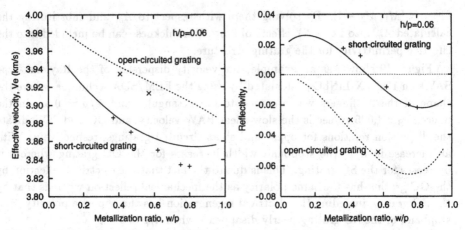

Fig. 20. w/p dependence of effective SAW velocity V_e and reflectivity r. Lines: theory, and ×, +: experiment.

agrees with the prediction by the conventional perturbation theory.[33] However, the perturbation theory suggests that the difference in r due to h could become zero as w/p approaches either unity or zero, whereas the present result suggests that the finite difference could exist when w/p becomes unity.

In the perturbation theory, $w/p \to 1$ means that the substrate is totally covered by a metallic film which is regarded as a physically and mechanically continuous medium. As shown in Fig. 21, however, the present analysis assumes that infinitesimally narrow slits exist in the metallic film even when $w/p \to 1$. Since each slit between two adjacent fingers makes the acoustic fields near the surface discontinuous, r may possess the h dependence even when $w/p \to 1$.

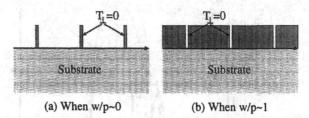

Fig. 21. Illustration for grating fingers for (a) $w/p \to 0$ and (b) $w/p \to 1$.

Figure 22 shows convergence of the calculation error as a function of M_2. The error was estimated from fractional change in the SAW velocity from the value when M_2 is sufficiently large. In the calculation, $M_1 = -M_2$ and $fp/V_B = 0.45$. Although the error becomes worse with an increase in h/p, its decrease with M_2 is still steep and monotonical. From the figure, $M_2 = 10$ seems sufficient for most cases.

Figure 23 shows the convergence of the calculation error as a function of the numbers N_x and N_z of FEM subdivisions in the width and thickness directions,

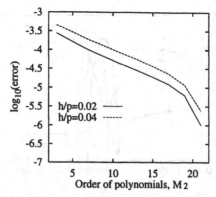

Fig. 22. Calculation error in SAW velocity as a function of M_2.

respectively. Here the discretization was performed as shown in Fig. 18, where equidistance sampling was applied for the thickness direction whereas the sampling was made dense at the finger edges so as to take the stress concentration into account.

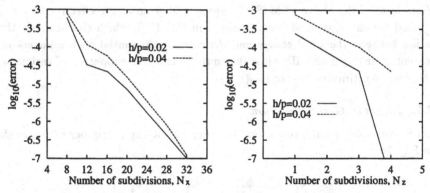

Fig. 23. Calculation error in SAW velocity as a function of N_x and N_z.

It is seen that the error decreases monotonically with an increase in N_x and N_z, and $N_x \cong 10$ and $N_z \cong 4$ seem to be enough. It is clear that N_z should be increased with an increase in h/p.

Figure 24 shows the calculated phase velocity and the attenuation of the Rayleigh-type SAW on $(0, 47.3°, 90°)$ $Li_2B_4O_7$ with Al grating of 4% p thickness as a function of fp/V_B where $V_B = 3,347$ m/sec is the velocity for the slow-shear BAW. The dependence seems similar to that for 128°YX-LiNbO$_3$ shown in Fig. 19. However, it is seen that stopband edges for the OC grating separate from those for the SC grating in this case. This is due to the asymmetry in the mechanical reflection characteristics caused by substrate anisotropy, and indicates that even solid-type IDTs exhibit directionality in the excitation characteristics. This property is called natural unidirectionality.[34]

It should be noted that this substrate is known to support the longitudinal leaky

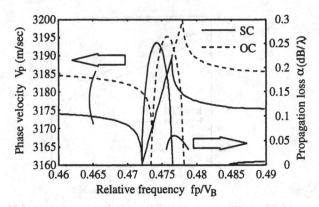

Fig. 24. Phase velocity and attenuation of Rayleigh-type SAW on (0,47.3°,90°) $Li_2B_4O_7$.

SAW,[35] and exhibits good natural unidirectionality for the Rayleigh-type SAW.[36]

6. Wave Excitation and Propagation in Grating Structures

Final form of the above FEM/SDA approach is given as the effective permittivity $\epsilon_g(s,\omega)$ for the grating structure defined in Eq. (52), which characterizes the relation between the total charge and the electrical potential on the fingers in the wavenumber β domain. Here it is shown that excitation properties of various waves can also be estimated by using $\epsilon_g(s,\omega)$.

6.1. *Discrete Green function*

Let us consider the infinite grating structure with a single strip per period as shown in Fig. 25.

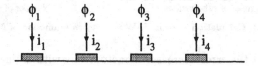

Fig. 25. Infinite grating structure.

Define the following new variables $Q(s)$ and $\Phi(s)$ by

$$Q(s) = \sum_{n=-\infty}^{+\infty} q_n \exp(+2\pi jns), \tag{70}$$

$$\Phi(s) = \sum_{n=-\infty}^{+\infty} \phi_n \exp(+2\pi jns), \tag{71}$$

where q_n and ϕ_n are the total charge and electrical potential on the n-th finger, respectively.

Since q_n and ϕ_n can be regarded as the Fourier expansion coefficients of $Q(s)$ and $\Phi(s)$, respectively, one can rewrite Eqs. (70) and (71) in the form of

$$q_n = \int_0^1 Q(s)\exp(-2\pi jns)ds, \qquad (72)$$

$$\phi_n = \int_0^1 \Phi(s)\exp(-2\pi jns)ds. \qquad (73)$$

Equations (72) and (73) show that q_n and ϕ_n are expressed as a sum of contributions of various grating modes having wavenumbers of $2\pi s/p$, where $0 \leq s \leq 1$. From the result described in Section 5, it is clear that $Q(s)$ and $\Phi(s)$ are not independent, and Eq. (51) indicated that their relation is given by

$$Q(s)/\Phi(s) = 2\sin(s\pi)\epsilon_g(s) = Y(s,\omega)/j\omega. \qquad (74)$$

Figure 26 shows calculated $\epsilon_g(s,f)$ on 128°YX-LiNbO$_3$ with Al grating of 2% p thickness at $f = 0.45 V_B/p$, where $V_B = 4,025$ m/sec is the slow-shear BAW velocity.

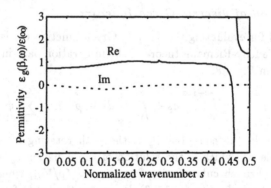

Fig. 26. Effective permittivity for grating structure on 128°YX-LiNbO$_3$ substrate at $h = 0.02p$ and $fp/V_B = 0.45$). Note $\epsilon_g(1-s) = \epsilon(s)$.

At $s \cong 0.46$, there is a pole corresponding to the radiation of the Rayleigh-type SAW on the SC grating. On the other hand, $\Im[\epsilon_g(s,f)]$ becomes complex when $s < 0.27$. This is due to the radiation of longitudinal BAW.

Substitution of Eq. (74) into Eq. (72) gives

$$q_n = \frac{-j}{\omega}\int_0^1 Y(s,\omega)\Phi(s)\exp(-2\pi jns)ds. \qquad (75)$$

One may then obtain the following relation by substituting Eq. (71) into Eq. (75):

$$q_k = \frac{-j}{\omega}\int_0^1 Y(s,\omega)\sum_{\ell=-\infty}^{+\infty}\phi_\ell\exp\{-2\pi js(k-\ell)\}ds \equiv \sum_{\ell=-\infty}^{+\infty}\phi_\ell G_{k-\ell}, \qquad (76)$$

where G_k is the discrete Green function[19] given by

$$G_k = \frac{-j}{\omega} \int_0^1 Y(s,\omega) \exp(-2\pi j k s) ds. \qquad (77)$$

As shown in Fig. 27, G_k represents the induced charge q_k when a unit potential is applied on the 0-th finger while the potential on other fingers is zero. Because of the reciprocity, $G_k = G_{-k}$ holds, and this gives the identity $\epsilon_g(1-s) = \epsilon_g(s)$.

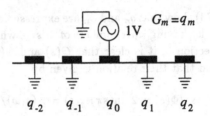

Fig. 27. Discrete Green function.

6.2. Evaluation of discrete Green function

Here a method for evaluating the discrete Green function G_n is described.[19]

From the Cauchy-Riemann theorem, the integration path in Eq. (77) can modified as shown in Fig. 28:

$$\int_0^1 ds = \int_0^{0-j\infty} ds + \int_{1-j\infty}^1 ds + \oint_{\Gamma_g} ds + \sum_{i=1}^3 \oint_{\Gamma_{Bi}} ds, \qquad (78)$$

for positive k. In the equation, Γ_g is the path rotating in a clockwise manner around the pole s_g, which is the solution of the equation $\epsilon_g^{-1}(s_g)=0$. Γ_{Bi} is the path along the branch cuts starting from $s_{Bi} = fp/V_{Bi}$, where V_{Bi} is the BAW velocity. Here, the subscript $i(=1,2,3)$ designates the type of BAWs, that is, slow-shear, fast-shear and longitudinal BAWs, respectively. The integration along the path of $s = [-j\infty, 1 - j\infty]$ vanishes because the integrand is zero.

The first two terms in the right-hand side in Eq. (78) give the contribution of electrostatic coupling. The third term represents the contribution of grating-mode radiation, and the fourth term comes from BAW radiation.

Here the contribution of the BAW radiation is ignored because it is usually not so significant for conventional SAW devices employing the Rayleigh-type SAWs. It should be noticed that the BAW radiation is important in devices employing SH-type SAWs.[37]

The contribution G_{en} of the electrostatic coupling in G_n is given by

$$G_{en} = 4 \int_0^{-j\infty} \sin(s\pi) \epsilon_g(s) \exp(-2\pi j n s) ds$$

$$\cong 4\epsilon_g(0) \int_0^{-j\infty} \sin(s\pi) \exp(-2\pi j n s) ds \cong -\epsilon(0)/\{\pi(n^2 - 1/4)\}. \qquad (79)$$

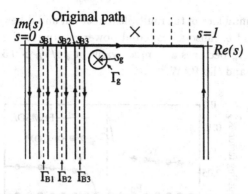

Fig. 28. Integration path for discrete Green function.

Applying the residue theorem to Eq. (77), one obtains the contribution G_{gn} of the grating-mode radiation as

$$G_{gn} = -j\frac{4}{\pi}\epsilon(\infty)K_g^2 \exp(-2\pi j s_g |n|). \qquad (80)$$

In the equation, K_g^2 is the effective electromechanical coupling factor of the grating-mode[19] given by

$$K_g^2 = \pi^2 \sin(s_g \pi) \left[\epsilon(\infty) \frac{\partial \epsilon_g(s)^{-1}}{\partial s} \Big|_{s=s_g} \right]^{-1}. \qquad (81)$$

Figure 29 shows the calculated dispersion relation of the Rayleigh-type SAW on 128°YX-LiNbO$_3$ with a SC Al grating of 2% p thickness. There is a stopband due to the Bragg reflection of the Rayleigh-type SAW at $0.482 < fp/V_B < 0.486$, and backscattering to the BAW occurs when $fp/V_B > 0.5$. In the figure, the dispersion curve for the OC grating is also shown for comparison.

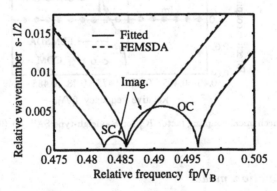

Fig. 29. Dispersion curves of the Rayleigh-type SAW on 128°YX-LiNbO$_3$.

Figure 30 shows the change in K_g^2 as a function of frequency. It is seen that K_g^2 changes very smoothly except at the stopband. Near the stopband, K_g^2 changes rapidly due to the influence of the multiple reflection, and it becomes pure imaginary within the stopband because all radiated power returns to the excitation point. It is interesting that K_p^2 becomes a complex value at $fp/V_B > 0.5$ due to the coupling between the SAW and the BAW.[37]

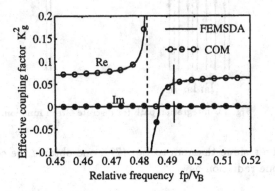

Fig. 30. Electromechanical coupling factor K_g^2 for Rayleigh-type SAW as a function of fp/V_B.

For comparison, Fig. 31 shows K_g^2 of the Rayleigh-type SAW on $(0,47.3°,90°)$ $Li_2B_4O_7$ with Al grating of 4% p thickness as a function of fp/V_B. It is seen that K_g^2 becomes infinite at both stopband edges and crosses zero. This reflects the existence of the natural unidirectionality.[34]

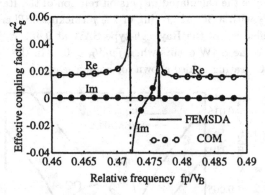

Fig. 31. Electromechanical coupling factor K_g^2 for Rayleigh-type SAW on $(0,47.3°,90°)$ $Li_2B_4O_7$.

6.3. Delta function model

Assume a structure where the propagation surface is completely covered with a SC grating. If the width and the spacing of the grating fingers are the same as those of IDT fingers, the discrete Green function analysis described in Sections 6.1 and 6.2 is applicable.

If there exist no floating fingers, the input admittance Y_{kk} of the k-th IDT consisting of M_k fingers is simply given by

$$Y_{kk} = j\omega \sum_{m=0}^{M_k-1} \sum_{n=0}^{M_k-1} S_{km} S_{kn} G_{m-n}. \tag{82}$$

The transfer admittance $Y_{k\ell}$ between the k-th and ℓ-th IDTs is given by

$$Y_{k\ell} = j\omega \sum_{m=0}^{M_k-1} \sum_{n=0}^{M_\ell-1} S_{km} S_{\ell n} G_{D_{k\ell}+m-n}, \tag{83}$$

where $D_{k\ell}$ is the normalized distance between the two IDTs, and

$$S_k = \begin{cases} 1, & \text{for hot-finger} \\ 0, & \text{otherwise} \end{cases}. \tag{84}$$

Thus, if S_k is specified, electrical characteristics of arbitrary IDT configurations can be determined numerically.

Although Eqs. (82) and (83) are completely identical with those assumed in the simple delta function model analysis,[38] the present analysis takes account of the effects of electrical and acoustic interactions amongst the fingers. Note that the structure is equivalent to the resonator at the Bragg frequency because of the infinite length.

If there exist floating fingers, IDT behaviors are characterized by the following procedure; (i) determination of the voltages of the floating fingers by applying a unit potential to the hot-fingers of an input IDT and by short-circuiting an output IDT, (ii) calculation of the total charges induced on the hot-fingers by substituting the voltages of the floating fingers into Eq. (82), and (iii) determination of input and transfer admittances by summing up the induced charges on input and output IDT fingers individually and multiplying by $j\omega$.

By applying the above mentioned procedure, the input admittance of a floating-electrode-type unidirectional transducer (FEUDT)[39] shown in Fig. 32 was analyzed. The FEUDT consists 40 finger-pairs with the aperture of 30λ, and 128°YX-LiNbO$_3$ was chosen as the substrate.

Fig. 32. Floating-electrode-type unidirectional transducer (FEUDT).

Figure 33 shows the calculated input admittance Y as a function of frequency. The result is in good agreement with the experiment.[39]

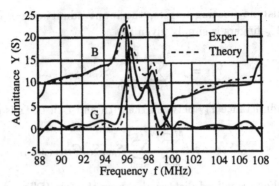

Fig. 33. Admittance Characteristics of FEUDT on 128°YX-LiNbO$_3$. Solid-line: experiment,[39] and broken-line: present analysis.

6.4. Input admittance of infinitely-long IDTs

Let us consider the grating structure shown in Fig. 25. When $\phi_n = (-1)^n v_0/2$, the structure is equivalent to the infinitely long single-electrode IDT with periodicity $p_I(=2p)$, and v_0 corresponds to the applied voltage.

From Eq. (71), $\Phi(\beta) = 2\pi v_0/p_I \times \delta(\beta - 2\pi/p_I)$. Then substitution of this relation into Eq. (75) gives

$$i_n = 2^{-1}(-1)^n Y(2\pi/p_I, \omega) v_0.$$

This means that the input admittance $\hat{Y}(\omega)$ per period for single-electrode IDTs with infinite length is given by[9]

$$\hat{Y}(\omega) = 2^{-1} Y(2\pi/p_I, \omega). \qquad (85)$$

Figure 34 shows calculated $\hat{Y}(\omega)/\omega\epsilon(\infty)$ on (0,47.3°,90°) Li$_2$B$_4$O$_7$ with a Al grating of 4% p thickness as a function of the relative frequency fp/V_B, where $V_B = 3,347$ m/sec is the velocity for the slow-shear BAW. Two resonances and antiresonances for the Rayleigh-type SAW are clearly seen. In this frequency region, $\Re[\hat{Y}(\omega)] = 0$ because of the cutoff for the BAW radiation. Comparison with Fig. 31 indicates that frequencies giving these resonances and antiresonances correspond to the stopband edges for the SC and OC gratings, respectively.

The response due to the longitudinal leaky SAW appears in a much higher frequency range.

6.5. COM parameter extraction

For the simulation of SAW devices, the COM theory is widely used. Since accuracy of the simulation is critically dependent upon that of parameters employed for the analysis. Their determination is the most important application for the present numerical techniques. Let us discuss applicability of these numerical methods to determine the parameters required for the COM analysis.

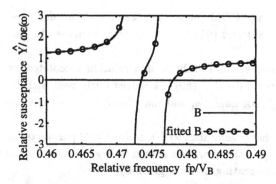

Fig. 34. Input admittance of infinitely long IDT on $(0, 47.3°, 90°)$ $Li_2B_4O_7$.

Let us discuss SAW propagation and excitation by using the COM equations shown below:

$$\frac{\partial U_+(X_1)}{\partial X_1} = -j\theta_u U_+(X_1) - j\kappa U_-(X_1) + j\zeta v, \tag{86}$$

$$\frac{\partial U_-(X_1)}{\partial X_1} = +j\kappa^* U_+(X_1) + j\theta_u U_-(X_1) - j\zeta^* v, \tag{87}$$

$$\frac{\partial I(X_1)}{\partial X_1} = -4j\zeta^* U_+(X_1) - 4j\zeta U_-(X_1) + j\omega C v, \tag{88}$$

where $U_\pm(X_1)$ is the mode amplitude propagating toward $\pm X_1$ direction, $I(X_1)$ is the current on the bus-bar, $\theta_u = \beta_u - 2\pi/p_I$, β_u is the wavenumber of an "unperturbed" mode under the SC grating, κ is the mutual coupling coefficient, ζ is the transduction coefficient, C is the static capacitance per unit length, and v is the applied voltage.

As an eigen value of Eqs. (86) and (87), one may obtain the wavenumber β_p of the "perturbed" mode under the SC grating as

$$\beta_p = 2\pi/p_I \pm \sqrt{\theta_u^2 - |\kappa|^2}. \tag{89}$$

For simplicity, here let us assume that β_u is proportional to ω. Then by using Eq. (89), the frequencies $\omega_{s\pm}$ at the stopband edges are given by

$$\omega_{s\pm}/V_e - 2\pi/p_I = \pm|\kappa|^2, \tag{90}$$

where $V_e = \omega/\beta_u$ is the effective SAW velocity.

Similarly, for the case where $dI(X_1)/X_1 = 0$ corresponding to the OC grating, the wavenumber β_o of the "perturbed" mode is given by

$$\beta_o = 2\pi/p_I \pm \sqrt{(\theta_u - 4|\zeta|^2/\omega C)^2 - |\kappa - 4\zeta^2/\omega C|^2}, \tag{91}$$

Under the above mentioned assumption, frequencies $\omega_{o\pm}$ at the stopband edges are given by

$$\omega_{o\pm}/V_e - 2\pi/p_I - 4|\zeta|^2/\omega C = \pm|\kappa - 4\zeta^2/\omega C|. \tag{92}$$

Thus, all necessary COM parameters can be determined by applying numerical fitting of Eqs. (89) and (91) to the dispersion curves obtained by the FEM/SDA analysis.

Figure 29 also shows the dispersion relation calculated by substituting determined parameters into Eq. (89). It agrees fairly well with that obtained directly from the FEM/SDA analysis, and the effectiveness of the COM analysis is demonstrated.

There are other possibilities to determine COM parameters. That is, by applying the COM theory to the structure shown in Fig. 33, the electromechanical coupling factor K_g^2 for the grating mode is given by

$$K_g^2 = \frac{\pi|\zeta|^2 p_I}{\omega C} \frac{\theta_u - |\kappa|\cos\psi}{\theta_p}, \qquad (93)$$

where $\psi = \angle(\kappa^*\zeta/\zeta^*)$. On the other hand, by applying the COM theory to the structure shown in Fig. 27, the input admittance $\hat{Y}(\omega)$ of the infinitely long IDT is given by

$$\hat{Y}(\omega) = j\omega C p_I \left(1 - \frac{8|\zeta|^2}{\omega C} \frac{\theta_u - |\kappa|\cos\psi}{\theta_u^2 - |\kappa|^2}\right). \qquad (94)$$

Thus, the COM parameters are determined by applying numerical fitting of Eqs. (93) or (94) to K_g^2 and $\hat{Y}(\omega)$, respectively, calculated directly by using the FEMSDA.

Figures 30, 31 and 34 also show K_g^2 and $\hat{Y}(\omega)$ calculated by using Eqs. (93) and (94). The agreement is excellent.

It should be noted that, comparison of Eq. (94) with Eqs. (89) and (91) indicates that the two resonance and two antiresonance frequencies in $\hat{Y}(\omega)$ coincide with the frequencies at the stopband edges for the SC and OC gratings, respectively.

For substrates without natural unidirectionality, only one pole and one zero appear in $\hat{Y}(\omega)$. Namely, the other pole and zero are degenerated. The location of degenerated zero and pole can be found by adding a tiny directivity. This is realized by giving a slight off-angle for the propagation direction from the crystal principle axes.[40]

By using derived COM parameters, the one-port SAW resonator on 128°YX-LiNbO3 was analyzed. The result is shown in Fig. 35. In the calculation, $h/p_I = 0.88\%$. The simulation agrees well with the experiment.* The response indicated by the arrow in the figure may be due to the coupling with the thickness vibration, which is not taken into account for the simulation.

7. Analysis of Multi-Finger Gratings

For the development of low loss IF SAW filters with a small group delay distortion, single-phase unidirectional transducers (SPUDT) are paid much attention. Usual SPUDTs possess multi-finger structures where multiple fingers with various widths

---

*This experimental result was given by Mr. Bungo of Mitsubishi Material Co. Ltd.

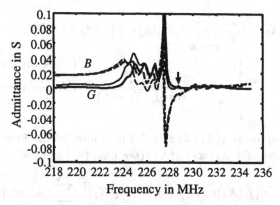

Fig. 35. Admittance characteristics of one-port SAW resonator on 128°YX-LiNbO₃. Bold lines: experiment, and thin lines: simulation.

are aligned within a unit period. Since SPUDT characteristics are critically dependent upon the finger arrangement, development of their simulation techniques is important.

Here ideas used in the FEM/SDA analysis are applied to multi-finger structures.[14] In this case, Aoki's theory[3,4] is employed instead of that by Bløtekjær in order to take charge concentration at the finger edges properly into account.

7.1. Aoki's theory for double-finger gratings

Figure 36 shows the periodic metallic grating of infinite acoustic length, where two types of fingers with widths w_1 and w_2, respectively, are aligned within one periodic length p. The analysis assumes that SAWs propagate toward the X_1 direction.

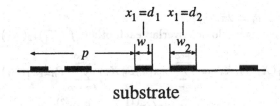

Fig. 36. Configuration used for analysis.

Define the following functions:

$$f_i(X_1) = \frac{\sqrt{2}}{2} \sum_{\ell=-\infty}^{+\infty} P_\ell(\Omega_i) \exp\{-j\beta_g(\ell+1/2)(X_1-d_i)\}$$

$$= \begin{cases} \dfrac{1}{\sqrt{\cos\{\beta_g(X_1-d_i)\}-\Omega_i}} & (|X_1-d_i| < w_i/2) \\ 0 & (|X_1-d_i| > w_i/2) \end{cases}, \quad (95)$$

$$g_i(X_1) = \frac{j\sqrt{2}}{2} \sum_{\ell=-\infty}^{+\infty} S_\ell P_\ell(\Omega_i) \exp\{-j\beta_g(\ell+1/2)(X_1-d_i)\}$$

$$= \begin{cases} 0 & (|X_1-d_i|<w_i/2) \\ \dfrac{\text{sgn}(X_1-d_i)}{\sqrt{\Omega_i-\cos\{\beta_g(X_1-d_i)\}}} & (|X_1-d_i|>w_i/2) \end{cases}, \quad (96)$$

where $\Omega_i = \cos(\pi w_i/p)$.

Then we represent $q(X_1)$ and $e(X_1)$ in a form of the Fourier transform with weighting factors of $f(X_1)$ and $g(X_1)$ (see Eqs. (95) and (96)):

$$q(X_1) = \{f_1(X_1)g_2(X_1) + g_1(X_1)f_2(X_1)\} \sum_{m=M_1}^{M_2} A_m \exp(-j\beta_m X_1)$$

$$= \sum_{n=-\infty}^{+\infty} \sum_{m=M_1}^{M_2} A_m F_{n-m} \exp(-j\beta_n X_1), \quad (97)$$

$$e(X_1) = g_1(X_1)g_2(X_1) \sum_{m=M_1}^{M_2} B_m \exp(-j\beta_m X_1)$$

$$= \sum_{n=-\infty}^{+\infty} \sum_{m=M_1}^{M_2} B_m G_{n-m} \exp(-j\beta_n X_1). \quad (98)$$

In Eqs. (97) and (98), A_m is the unknown coefficient,

$$F_n = \frac{j}{2} \sum_{\ell=-\infty}^{+\infty} P_{n-\ell-1}(\Omega_1) P_\ell(\Omega_2)(S_\ell + S_{n-\ell-1}) \exp(j\eta^{(1)}_{n-\ell-1/2} + j\eta^{(2)}_{\ell+1/2}), \quad (99)$$

$$G_n = -\frac{1}{2} \sum_{\ell=-\infty}^{+\infty} P_{n-\ell-1}(\Omega_1) P_\ell(\Omega_2) S_{n-\ell-1} S_\ell \exp(j\eta^{(1)}_{n-\ell-1/2} + j\eta^{(2)}_{\ell+1/2}), \quad (100)$$

where $\eta^{(i)}_n = n\beta_g d_i = 2\pi n d_i/p$.

Because the fingers do not overlap each other, $f_1(X_1)f_2(X_1) = 0$. Since

$$f_1(X_1)f_2(X_1) = \sum_{n=-\infty}^{+\infty} \exp(j\beta_n X_1) \sum_{\ell=-\infty}^{+\infty} P_{n-\ell-1}(\Omega_1) P_\ell(\Omega_2)$$
$$\times \exp(j\eta^{(1)}_{n-\ell-1/2} + j\eta^{(2)}_{\ell+1/2}) \quad (101)$$

from Eq. (95), the following relation holds for arbitrary n:

$$\sum_{\ell=-\infty}^{+\infty} P_{n-\ell-1}(\Omega_1) P_\ell(\Omega_2) \exp(j\eta^{(1)}_{n-\ell-1/2} + j\eta^{(2)}_{\ell+1/2}) = 0. \quad (102)$$

This relation simplifies Eqs. (99) and (100) as

$$G_n = -\sum_{\ell=0}^{n-1} P_{n-\ell-1}(\Omega_1) P_\ell(\Omega_2) \exp(j\eta^{(1)}_{n-\ell-1/2} + j\eta^{(2)}_{\ell+1/2}), \quad (103)$$

for positive n, and
$$F_n = -jS_nG_n. \tag{104}$$
Note that $G_0 = 0$ and $G_{-n} = G_n^*$. Since the summation in Eq. (103) has only to be done over finite ℓ, Eqs. (103) and (104) enable us to calculate F_n and G_n much faster than Eqs. (99) and (100).

From Eqs. (97) and (98), the Fourier transforms $Q(\beta_n)$ and $E(\beta_n)$ of $q(X_1)$ and $e(X_1)$, respectively, are given by
$$Q(\beta_n) = -j \sum_{m=M_1}^{M_2} A_m S_{n-m} G_{n-m}, \tag{105}$$
$$E(\beta_n) = \sum_{m=M_1}^{M_2} B_m G_{n-m}. \tag{106}$$
Substitution of Eqs. (105) and (106) into Eq. (43) gives
$$\sum_{m=M_1}^{M_2} G_{n-m}\{B_m - A_m S_{n-m} S_n/\epsilon(\beta_n/\omega)\} = 0. \tag{107}$$

It should be noted that $\epsilon(\beta_n/\omega) \to \epsilon(\infty)$ when $|n| \to \infty$, and that the relation $\epsilon(\beta_n/\omega) \cong \epsilon(\infty)$ holds if $|\omega/\beta_n|$ is not very close to the SAW velocity V_S. So if it is assumed that $\epsilon(\beta_n/\omega) = \epsilon(\infty)$ for $n < N_1$ or $N > N_2$, $B_m = A_m/\epsilon(\infty)$ so that Eq. (107) holds for $n < M_1$ or $n \geq M_2$. Thus Eq. (107) becomes
$$\sum_{m=M_1}^{M_2} A_m G_{n-m}\{1 - S_{n-m} S_n \epsilon(\infty)/\epsilon(\beta_n/\omega)\} = 0 \tag{108}$$
for $n = [M_1 + 1, M_2 - 1]$.

By applying the Floquet theorem of Eq. (1), the potential $\Phi^{(\ell)}(s)$ of the ℓ-th finger is given by
$$\Phi^{(\ell)}(s) = -\int_{-\infty}^{d_\ell - w_\ell/2} e(X_1)dX_1 = -\int_{-p+d_\ell+w_\ell/2}^{d_\ell - w_\ell/2} \frac{e(X_1)dX_1}{1 - \exp(2\pi js)}. \tag{109}$$
Substitution of Eq. (98) into Eq. (109) gives
$$\Phi^{(1)}(s) = -j \frac{\exp(-\pi js)}{2\sin(2\pi s)} \sum_{m=M_1}^{M_2} A_m \{V_m^{(1)} + V_m^{(2)} \exp(2\pi js)\}, \tag{110}$$
$$\Phi^{(2)}(s) = -j \frac{\exp(-\pi js)}{2\sin(2\pi s)} \sum_{m=M_1}^{M_2} A_m \{V_m^{(1)} + V_m^{(2)}\}, \tag{111}$$
where
$$V_m^{(1)} = \int_{-p+d_2+w_2/2}^{d_1 - w_1/2} g_1(X_1)g_2(X_1)\exp(-j\beta_m X_1)dX_1, \tag{112}$$
$$V_m^{(2)} = \int_{d_1+w_1/2}^{d_2 - w_2/2} g_1(X_1)g_2(X_1)\exp(-j\beta_m X_1)dX_1. \tag{113}$$

The current $I^{(k)}(s)$ flowing into the k-th finger is given by

$$I^{(k)}(s) = j\omega \int_{d_k-w_k/2}^{d_k+w_k/2} q(X_1)dX_1 = \sum_{m=M_1}^{M_2} A_m W_m^{(k)}(s), \qquad (114)$$

where

$$W_m^{(k)}(s) = j\omega \int_{d_k-w_k/2}^{d_k+w_k/2} g_k(X_1)f_{3-k}(X_1)\exp(-j\beta_m X_1)dX_1. \qquad (115)$$

The numerical analysis is carried out by the following procedure. After calculating $V_m^{(i)}$ and $W_m^{(i)}$ by numerical integration for given values of s and ω, the linear equations composed of Eqs. (108), (110) and (111) are solved with respect to A_m. This gives A_m in the form of

$$A_m = p_{m1}(s)\Phi^{(1)}(s) + p_{m2}(s)\Phi^{(2)}(s). \qquad (116)$$

Substitution of Eq. (116) into Eq. (114) gives

$$I^{(k)}(s) = \sum_{\ell=1}^{2} Y_{k\ell}(s,\omega)\Phi^{(\ell)}(s), \qquad (117)$$

where $k = 1$ or 2, and

$$Y_{k\ell}(s,\omega) = \sum_{m=M_1}^{M_2} W_m^{(k)}(s)p_{m\ell}(s), \qquad (118)$$

is the transfer admittance matrix[15] for the grating structure.

Effects of the finite film thickness can be introduced by using the technique described in Section 5.2. That is, since Eqs. (67) and (68) hold independent of the number of fingers per unit period, Eqs. (67) should be used instead of Eq. (43) for the derivation of Eq. (107). Then the simultaneous linear equations of Eq. (108) to be solved become

$$\sum_{m=M_1}^{M_2} A_m \left[G_{n-m} - S_n \epsilon(\infty) \sum_{\ell=-\infty}^{+\infty} S_{\ell-m} H_{n\ell} G_{\ell-m} \right] = 0. \qquad (119)$$

$$\text{for } n = [M_1 + 1, M_2 - 1].$$

The authors has extended Aoki's theory for triple-finger gratings (three fingers in one periodic section).[15] The formalism is given in Appendix A. The results are quite similar to the two-finger case described here, and effects of finite film thicknesses can also be taken into account.[14] These ideas were implemented into the software MULTI.[16,17]

As a demonstration, the SAW propagation characteristics for the triple-finger grating shown in Fig. 37 were analyzed. Al and 128°YX-LiNbO$_3$ are assumed as the electrode and substrate materials. Note that the grating periodicity p is equal to the IDT periodicity p_I in this case.

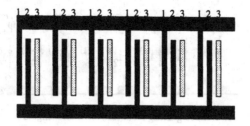

Fig. 37. Triple finger grating.

Figure 38 shows the dispersion relation near $fp/V_B \cong 1$, where $h_1 = h_2 = h_3 = 0.04p$, $w_1 = w_2 = w_3 = p/8$, $d_1 = -p/4$, $d_2 = 0$, and $d_3 = p/4$. To extract the COM parameters, the following two cases were calculated, that is, 1) short-circuited (SC) case: fingers 1 and 2 are connected with the bus-bar, and 2) open-circuited (OC) case: all fingers are isolated from each other. It is seen that the stopband edges for

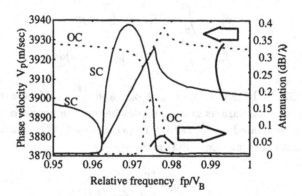

Fig. 38. SAW dispersion relation in triple finger grating.

the SC grating do not coincide with those for the OC grating. This is because the IDT possesses directionality. From the location of these stopband edges, we can determine the COM parameters by using the procedure described in Section 6.5.

7.2. Analysis of excitation properties of multi-finger grating

In Section 6, we have shown various techniques for the characterization of SAW propagation and excitation for "single-finger" gratings. Of course, these ideas can be extended for the multi-finger structures.[15,37]

Let us consider the finger configuration shown Fig. 39.

Define $\Phi^{(k)}(s)$ and $I^{(k)}(s)$ associated with potential $\phi_n^{(k)}$ and current $i_n^{(k)}$ of the k-th strip in the n-th period, respectively, i.e.,

$$\Phi^{(k)}(s) = \sum_{n=-\infty}^{+\infty} \phi_n^{(k)} \exp(+2\pi j n s), \qquad (120)$$

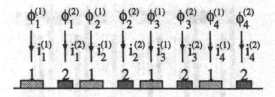

Fig. 39. Infinite grating structure with two fingers per period.

$$I^{(k)}(s) = \sum_{n=-\infty}^{+\infty} i_n^{(k)} \exp(+2\pi j n s), \tag{121}$$

where $s = \beta p/2\pi$. Since $\phi_n^{(k)}$ and $i_n^{(k)}$ can be regarded as the Fourier expansion coefficients of $\Phi^{(k)}(s)$ and $I^{(k)}(s)$, respectively, Eqs. (120) and (121) can be rewritten as

$$\phi_n^{(k)} = \int_0^1 \Phi^{(k)}(s) \exp(-2\pi j n s) ds, \tag{122}$$

$$i_n^{(k)} = \int_0^1 I^{(k)}(s) \exp(-2\pi j n s) ds. \tag{123}$$

Equations (122) and (123) suggest that $\phi_n^{(k)}$ and $i_n^{(k)}$ could be expressed as a sum of contributions from various grating modes having the normalized wavenumber s, where $\Phi^{(k)}(s)$ is related to $I^{(k)}(s)$ by $Y_{k\ell}(s)$ in Eq. (117). Then, substitution of Eqs. (117) and (120) into Eq. (123) gives

$$i_n^{(k)} = j\omega \sum_{n=-\infty}^{+\infty} \sum_{\ell=1}^{L} G_{n-\ell}^{(k\ell)} \phi_n^{(\ell)}, \tag{124}$$

where $k = [1,2]$ and $k = [1,3]$, respectively, for the double- and triple-finger gratings, and $G_n^{(k\ell)}$ is the discrete Green function given by

$$G_n^{(k\ell)} = \frac{-j}{\omega} \int_0^1 Y_{k\ell}(s,\omega) \exp(-2\pi j n s) ds. \tag{125}$$

Let us consider the IDT with triple-fingers per period shown in Fig. 37. In this case, $p = p_I$, $\Phi^{(1)}(\beta) = 2\pi v_0/p_I \times \delta(\beta - 2\pi/p_I)$, $\Phi^{(2)}(\beta) = 0$ and $I^{(3)}(\beta) = 0$. Then the input admittance $\hat{Y}(\omega)$ per period for the IDTs with infinite length is given by

$$\hat{Y}(\omega) = Y_{11}(2\pi/p_I, \omega) - Y_{13}(2\pi/p_I, \omega)Y_{31}(2\pi/p_I, \omega)/Y_{33}(2\pi/p_I, \omega). \tag{126}$$

Figure 40 shows calculated $\hat{Y}(\omega)/\omega\epsilon(\infty)$ as a function of normalized frequency fp_I/V_B for the structure shown in Fig. 37 on 128°YX-LiNbO$_3$ with Al fingers of 4% p_I thickness. Due to the directivity caused by the asymmetry of the IDT structure, two resonances and antiresonances are clearly seen. In the figure, $\hat{Y}(\omega)$

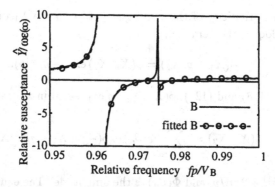

Fig. 40. Input admittance of infinite triple-finger IDT on 128°YX-LiNbO$_3$.

calculated by using COM is also shown, and it agrees well with the result of the direct computation.

8. Analysis of Oblique SAW Propagation

For the development of low loss IF SAW filters with superior out-of-band rejection, resonator SAW filters are widely used. Presently, the behavior of transverse modes in the grating structures is investigated extensively so as to achieve controlled out-of-band rejection and passband shape.[41,42]

Here the theory proposed by Wagner and Männer[43] is described, aiming at analyzing oblique SAW propagation under the metallic grating, and the theory is extended so as to include effects of the finite film thickness.[44]

Consider plane SAWs obliquely propagating under the metallic grating shown in Fig. 41. Assuming the wavenumbers of SAWs propagating toward the X_1 and

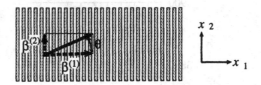

Fig. 41. Obliquely propagating SAWs.

X_2 directions to be $\beta^{(1)}$ and $\beta^{(2)}$, respectively, one can relate these wavenumbers with the propagation angle θ with respect to the X_1 axis as

$$\beta^{(2)} = \beta^{(1)} \tan \theta. \tag{127}$$

Since the metallic grating is uniform in the X_2 direction, the SAW field $\phi(X_1, X_2)$ on the surface of the substrate varies according to,

$$\phi(X_1, X_2) \propto \exp(-j\beta^{(2)} X_2). \tag{128}$$

Because of the periodicity of the grating structure in the X_1 direction, $\phi(X_1, X_2)$ satisfies the Floquet theorem, i.e.,

$$\phi(X_1 + p, X_2) = \phi(X_1, X_2)\exp(-j\beta^{(1)}p). \tag{129}$$

From Eqs. (128) and (129), $\phi(X_1, X_2)$ is expressed in a form of

$$\phi(X_1, X_2) = \sum_{n=-\infty}^{+\infty} \Phi(\beta_n)\exp(-j\beta_n X_1 - j\beta^{(2)} X_2), \tag{130}$$

where $\beta_n = \beta^{(1)} + 2\pi n/p$, and $\Phi(\beta_n)$ is the amplitude. The equation suggests that $\phi(X_1, X_2)$ under the grating structure could be expressed as a sum of various plane waves having wavenumbers β_n and $\beta^{(2)}$ toward the X_1 and X_2 directions, respectively. Hence, although the SAW field generally has three dimensional variation, it is reduced to two dimensional problem. Then $\phi(X_1, X_2)$ can be analyzed by specifying the angular frequency ω and $\beta^{(2)}$ as parameters. By using this idea, the FEM/SDA analysis described in Section 5 is applicable to the oblique incident case. This technique was implemented into the software OBLIQ.[16,17,44]

Note that since the grating fingers are assumed to be of infinite length in the X_2 direction, the electric field associated with the obliquely propagating SAWs is always short-circuited, independent of the electrical connection; the OC grating also behaves like the SC grating except for the normal incident case.

Figure 42 shows the frequency dispersion of the velocity of the Rayleigh-type SAW with the propagation angle θ (see Fig. 41) as a parameter. As a substrate, 128°YX-LiNbO$_3$ was chosen, and for the metallic grating, Al fingers are assumed to be rectangular with $h/p = 0.024$ and $w/p = 0.5$.

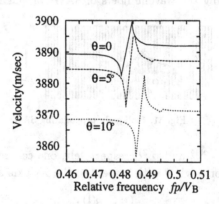

Fig. 42. θ dependence of phase velocity of Rayleigh-type SAW velocity on 128°YX-LiNbO$_3$.

With an increase in θ, the stopband at $fp/V_B \cong 0.48 - 0.49$ moves toward higher frequencies. This originates from the convex shape of the SAW slowness curve around $\theta \cong 0$ on the $X_1 - X_2$ plane of 128°YX-LiNbO$_3$.

9. Analysis of SH-type SAW

The numerical techniques described above are also applicable to SH-type SAWs, which are now widely used in the RF section for mobile communication systems. The behavior of the SH-type SAWs is similar to that of the Rayleigh-type SAWs at first look, and direct application of the above techniques looks fine at first trial. However, if better understanding and/or precise simulation are needed, attention must be paid to the following for their characterization.

Figure 43 shows the change in the SH-type SAW velocity on $36°$YX-LiTaO$_3$ with Al finger thickness h, where $V_B = 4,226$ m/s is the fast-shear BAW velocity. With an increase in h, the stopband width for the SC grating increases, and at $h/p > 0.05$, its upper edge coincides with the cut-off frequency for the back-scattering of the fast-shear BAW. In addition, the amplitude of the back-scattered BAW also increases with an increase in h. It should be noted that $\Im(\beta) \neq 0$ at frequencies lower than the stopband. This is due to the intrinsic leaky nature of the SH-type SAW on $36°$YX-LiTaO$_3$.

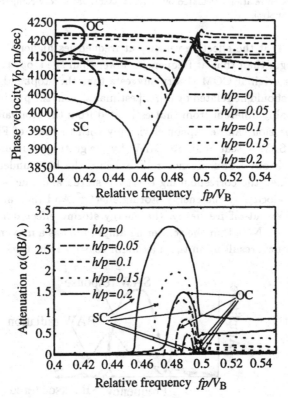

Fig. 43. Dispersion in phase velocity V_p and attenuation α of SH-type SAW on $36°$YX-LiTaO$_3$ with Al film thickness h.

Figure 44 compares the dispersion relation obtained by the FEM/SDA analysis with that calculated by the COM analysis described in Section 6.5. In the calculation, Al metallic grating with 10% p thickness is assumed to be short-circuited. Comparing with the result for the Rayleigh-type SAW shown in Fig. 29, the agreement is rather poor, particularly, near the upper edge of the stopband.

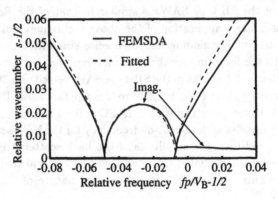

Fig. 44. Dispersion relation calculated by conventional COM theory compared with that obtained by FEM/SDA analysis.

It is seen that the effective velocity at frequencies higher than the stopband is relatively larger than that at frequencies lower than the stopband. In other words, although the original COM theory suggests that the stopband width and out-of-band ripples should be related by $|\kappa|$, $|\kappa|$ estimated from the passband characteristics was considerably different from that estimated from the stopband characteristics.

This dispersion can be explained visually with the help of Fig. 45.[45] Although the SH-type SAW is scattered into BAWs by the grating structure, scattered BAWs cancel to each other and are not radiated into the bulk under the BAW cut-off frequency. Then the corresponding energy is stored and causes a reduction of the leaky SAW velocity by the energy storing effect.[23] And then at frequencies higher than the BAW cut-off frequency, the energy storing effect diminishes due to the BAW radiation. Note that the scattering amplitude has a maximum near the BAW cut-off frequency, resulting in the maximum velocity reduction.

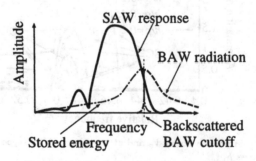

Fig. 45. Origin of dispersion near stopband.

It should be noted that the cut-off frequency is close to the resonance frequency for the SH-type SAWs. Then the variation in the energy storing effect considerably affects the resonant characteristics of such grating structures, such as reflectors and IDTs.

Extended COM models were proposed so as to include this effect.[46,47]

As for the conventional Rayleigh-type SAWs, since they are considerably slower than the BAWs on the same substrate, effects of the dispersion in the energy storing effect are not significant.

It should be noted that the BAWs significantly influence IDT characteristics even for resonators.[37] That is, BAWs launched by an IDT are detected by the same IDT as the SAWs generated by the scattering (reflection) at the grating structures. Since the scattering amplitude is dependent upon frequency as described above, influence of the BAW radiation is very complicated and has not been well understood yet.

10. Conclusions

This paper reviewed numerical techniques used for the analysis of excitation and propagation of the SAW in the metallic grating structures.

First, the FEM, BEM and SDA were compared for the use of SAW analysis. It is shown that

1. The FEM can be applied directly to arbitrary structures.
2. Although applicability of the BEM and the SDA is relatively limited, they can reduce program size significantly. These techniques are applicable to the metallic grating structures on flat piezoelectric substrates.
3. For rapid calculation, the charge concentration at the finger edges must be taken into account properly. Since continuity of the field variable is assumed in traditional FEM and BEM, they are less efficient for the purpose.
4. By combining with the FEM analysis for the finger region, the BEM and the SDA can take the effects of finite finger thickness into consideration.
5. The calculation procedure for the SDA can be significantly simplified for infinitely long gratings. Very rapid calculation can be realized by the combination with Bløtekjær's theory, which takes the charge concentration at finger edges into account efficiently.

It was also shown how skillfully excitation and propagation properties are characterized by using the FEM/SDA technique. Furthermore, extended FEM/SDA theories were also given for the analysis of multi-finger grating structures and oblique SAW propagation.

Finally, a brief discussion was given for the application of these techniques toward the analysis of SH-type SAWs.

The numerical techniques are quite powerful as described in this paper. However, there still remains several obstacles which must be overcome in order to realize precise and fast simulation and design tools. For example, the lateral guiding effect of leaky SAWs[41] and their propagation at discontinuities[48] are of current interest.

Progress in SAW technologies will stimulate further progress of the simulation and design technologies.

Acknowledgments

The authors express their thanks to the staff members and present and former students in the laboratory for their tireless efforts and innovative ideas.

References

1. K. Bløtekjær, K. A. Ingebrigtsen and H. Skeie, "A Method for Analysing Waves in Structures Consisting of Metallic Strips on Dispersive media", *IEEE Trans. Electron. Devices*, **ED-20** (1973) pp. 1133–1138.
2. K. Bløtekjær, K. A. Ingebrigtsen and H. Skeie, "Acoustic Surface Wave in Piezoelectric Materials with Periodic Metallic Strips on the Substrate", *IEEE Trans. Electron. Devices*, **ED-20** (1973) pp. 1139–1146.
3. T. Aoki and K. A. Ingebrigtsen, "Equivalent Circuit Parameters of Interdigital Transducers Derived from Dispersion Relation for Surface Acoustic Waves in Periodic Metal Gratings", *IEEE Trans. Sonics and Ultrason.*, **SU-24** (1977) pp. 167–178.
4. T. Aoki and K. A. Ingebrigtsen, "Acoustic Surface Waves in Split Strip Metal Gratings on a Piezoelectric Surface", *IEEE Trans. Sonics and Ultrason.*, **SU-24** (1977) pp. 179–193.
5. B. A. Auld, *Acoustic Fields and Waves in Solids*, Vol. II, Chapter 12, Wiley-Interscience, New York, 1973, pp. 271–332.
6. B. A. Auld, *Acoustic Fields and Waves in Solids*, Vol. II, Chapter 13, Wiley-Interscience, New York, 1973, pp. 333–378.
7. M. Koshiba and K. Ohbuchi, "An Analysis of Surface-Acoustic-Wave Devices Using Coupling-of-Mode Theory and Finite-Element Method", *Jpn. J. Appl. Phys.*, **30**, Suppl. 30-1 (1991) pp. 140–142.
8. H. P. Reichinger and A. R. Baghai-Wadji, "Dynamic 2D analysis of SAW devices including massloading", *Proc. IEEE Ultrason. Symp.*, 1972, pp. 7–10.
9. P. Ventura, J.M. Hodé and M. Solal, "A New Efficient Combined FEM and Periodic Green's Function Formalism for Analysis of Periodic SAW Structures", *Proc. IEEE Ultrason. Symp.*, 1995, pp. 263–268.
10. K. Hashimoto and M. Yamaguchi, "Derivation of Coupling-of-Modes Parameters for SAW Device Analysis by Means of Boundary Element Method", *Proc. IEEE Ultrason. Symp.*, 1991, pp. 21–26.
11. P. Ventura, J. M. Hodé and B. Lopez, "Rigorous Analysis of Finite SAW Devices with Arbitrary Electrode Geometries", *Proc. IEEE Ultrason. Symp.*, 1995, pp. 257–262.
12. P. Ventura, J. M. Hodé, M. Solal, J. Desbois and J. Ribbe, "Numerical Methods for SAW Propagation Characterization", *Proc. IEEE Ultrason. Symp.*, 1998, pp. 175–186.
13. G. Endoh, K. Hashimoto and M. Yamaguchi, "SAW Propagation Characterisation by Finite Element Method and Spectral Domain Analysis", *Jpn. J. Appl. Phys.*, **34**, 5B (1995) pp. 2638–2641.
14. K. Hashimoto, G. Q. Zheng and M. Yamaguchi, "Fast Analysis of SAW Propagation under Multi-Electrode-Type Gratings with Finite Thickness", *Proc. IEEE Ultrason. Symp.*, 1997, pp. 279–284.
15. K. Hashimoto and M. Yamaguchi, "Discrete Green Function Theory for Multi-Electrode Interdigital Transducers", *Jpn. J. Appl. Phys.*, **34**, 5B (1995) pp. 2632–2637.
16. K. Hashimoto and M. Yamaguchi, "Free Software Products for Simulation and Design

of Surface Acoustic Wave and Surface Transverse Wave Devices", *Proc. IEEE International Frequency Control Symp.*, 1996, pp. 300-307.
17. http://www.sawlab.te.chiba-u.ac.jp/users/ken/freesoft.html
18. Y. Zhang, J. Desbois and L. Boyer, "Characteristic Parameters of Surface Acoustic Waves in a Periodic Metal Grating on a Piezoelectric Substrate", *IEEE Trans. Ultrason., Ferroelec. and Freq. Cont.*, **UFFC-40**, 3 (1993) pp. 183-192.
19. K. Hashimoto and M. Yamaguchi, "Analysis of Excitation and Propagation of Acoustic Waves Under Periodic Metallic-Grating Structure for SAW Device Modeling", *Proc. IEEE Ultrason. Symp.*, 1993, pp. 143-148.
20. P. Ventura, J. M. Hodé, M. Solal and L. Chommeloux, "Accurate Analysis of Pseudo-SAW Devices", *Proc. 9th European Time and Frequency Forum*, 1995, pp. 200-204.
21. R. E. Collin, *Field Theory of Guided Waves: 2nd Ed.*, 9.1, IEEE Press, Piscataway, NJ, 1991, pp. 605-608.
22. K. Hashimoto, M. Yamaguchi and H. Kogo, "Interaction of High-Coupling Leaky SAW with Bulk Waves under Metallic-Grating Structure on 36°YX-LiTaO$_3$", *Proc. IEEE Ultrason. Symp.*, 1985, pp. 16-21.
23. R. C. M. Li and J. Melngailis, "The Influence of Stored Energy at Step Discontinuities on the Behaviour of Surface-Wave Gratings", *IEEE Trans. Sonics and Ultrason.*, **SU-22** (1977) pp. 189-198.
24. R. E. Collin, *Field Theory of Guided Waves 2nd Ed.*, 6.2, IEEE Press, Piscataway, NJ, 1991, pp. 419-430.
25. http://www.asci.fr
26. K. Koshiba, "Theoretical Determination of Equivalent Circuit Parameters for Interdigital Surface-Acoustic-Wave Transducers", No. 18 Meeting Report, No. 150 Committee in Japan Society of Promotion of Science, (1989.4.27) in Japanese.
27. R. F. Milsom, N. H. C. Reilly and M. Redwood, "Analysis of generation and detection of surface and bulk acoustic waves by interdigital transducers", *IEEE Trans. Sonics and Ultrason.*, **SU-24** (1977) pp. 147-166.
28. R. F. Milsom, N. H. C. Reilly and M. Redwood, "The interdigital transducer", in *Surface Wave Filters*, edited by H.Matthews, Wiley, 1977, pp. 55-108.
29. V. P. Plessky and T. Thorvaldsson, "Rayleigh Waves and Leaky SAW in Periodic Systems of Electrodes: Periodic Green's Function Analysis", *Proc. IEEE Ultrason. Symp.*, 1992, pp. 461-464.
30. K. Hashimoto, Y. Watanabe, M. Akahane and M. Yamaguchi, "Analysis of Acoustic Properties of Multi-Layered Structures by Means of Effective Acoustic Impedance Matrix", *Proc. IEEE Ultrasonics Symp.*, 1990, pp. 937-942.
31. S. Jen and C. S.Hartmann, "An Improved Model for Chirped Slanted SAW Devices", *Proc. IEEE Ultrason. Symp.*, 1989, pp. 7-14.
32. K.Ibata, T.Omori, K.Hashimoto and M.Yamaguchi, "Polynomial Expression for SAW Reflection by Aluminium Gratings on 128°YX-LiNbO$_3$", *Proc. IEEE Ultrason. Symp.*, 1998, pp. 193-197.
33. Y. Suzuki, and H. Shimizu, "Reflection of surface acoustic waves from gratings on the surface of piezoelectric substrates", *Trans. IEICE*, **J69-A**, 6 (1986) pp. 764-774 (in Japanese).
34. P. V. Wright, "Natural Single-Phase Unidirectional Transducer", *Proc. IEEE Ultrason. Symp.*, 1985, pp. 58-63.
35. T. Sato and H. Abe, "Longitudinal Leaky Surface Waves for High Frequency SAW Device Application", *Proc. IEEE Ultrason. Symp.*, 1995, pp. 305-315.
36. M. Takeuchi, H. Odagawa and K. Yamanouchi, "Crystal Orientations for Natural Single Phase Unidirectional Transducers (NSPUDT) on Li$_2$B$_4$O$_7$", *Electron. Lett.*, **30**, 24 (1994) pp. 2081-2082.

37. K. Hashimoto, M. Yamaguchi, G. Kovacs, K. C. Wagner, W. Ruile and R. Weigel, "Effects of Bulk Wave Radiation on IDT Conductances on 42°YX-LiTaO$_3$", *IEEE Trans. Ultrasonics, Ferroelec., and Freq. Contr.*, under review.
38. R. H. Tancrell and M. G. Holland, "Acoustic Surface Wave Filters", *Proc. IEEE*, **59** (1971) pp. 393–409.
39. M. Takeuchi and K. Yamanouchi, "Field Analysis of SAW Single-Phase Unidirectional Transducers Using Internal Floating Electrodes", *Proc. IEEE Ultrason. Symp.*, 1988, pp. 57–61.
40. K. Hashimoto, J. Koskela and M. M. Salomaa, "Fast Determination of Coupling-Of-Modes Parameters Based on Strip Admittance Approach", *Proc. IEEE Ultrason. Symp.*, 1999, to be published.
41. M. Solal, J. Knuuttila and M. M. Salomaa, "Modelling and Visualization of Diffraction Like Coupling in SAW Transversely Coupled Resonator Filters", *Proc. IEEE Ultrason. Symp.*, 1999, to be published.
42. K. Hirota and K. Nakamura, "Analysis of SAW Grating Waveguides Considering Velocity Dispersion Caused by Reflectivity", *Proc. IEEE Ultrason. Symp.*, 1999, to be published.
43. K. C. Wagner and O. Männer, "Analysis of Obliquely Propagating SAWs in Periodic Arrays of Metallic Strips", *Proc. IEEE Ultrason. Symp.*, 1992, pp. 427–431.
44. K. Hashimoto, G. Endoh, M. Ohmaru and M. Yamaguchi, "Analysis of Surface Acoustic Waves Obliquely Propagating under Metallic Gratings with Finite Thickness", *Jpn. J. Appl. Phys.*, **35**, Part 1, 5B (1996) pp. 3006–3009.
45. K. Hashimoto and M. Yamaguchi, "Coupling-Of-Modes Modelling for Fast and Precise Simulation of Leaky Surface Acoustic Wave Devices", *Proc. IEEE Ultrason. Symp.*, 1995, pp. 251–256.
46. V. Plessky, "Two Parameter Coupling-of-Modes Model for Shear Horizontal Type SAW Propagation in Periodic Gratings", *Proc. IEEE Ultrason. Symp.*, 1993, pp. 195–200.
47. B. P. Abbott and K. Hashimoto, "A Coupling-of Modes Formalism for Surface Transverse Wave Devices", *Proc. IEEE Ultrason. Symp.*, 1995, pp. 239–245.
48. Y. Sakamoto, K. Hashimoto and M. Yamaguchi, "Behaviour of LSAW Propagation at Discontinuous Region of Periodic Grating", *Jpn. J. Appl. Phys.*, **37**, 5B (1998) pp. 2905–2908.

Appendix A Extension to Triple-Finger Gratings

Here, as an extension of the theory described in Section 7.1, let us discuss triple-finger gratings, which consist of three fingers with widths w_1, w_2 and w_3 in one periodic length p.

Let $q(X_1)$ and $e(X_1)$ be given by

$$q(X_1) = \{f_1(X_1)g_2(X_1)g_3(X_1) + g_1(X_1)f_2(X_1)g_3(X_1)$$
$$+ g_1(X_1)g_2(X_1)f_3(X_1)\} \sum_{m=M_1}^{M_2} A_m \exp(-j\beta_{m-1/2}X_1)$$

$$= \sum_{n=-\infty}^{+\infty} \sum_{m=M_1}^{M_2} A_m F_{n-m} \exp(-j\beta_n X_1), \qquad (A.1)$$

$$e(X_1) = g_1(X_1)g_2(X_1)g_3(X_1) \sum_{m=M_1}^{M_2} B_m \exp(-j\beta_{m-1/2}X_1)$$

$$= \sum_{n=-\infty}^{+\infty} \sum_{m=M_1}^{M_2} B_m G_{n-m} \exp(-j\beta_n X_1), \qquad (A.2)$$

where

$$G_n = -j\sqrt{2} \sum_{\ell=1}^{n} \sum_{k=0}^{n-\ell} P_k(\Omega_1) P_{\ell-1}(\Omega_2) P_{n-k-\ell}(\Omega_3)$$
$$\times \exp(j\eta_{k+1/2}^{(1)} + j\eta_{\ell+1/2}^{(2)} + j\eta_{n-k-\ell-1/2}^{(3)}) \qquad (A.3)$$

for positive n, $G_0 = 0$, $G_{-n-1} = G_n^*$, and

$$F_n = -j S_n G_n. \qquad (A.4)$$

As can be seen, Eq. (A.4) is the same as Eq. (104). This implies that Eqs. (105) to (107) developed for double-finger gratings also hold for triple-finger gratings. Then the equation to be solved is exactly the same as Eq. (108) but the range of n must be changed to $[M_1 + 1, M_2 - 2]$ so as to make the number of linear equations equal to the number of unknowns A_m.

Once A_m are determined, the potential $\Phi^{(\ell)}(s)$ of the ℓ-th finger is given by

$$\Phi^{(1)}(s) = -j\frac{\exp(-j\pi s)}{2\sin(2\pi s)} \sum_{m=M_1}^{M_2} A_m \{V_m^{(1)} + (V_m^{(2)} + V_m^{(3)}) \exp(2j\pi s)\}, \qquad (A.5)$$

$$\Phi^{(2)}(s) = -j\frac{\exp(-j\pi s)}{2\sin(2\pi s)} \sum_{m=M_1}^{M_2} A_m \{V_m^{(1)} + V_m^{(2)} + V_m^{(3)} \exp(2j\pi s)\}, \qquad (A.6)$$

$$\Phi^{(3)}(s) = -j\frac{\exp(-j\pi s)}{2\sin(2\pi s)} \sum_{m=M_1}^{M_2} A_m \{V_m^{(1)} + V_m^{(2)} + V_m^{(3)}\}, \qquad (A.7)$$

where

$$V_m^{(1)} = \int_{-p+d_3+w_3/2}^{d_1-w_1/2} g_1(X_1) g_2(X_1) g_3(X_1) \exp(-j\beta_{m-1/2} X_1) dX_1, \qquad (A.8)$$

$$V_m^{(2)} = \int_{d_1+w_1/2}^{d_2-w_2/2} g_1(X_1) g_2(X_1) g_3(X_1) \exp(-j\beta_{m-1/2} X_1) dX_1, \qquad (A.9)$$

$$V_m^{(3)} = \int_{d_2+w_2/2}^{d_3-w_3/2} g_1(X_1) g_2(X_1) g_3(X_1) \exp(-j\beta_{m-1/2} X_1) dX_1. \qquad (A.10)$$

In addition, the current $I^{(k)}(s)$ flowing into the k-th finger is given by

$$I^{(k)}(s) = \sum_{m=M_1}^{M_2} A_m W_m^{(k)}(s), \qquad (A.11)$$

where

$$W_m^{(1)}(s) = j\omega \int_{d_1-w_1/2}^{d_1+w_1/2} f_1(X_1) g_2(X_1) g_3(X_1) \exp(-j\beta_{m-1/2} X_1) dX_1, \qquad (A.12)$$

$$W_m^{(2)}(s) = j\omega \int_{d_2-w_2/2}^{d_2+w_2/2} g_1(X_1)f_2(X_1)g_3(X_1)\exp(-j\beta_{m-1/2}X_1)dX_1, \quad \text{(A.13)}$$

$$W_m^{(3)}(s) = j\omega \int_{d_3-w_3/2}^{d_3+w_3/2} g_1(X_1)g_2(X_1)f_3(X_1)\exp(-j\beta_{m-1/2}X_1)dX_1. \quad \text{(A.14)}$$

HIGH-PERFORMANCE SURFACE TRANSVERSE WAVE RESONATORS IN THE LOWER GHz FREQUENCY RANGE

IVAN D. AVRAMOV

Institute of Solid State Physics
72, Tzarigradsko Chaussee Blvd., 1784 Sofia, Bulgaria

Since the first successful surface transverse wave (STW) resonator was demonstrated by Bagwell and Bray in 1987, STW resonant devices on temperature stable cut orientations of piezoelectric quartz have enjoyed a spectacular development. The tremendous interest in these devices is based on the fact that, compared to the widely used surface acoustic waves (SAW), the STW acoustic mode features some unique properties which makes it very attractive for low-noise microwave oscillator applications in the 1.0 to 3.0 GHz frequency range in which SAW based or dielectric resonator oscillators (DRO) fail to provide satisfactory performance. These STW properties include: high propagation velocity, material Q-values exceeding three times those of SAW and bulk acoustic waves (BAW) on quartz, low propagation loss, unprecedented 1/f device phase noise, extremely high power handling ability, as well as low aging and low vibration sensitivity. This paper reviews the fundamentals of STW propagation in resonant geometries on rotated Y-cuts of quartz and highlights important design aspects necessary for achieving desired STW resonator performance. Different designs of high- and low-Q, low-loss resonant devices and coupled resonator filters (CRF) in the 1.0 to 2.5 GHz range are characterized and discussed. Design details and data on state-of-the-art STW based fixed frequency and voltage controlled oscillators (VCO) with low phase noise and high power efficiency are presented. Finally, several applications of STW devices in GHz range data transmitters, receivers and sensors are described and discussed.

1. Introduction

Surface transverse wave (STW) based microwave resonators have enjoyed considerable interest among the frequency control community over the last decade. This is because STW resonators have demonstrated unprecedented characteristics in terms of low device loss, high Q, low 1/f noise, high radio frequency (RF) power handling ability, low aging and low vibration sensitivity. This has allowed significant improvement of microwave oscillator performance in the 1.0 to 3.0 GHz frequency range in which it is very difficult to achieve low phase-noise performance at a reasonable cost using other physical principles. Currently, STW-based resonators and oscillators are being used increasingly in high-performance navigation and weapon guiding systems, microwave test and measurement equipment, modern communications, wireless remote sensing as well as in chemical and biological sensor systems.

STW resonators are based on the guided shear horizontal (SH) wave which was first described in 1911 by A. E. Love in an attempt to explain seismic data.[1] Independently from each other, in 1968 and 1969, respectively, Bleustein and Gulyaev showed that this wave can exist and propagate in nonperiodically surface corrugated piezoelectric substrates. Furthermore, it is nondispersive and features some attractive properties for applications in acoustic wave devices, such as high propagation velocity, low acoustic loss and good temperature stability.[2,3] Unfortunately, more than a decade passed before the Bleustein-Gulyaev wave (BGW) could find a practical application because at that time, nobody invented an appropriate transducer for efficient excitation and detection of BGW. The turning point came when, in 1977, Lewis and Browning demonstrated efficient excitation and detection of SH acoustic waves on piezoelectric substrates by means of interdigital transducers (IDTs) which are widely used in surface acoustic wave (SAW) devices.[4,5] The authors called this wave mode surface skimming bulk wave (SSBW), identified its advantages for potential applications and suggested temperature compensated cut orientations on piezoelectric quartz which were very attractive for stable oscillator applications above 1 GHz. One of these temperature stable orientations was close to the widely used for bulk acoustic wave (BAW) devices AT-cut and offered 60% higher propagation velocity than SAW on ST-cut quartz, meaning 1.6 times higher device frequency at the same lithographic resolution.[4] In addition, SSBWs were found to have lower propagation loss than their SAW counterparts.[4]

A serious problem in all SSBW devices was that they suffered substantial losses due to radiation of wave energy into the bulk of the substrate and these losses increase with the distance between the IDTs.[4,5] Thus, designing a low-loss SSBW device with reasonably high group delay for low oscillator noise, was a matter of impossibility. That is why, a lot of efforts were dedicated towards reducing device loss by converting the IDT excited SSBW into a guided BGW which propagates between the IDTs with low bulk radiation. As shown in Refs. /7/, /8/ and /9/, several techniques including recessed transducers, periodic grooved and metal strip gratings as well as uniform metal layers between the IDTs, were implemented in a variety of fundamental mode and overtone devices, mainly filters and delay lines for oscillator applications in the 1.0 to 4.0 GHz range.[10-18] Despite all efforts, radiation into the bulk was difficult to control even in short nonresonant SSBW devices and, therefore, they rarely demonstrated less than 20 dB of unmatched insertion loss. In fact, typical untuned device loss values were in the 20 to 40 dB range and it was impossible to achieve sufficiently low oscillator phase noise at such high insertion loss and moderate group delay values. At that time, single-port and two-port resonators based on the standing wave principle and using the Rayleigh type of SAW, were well known for their high Q and low loss and were the best choice for designing high-performance low-noise SAW oscillators.[19,20] Unfortunately, due to fairly high propagation and ohmic losses, the practical frequency limit of SAW resonators was only about 1.5 GHz.[21] It was quite obvious that, the only way to move to higher frequencies, was to design a high-Q SH-wave resonator, similar or analogous to its SAW counterpart. The problem was how to confine the SH-wave energy sufficiently well to the substrate surface where the acoustic wave can propagate with low radiation into the bulk of the substrate and can efficiently be excited and detected using IDTs, as this is done in SAW resonators. Although there were several attempts to build SH-wave resonators, most of them had limited success because it seemed to be impossible to design an efficient reflector for the shear horizontal acoustic wave mode.[8,10,22] This situation changed dramatically when Bagwell and Bray demonstrated the first high-performance STW resonators in 1987.[23] In their devices they successfully applied the phenomenon of energy trapping by means of reflective periodic metal-strip gratings, that had been investigated by Auld and Thompson in earlier work with respect to guided SH-waves which they called surface transverse waves.[24,25] Bagwell and Bray succeeded in designing an efficient periodic metal-strip wave guide placed inside the resonant cavity between both IDT, which provided excellent energy trapping of the STW to the surface of the quartz substrate. The result was a single mode 1.7 GHz two-port STW resonator with an unloaded Q of 5600 which is as high as 92% of the material Q-limit for SAW's.[23] This remarkable result inspired many scientists, including the author of this paper, to start extensive systematic studies on STWs in quartz and explore their advantages in a variety of applications.

This paper will review the fundamentals of STW propagation under periodic metal-strip gratings and will concentrate on the design and characterization of high-performance single- and two-port resonators and narrowband filters for oscillator applications in the 1.0 to 3.0 GHz frequency range. The longitudinal mode behavior and its dependence on mass loading and spacer variation in STW-resonant geometries will be discussed and verified by experimental data to illustrate peculiarities of STW in such geometries and to highlight important design rules. It will be shown that, with a careful control over the longitudinal modes, the resonant Q can be controlled within wide limits to achieve, for example, a desired voltage controlled oscillator (VCO) tuning range. Data on state-of-the-art STW resonators at frequencies up to 2.5 GHz will be presented and discussed.

The second part of the paper will present the design and performance of stable low-noise microwave oscillators for a variety of applications. It will be shown that in the frequency range of 1.0 to 2.5 GHz, the unique features of STW can greatly improve oscillator performance in terms of close-to carrier phase noise, thermal noise floor, output power, RF/d.c. efficiency and tuning range.

Finally, a few system applications using STW resonant devices, will be presented. These will cover simple but highly efficient GHz range data transmitters and receivers for local-area radio networks and wireless sensor systems. A simple STW based high-resolution humidity sensor will highlight the potential of polymer coated STW resonators in chemical and biological sensor systems.

2. STW Propagation under Periodic Metal-Strip Gratings

In practical devices SH-types of acoustic waves are excited by means of IDTs in a direction perpendicular to the X-axis on selected temperature compensated rotated Y-cut orientations of piezoelectric quartz, such as AT-cut, to provide sufficient device temperature stability.[4,5] If the IDTs are separated by a free surface from each other, as shown in Fig. 1 a), then the SH-wave is a SSBW. In this case, the power flow is radiated at the angle Θ with respect to the surface and diffracts into the bulk of the crystal. This results in an increased insertion loss, especially when the IDTs are separated by a large distance from each other.[4,5] If a metal strip grating with a period equal to that of the IDT is deposited between the IDTs, as shown in Fig. 1 b), it slows the wave down, converting it from a SSBW into a STW.[24,25] In this case, the slowing effect of the grating confines the wave energy close to the surface, preventing it from dissipation into the bulk. This phenomenon is also called energy trapping and strongly depends on the mass loading parameters of the grating: film thickness h and mark-to-period *(m/p)* ratio, defined in Fig. 1 c). Energy trapping causes the k-vector to become parallel to the substrate surface drastically reducing the device loss. As shown in Refs. /24/ and /25/, the slowing effect of the grating is two-fold: first, the trapping mechanism is similar to that of a Love wave where addition of a slow smooth surface layer slows the SSBW down, creating a decay of the wave function into the depth of the substrate. Second, additional slowing occurs due to multiple reflections from the strips of the grating which create a stopband in the *sin(x)/x* transfer function *S21* of the IDTs as shown in Fig. 2. The behavior of the transfer function close to the stopband has three characteristic regions which are essential for the design of practical resonant devices. Below and at the lower edge of the stopband, the SH-wave is well trapped to the surface and is, therefore, a STW. We will call this region the "STW-region". If a Fabri-Perot resonator is designed in such manner that its resonance is positioned in this region, the device will have low loss and high Q. As it will be shown later in this paper, practical resonators designed to operate in the STW-region achieve Q-values greatly exceeding the material limit for SAW and approaching the material Q for SH-waves.

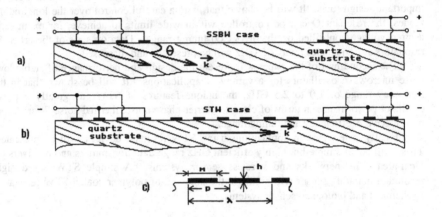

Fig. 1. Propagation of IDT excited SH-waves on rotated Y-cuts of quartz: a) SSBW case; b) STW case; c) definition of the metallisation parameters h and (m/p), and the acoustic wavelength λ, (IEEE© 1993).

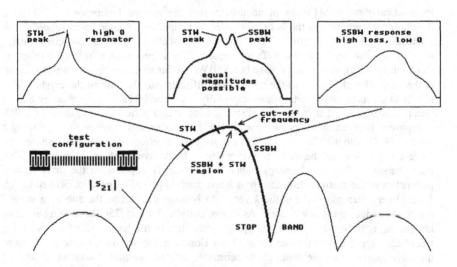

Fig. 2. Transfer function of a test device consisting of two IDT separated by a reflective metal-strip grating, (IEEE© 1993).

The second characteristic part of the transfer function is on both sides of the so called "cut-off frequency" that is the frequency at which the stopband behavior sets on.[9] In 1980 it was shown by D. L. Lee that here both waves, the STW and the SSBW can exist and their magnitude and difference in propagation velocities will strongly depend on the metallization thickness and the number of active finger pairs in the IDT.[9] In this part which we will call "STW/SSBW" region, the close-to-resonance device frequency response is formed by two adjacent longitudinal modes, as shown in Fig. 2. The left one is a STW-mode which is well trapped to the surface and the right one is a SSBW-mode which is dispersive and dissipates part of its energy into the bulk. By careful control of the metallisation parameters and the length of the resonant cavity, it is possible to move the SSBW-mode further away or closer to the STW-mode. This is a powerful method for control of the resonant Q and was successfully applied to the design of Chebyshev-type combined mode resonator filters (CMRF) and low-Q resonators for VCO applications.[26,27]

Finally, the third part of the transfer function in Fig. 2, which we will call "SSBW-region", is on the left stopband slope where the SH-wave suffers strong diffraction into the bulk and propagates as a SSBW. In this region, it is impossible to achieve high Q and low device loss. That is why, the SSBW-region is of no importance to the design of practical devices. However, special care should be taken not to move the resonance into the SSBW-region because a device like this will never work satisfactorily.

3. Operation Principle and Basic Geometries for STW Resonant Devices

Once the wave energy is well trapped to the substrate surface, the STW behaves just like a SAW, although the shear horizontal particle motion is different from the elliptic motion in SAW devices. Therefore, Fabri-Perot STW based resonators use the same topology as SAW-based ones. The simplest implementation of such a resonator is the single-port device schematically shown in Fig. 3. It consists of an IDT and two metal-strip reflector gratings on both sides of it, which have the same periodicity as the IDT. Thus, the IDT with its internal reflections, is also part of the reflector structure. Since the former has bi-directional behavior, it will excite an acoustic wave propagating as an incident wave towards each of the gratings. A small part of wave energy, which is typically fraction of a percent, will be reflected at each metal strip and will return to the IDT, as shown in Fig. 3. Thus, the farther the incident wave penetrates into the reflector gratings, the more of its energy will be reflected back, until the entire wave energy returns back to the transducer. The interference between the incident and reflected wave will form standing waves along the resonant cavity. As a good approximation, we can assume that each grating has an effective center of reflection and acts, therefore, like a mirror. Since the distance between the mirrors is fixed, standing waves can occur only at discrete frequencies $f_n = f_r(1 \pm n/M)$ as illustrated in Fig. 3. In this notation, $n=0, 1, 2, ..., M$ is the penetration depth or the distance between the mirrors in half wavelengths and f_r is the frequency of the central longitudinal standing wave mode inside the cavity. If the fingers of the IDT are positioned in such manner that they coincide with the maxima of one of the standing wave modes, (f_r in Fig. 3), then this longitudinal mode will be detected with the highest amplitude, while for all the adjacent ones, which, due to the large penetration depth, are generally fairly close in frequency, the IDT fingers will be on the slopes. In the device frequency response, this will generate a sharp resonant peak at the central mode f_r and several lower-amplitude ripples on both sides of it which correspond to the adjacent longitudinal modes. This behavior is illustrated in Fig. 4 for the simple structure in Fig. 3.

It should be noted that two-port Fabri-Perot SAW and STW resonators operate according to the same principle. The only difference is, that the IDT is split into two sections to form a two-port device which may be more convenient to work with, especially at higher frequencies. Generally, single-port resonators are more appropriate for applications in negative resistance oscillators, while two-port ones are implemented in feedback-loop oscillators.

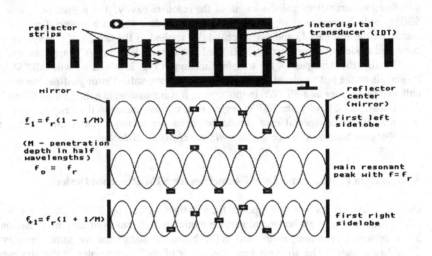

Fig. 3. Operation principle of a Fabri-Perot single-port resonator using SAW or STW.

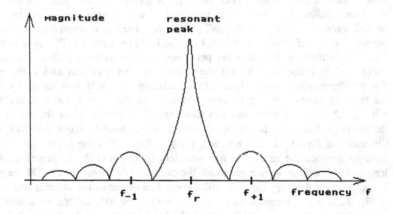

Fig. 4. Idealized frequency response of a SAW or STW based Fabri-Perot resonator using the geometry from Fig. 3.

3.1. Basic geometries of STW resonant devices

The basic single-port STW resonator geometry is shown in Fig. 3 and was discussed above. Here we will show and discuss basic topologies of two-port STW devices which are more indicative of the peculiarities of STW behavior in metal-strip resonant devices. Three of the most commonly used device geometries are shown in Fig. 5. They all use the synchronous IDT configuration, first proposed by Cross and Shreve, in which the IDTs are synchronously placed with the reflector gratings.[19] This arrangement is known for the fact that it positions the main resonance close to the lower stopband edge where the SH-wave is well trapped to the surface and propagates as a STW, (see Fig. 2). The simplest arrangement of this type is the short-cavity single-mode resonator shown in Fig. 5 a). Since the cavity length L between the IDT is just a few wavelengths long, by careful adjustment of the spacer S, the resonance behavior can be controlled in such a manner that

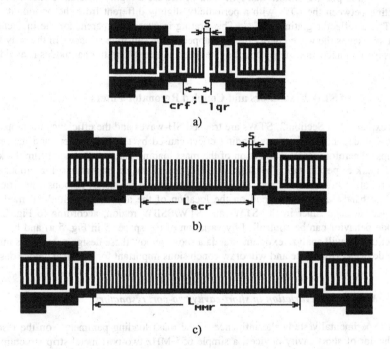

Fig. 5. Commonly used topologies of STW resonant devices: a) short-cavity geometry for low-Q resonators and coupled resonator filters; b) extended-cavity geometry for maximizing the Q in single-mode two-port resonators; c) long-cavity device for multimode high-Q resonators, (IEEE© 1993).

a low-Q resonator (LQR) or an in-line coupled resonator filter (CRF) response can be obtained.[26,27] As it will be shown later in this paper, this short-cavity device can be designed to also support a high-Q single-mode resonance. If the Q-factor of this resonance has to be maximized, then the best way to do this is to extend the cavity between the IDTs in such manner that the mirrors in Fig. 3 are moved as much apart as possible before multimoding occurs.[27] The extended cavity resonator (ECR) in Fig. 5 b) generates a high-Q single-mode resonance which can readily approach or exceed the material Q-limit for SAW on quartz.

If the cavity between the IDTs is made several hundred wavelengths long, then multiple resonances of very high Q will occur inside the stopband. The structure supporting such multimode resonator (MMR) operation is shown in Fig. 5 c). Although in a MMR the modes are difficult to control, taking advantage of the fact that the higher frequency modes dissipate energy into the bulk and have higher loss than the lower frequency trapped ones, it is still possible to design a MMR in which one resonance of very high Q is by up to 5 dB higher in magnitude than the adjacent ones.[27] This difference in magnitudes is sufficient to provide stable single-mode oscillation in a microwave oscillator. The MMR configuration is appropriate for oscillator applications in which maximum device Q is of primary importance.

It is important to note that ECR and MMR use a highly efficient center wave guide grating between the IDTs with a periodicity slightly different from the periodicity of the IDT and reflector gratings.[23,24,25] This grating is quasi transparent for the incident wave but still keeps the wave energy tightly trapped to the substrate surface. In this way, device loss is minimized and energy coupling between adjacent longitudinal modes is avoided.

4. Design of STW Resonators and Coupled Resonator Filters

As explained in Section 2, STWs are trapped SH-waves and the efficiency of trapping will strongly depend on the mass loading effect caused by the IDT fingers and the reflector strips. Therefore, careful selection of the mass loading parameters: metal film thickness h and mark-to-period ratio *(m/p)* in Fig. 1 c) are of major importance for proper device operation. On the other hand, the shape of the device frequency response, insertion loss, Q and bandwidth, will depend on the location of the adjacent longitudinal modes with respect to each other in the STW- and STW/SSBW region, according to Fig. 2. This mode behavior can be controlled by variation of the spacer S in Fig. 5 a) and b). In this Section, we will analyze experimental data showing how these design parameters influence the device performance and will draw conclusions important for practical device design.

4.1. *Mass loading variation in short-cavity two-port resonators*

To experimentally study the influence of the mass loading parameters on the resonance behavior of short-cavity devices, a simple 655 MHz two-port metal-strip structure, using the configuration in Fig. 5 a) was designed.[28] The cavity L_{lqr} between the IDTs was just a few wavelengths long. This device was fabricated on AT-cut quartz with 5 different aluminum (Al) film thickness values using the same photomask and the actual film thickness h and *m/p*-ratio were measured precisely. Since there was a slight variation of the *m/p*-ratio from device to device, we introduced the effective mass loading parameter M_{eff} which we calculated in percent for every individual device as:

$$M_{eff} = (h/\lambda)(m/p + 0.5) \qquad (1)$$

The frequency responses of the device for five different M_{eff} values are shown in Fig. 6 for comparison. The notch at the bottom right of the plot, which corresponds to the upper stopband edge, remains fixed for all five devices and does not move with changes in effective mass loading. We calculated the effective propagation velocity at this notch as $v_{eff} = \lambda f_{notch}$ and found it to be 5098 m/s which corresponds to the free SSBW velocity on this quartz cut orientation.[4] Therefore, the position of the upper stopband is determined only by the Bragg frequency of the grating and the free SSBW velocity. To a good approximation, the lower stopband edge in synchronous IDT resonant devices is located at the frequency of the main resonance.[19] As evident from the plots in Fig. 6, with increased mass loading, the stopband width increases by stretching the lower stopband edge towards lower frequencies.

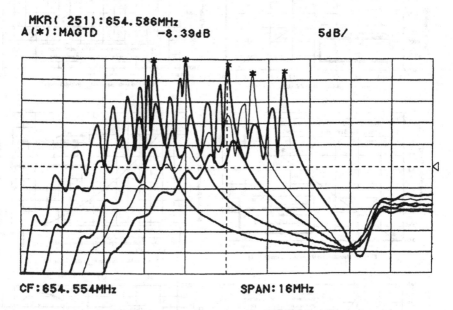

Fig. 6. Frequency responses of the short-cavity device at five different M_{eff} values: from right to the left: 1,242%; 1.397%; 1.612%; 1.732% and 1.944%.

In this study, we also investigated the behavior of other characteristic device parameters as a function of the effective mass loading. The results are shown in Fig. 7 a) through f). As shown in Fig. 7 a), b) and c), increased mass loading causes a linear decrease in device frequency and STW velocity at the frequency of main resonance, as well as linear increase in stopband width, respectively. The device sidelobe suppression and insertion loss decrease exponentially with increased mass loading, as evident from Fig. 7 d) and e), respectively. This is an excellent illustration of the fact that higher mass loading traps better not only the main resonance but also the lower order adjacent longitudinal modes at the lower stopband edge. Therefore, the STW coupling coefficient is not constant, but increases with mass loading. Another indication of that is evident also from Fig. 6 in which the oscillation magnitude of those modes increases with mass loading. Finally, Fig. 7 f) shows the behavior of the device loaded and unloaded Q which varies strongly with mass loading. In this particular device, maximum Q is obtained at $M_{eff} \approx 1.8\%$. As will be shown later, the device Q depends on the location of both adjacent longitudinal modes with respect to each other at the lower stopband edge. This implies that, in different device geometries, the mass loading value yielding maximum device Q, will be different.

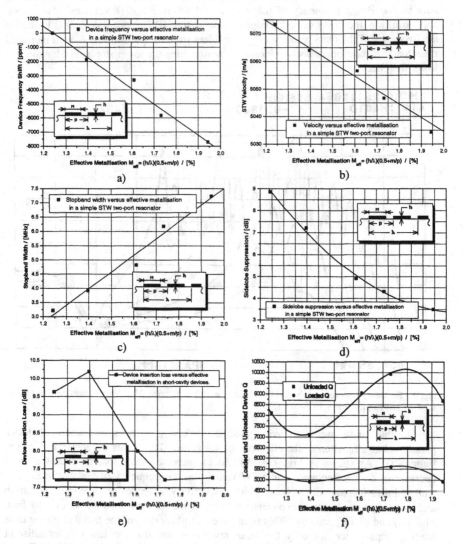

Fig. 7. Influence of the effective metallisation on the characteristics of short-cavity STW resonant devices.

4.2. Influence of the film thickness on the characteristics of extended-cavity devices

Extended cavity and multimode devices use large cavity lengths between the IDTs and their longitudinal modes are located closer to each other. Therefore, such devices are generally more sensitive to mass loading tolerances than short-cavity ones. The data in Fig. 8 compares the frequency and group delay responses of three 1.27 GHz ECR devices with identical geometry, fabricated with three different Al film thicknesses.[29] The spacer

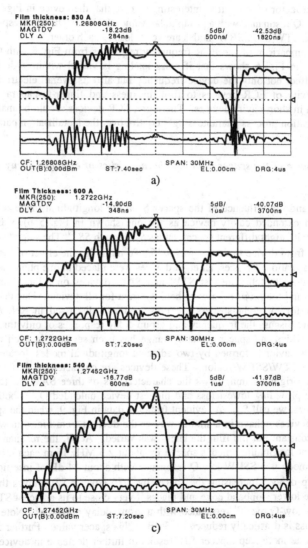

Fig. 8. Film thickness variation in 1.27 GHz ECR devices: a) 830 A; b) 600A; c) 540A; S=const, (IEEE© 1996).

S according to Fig. 5 b) was kept constant. It is evident that the device with the highest film thickness in Fig. 8 a) has several longitudinal modes with the same magnitude. There are several high-Q ripples in the STW-region at the lower stopband edge, while a low-Q SSBW resonance is located about 3 MHz higher in frequency in the STW/SSBW-region. As the film thickness decreases, the SSBW resonance moves fast towards the lower stopband edge until it coincides with one of the STW modes and forms a single high-Q resonance, shown in Fig. 8 c). In this process, the device group delay increases from 284 ns in Fig. 8 a) to 600 ns in Fig. 8 c) which means that the device loaded Q increases by more than a factor of two. It is interesting to note that the device in Fig. 8 b) has a very smooth low-Q resonance with a group delay value of 348 ns which corresponds to the case in which the STW and SSBW modes are very close to each other.

The comparison of the ECR frequency responses from Fig. 8 with the short cavity device data in Fig. 6 shows that if mass loading is varied within wide limits, this can seriously affect not only the device frequency but also its overall electrical performance. The sensitivity of ECR devices to film thickness and finger-to-gap ratio tolerances is higher than in short-cavity devices. Therefore, ECR devices should be manufactured only if the fabrication process guarantees tight control over the metallisation parameters.[30]

4.3. Influence of the spacer S on the characteristics of extended-cavity STW and SAW devices

To understand the influence of the spacer S on the longitudinal mode behavior in STW devices, an extended cavity device as the one shown in Fig. 5 b), was designed and fabricated with eight different spacer values $S1$ through $S8$.[29] The step at which the spacer was varied from device to device was $\lambda/16$. To minimize the effect of film thickness and finger-to-gap ratio tolerances, all eight devices were placed in one photomask and then the latter was used to fabricate test STW and SAW devices on AT-cut quartz. The film thickness h and the (m/p)-ratio were held constant for all devices. Detailed data from this experiment for all eight spacers are presented and discussed in Ref. /29/. Here we compare and discuss the frequency and group delay responses of only three characteristic devices in which the spacer variation range has been selected in such manner that the resonance behavior is formed by two adjacent longitudinal modes, located close to each other in the STW/SSBW-region. These devices are characterized in the left column of Fig. 9. The right column shows the characteristics of three SAW devices accordingly. The spacer providing lowest loss and highest device unloaded Q for both the STW and SAW devices, we call S_{opt}. As evident from the data in Fig. 9, when the spacer is altered, the STW devices behave in a similar way as the devices from Fig. 8 in which the spacer was kept constant but the film thickness was varied. At $S1$, the resonance is formed by two adjacent longitudinal modes spaced by about 2.5 MHz from each other. The higher frequency mode is a SSBW low-Q resonance with about 14 dB of insertion loss and 328 ns of group delay. Increasing the spacer by $\lambda/16$ (spacer $S2$), moves the SSBW mode towards the lower stopband edge and places it very close to its adjacent STW mode. This results in a low-Q overall resonance with a group delay of 344 ns. Note that the device insertion loss is drastically reduced by 5 dB at this spacer value. Further reduction of the spacer by one more step (spacer $S3$), results in further decrease in device loss to 7.3 dB while the loaded device Q is doubled (group delay is 620 ns). This is the optimum case

(S_{opt}) in which both adjacent longitudinal modes coincide with each other and form a high-Q single-mode resonance at the lower stopband edge.

The behavior of the SAW devices (right column in Fig. 9), is fairly different. Here the device always has a high-Q resonance and the lower-order longitudinal modes are suppressed by 8 to 10 dB. When the spacer is varied from S6 to S8, the main resonance moves towards lower frequencies but the insertion loss and group delay do not change significantly. It is interesting to note that S6 features the highest group delay value meaning that with respect to device loaded Q, SAWs behave in the opposite way to STWs.

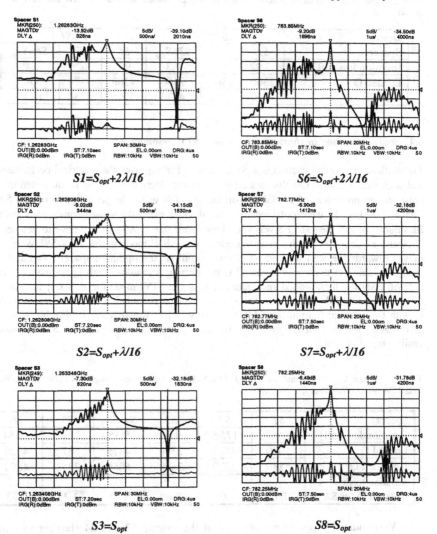

Fig 9. Spacer variation in STW- and SAW-based ECR devices. Mass loading parameters: m/p, $h=const$, (IEEE© 1996).

The optimum spacer value for SAWs is totally different from STWs propagating in such device geometries. As evident from Fig. 9, the difference in optimum spacers is $5\lambda/16$ which is consistent also with the data reported in Ref. /31/. Another illustration of the main differences between SAWs and STWs in extended cavity resonant devices is the data in Table 1 which compares spacer values, device loss, as well as loaded and unloaded device Q for the SAW and STW devices from Fig. 9.

Table 1. Comparison of optimum spacer ranges and electrical characteristics of the SAW and STW devices characterized in Fig. 9.

Device	STW			SAW		
Spacer	$S1$	$S2$	$S3=S_{opt}$	$S6$	$S7$	$S8=S_{opt}$
Insertion loss [dB]	13.92	9.02	**7.3**	9.2	6.9	**6.49**
Loaded Q	1300	1365	**2460**	4180	3470	**3540**
Unloaded Q	1630	2110	**4330**	6395	6330	**6725**

4.4. *Spacer variation in short-cavity two-port STW resonators*

The method of spacer variation is a powerful tool for adjusting the electrical performance and achieving a desired Q-value in a STW resonator. Here this method is illustrated in the optimization procedure of a 2 GHz short-cavity STW using the geometry from Fig. 5 a). The data in Fig. 10 were obtained from six devices using the same geometry and metal film thickness but the spacer S was varied by a step size of $\lambda/14$ from device to device. It is clearly evident how, with spacer variation, the SSBW mode in the STW/SSBW-region moves towards the well trapped STW mode until, at the optimum spacer $S4=S_{opt}$, both modes coincide at the lower stopband edge forming a low-loss high Q resonance. The data on insertion loss, loaded and unloaded Q at the STW mode for all six spacer values are compared in Table 2. The last row in this table is the relation of the device unloaded Q with the material Q-limit for SAW of 5240 which is theoretically achievable with a 2 GHz SAW device.[20] It is quite obvious that even short-cavity practical STW devices, easily exceed this limit.

Table 2. Insertion loss, loaded- and unloaded Q of the STW mode versus spacer variation in the short-cavity 2 GHz STW two-port resonator from Fig. 10.

Spacer	$S1$	$S2$	$S3$	$S4=S_{opt}$	$S5$	$S6$
Insertion loss [dB]	16.67	17.95	**9.06**	**9.18**	14.41	18.64
Loaded Q	5020	4470	**2350**	**4475**	5872	4320
Unloaded Q	5880	5320	**3630**	**6860**	7252	4890
Relation to the material Q for SAW	1.12	1.01	**0.69**	**1.31**	1.38	0.93

Very indicative results are obtained at the spacers S3 and S4 showing that with appropriate selection of the spacer value, the device loaded and unloaded Q can be varied by a factor of two in short cavity devices while keeping the device insertion loss nearly

Surface Transverse Wave Resonators in Lower GHz Frequency Range 749

constant and very low. This feature which is unique for STW devices allows precise optimization of the resonator loaded Q to achieve a desired tuning range in a VCO while keeping maximum phase noise suppression in the oscillator.

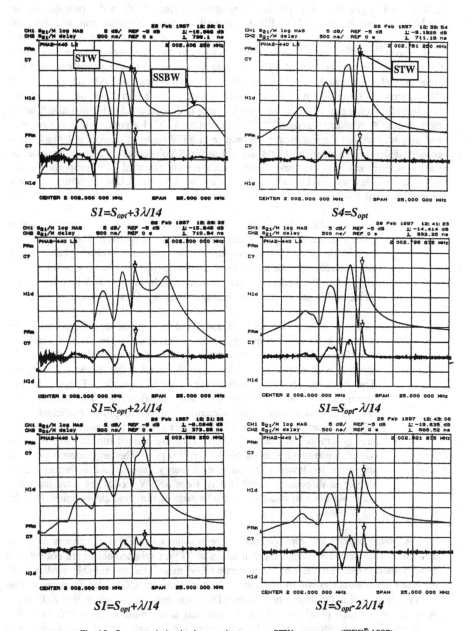

Fig. 10. Spacer variation in short-cavity two-port STW resonators, (IEEE© 1997).

4. 5. Simulation of STW resonant devices

Several authors have dedicated a significant amount of work to creating mathematical models and algorithms for precise simulation of the STW-mode propagation in fundamental and harmonic IDT filters, as well as periodic metal-strip and grooved reflective geometries, mainly for high-Q resonator based oscillator applications.[14,15,22,24,25,32,33,34,35] In the author's opinion, the most powerful simulation tool for modeling the performance of STW resonant devices is the coupling-of-modes (COM) algorithm which is widely used for predicting the performance of SAW devices. Since this algorithm is described in detail in Plessky's work elsewhere in this book, here we will briefly address only that part of effort that has been dedicated to adopting this algorithm to modeling STW resonant devices.[36,37,38,39,40] Creating a COM algorithm that can reliably and precisely predict the electrical performance of STW resonant devices is a much more complicated problem than the COM algorithm for SAW devices. This is because, as shown experimentally in the previous sections, there are essential STW-properties which must be modeled very accurately and then included in the COM algorithm. These properties are as follows:

1. Trapped shear horizontal surface acoustic waves, propagating under reflective metal-strip gratings, are strongly dispersive over the narrow bandwidth of the grating stopband. Not only their velocity but also the electromechanical coupling coefficient depends on the degree of energy trapping to the surface of the substrate which, in turn, is a function of the metallisation parameters film thickness h and mark-to-period (m/p) ratio.

2. Once h and m/p are selected and fixed in a given device geometry, the resonant behavior is formed by several longitudinal modes each of which has a different degree of energy trapping and dispersion, dependent on its location with respect to the lower stopband edge that is the STW-region. The lower frequency longitudinal modes are STW modes. They are located closer to the lower stopband edge and are, therefore, well trapped, less dispersive and feature higher electromechanical coupling coefficient. The higher frequency modes are SSBW modes because they radiate part of their energy into the bulk of the crystal. They are located further away from the lower stopband edge, in the STW/SSBW region. The larger their offset from that edge, the lower is their degree of energy trapping and the higher is their degree of dispersion. That is why, these SSBW modes are more lossy and more sensitive to small variations in mass loading and cavity length than the STW ones.

Although, significant improvement has been achieved in COM algorithms for STW devices over the last few years, the peculiarities of STW, explained above are not yet well enough understood. That is why, an accurate COM model that can reliably predict STW resonator performance is still to be created. Nevertheless, the COM algorithm can still be a very helpful tool in the design process of STW resonators. As shown in Ref. /27/, it can be used for a rough prediction of the longitudinal mode behavior in a certain device geometry. This is illustrated in Fig. 11 which compares experimental data, obtained from a 976 MHz extended cavity resonator, with simulation results obtained from a COM-algorithm used for modeling SAW devices. From that comparison it becomes evident that the agreement between theory and experiment is not very good, especially above the lower stopband edge, but the overall resonance and longitudinal-mode behavior is predicted with sufficient accuracy. Once the number of modes and their location with respect to the lower stopband edge are roughly determined by the COM algorithm, the actual device performance can be predicted using the experience and experimental data described in Section 4. For a fine adjustment of the device Q, insertion loss and side-lobe suppression, a spacer correction in a second iteration step may be necessary.

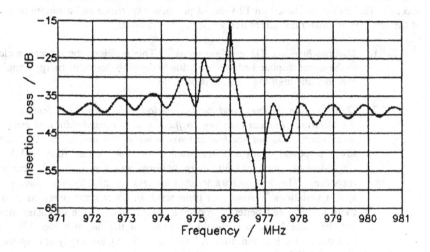

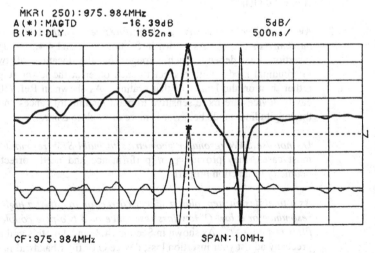

Fig. 11. COM-predicted frequency response (upper graph) and experimental frequency and group delay responses (lower graph) of a 975 MHz extended cavity STW resonator, (IEEE® 1993).

4.6. A few helpful rules for the practical design of successful metal-strip STW resonators

Although metal-strip STW-resonators use simple device geometries which are very similar to the geometry of a SAW resonator, designing a good STW resonator is not a trivial problem.[31] If the mass loading parameters and cavity length are not properly selected for a chosen device geometry, one can easily end up with high loss, low Q and very distorted device frequency responses. To avoid this, and to help the design engineer benefit from the remarkable properties of STW resonators, the author would like to suggest a few practical design rules which do not necessarily guarantee but will increase the chances of success. These rules are based on 12 years of practical experience and a variety of state-of-the-art STW resonator designs at frequencies up to 2.5 GHz.

1. *Use synchronous IDT configurations.*[19] This positions the resonance close to the lower stopband edge where, due to the high degree of energy trapping, low-loss and high-Q operation is possible.

2. *Optimum film thickness and mark-to-period ratio are chosen dependent on the electromechanical coupling coefficient, device bandwidth and insertion loss desired. Do not use higher mass loading than necessary.* Due to the lower stopband width and decreased dispersion, thinner films and lower finger-to-gap ratios provide more uniform and less distorted frequency responses. The device characteristics are less sensitive to mass loading and spacer variations.[30] Also, thin films result in lower static stress, as well as more uniform and better reproducible temperature device characteristics.[46] On the other hand, one should be aware of the fact that too thin films increase ohmic loss and, finally, limit the device Q, especially at frequencies above 2.0 GHz.[30]

3. *Avoid using very long wave guide gratings between the IDT in an attempt to maximize device Q.* As shown in Sections 4.3. and 4.4., the device Q and insertion loss depend much stronger on the location of two adjacent longitudinal modes with respect to each other at the lower stopband edge, rather than on the length of that grating. As shown in Ref. /30/, extended cavity devices are more sensitive to mass loading tolerances and, generally, the longitudinal mode behavior is more difficult to control.

4. *Do not use SAW resonator geometries to build STW resonant devices.* In most cases, they provide poor performance and need corrections of the geometry to function properly.[31]

5. *As a final design step, plan a spacer correction to adjust a high-Q, low-loss resonance or a low-Q, low-loss resonance or a two-pole coupled resonator filter response.*[29,30] As shown in Sections 4.3. and 4.4., this will allow you to precisely adjust your insertion loss, device Q or filter bandwidth.

5. Electrical Characterization of STW Resonators

The design of high-performance STW-resonator based oscillators requires precise extraction of the equivalent circuit parameters of STW two-port and single-port devices. The methods of parameter extraction are well known and widely used for the characterization of SAW resonators and narrowband filters. The only difference that has to be taken into account is that the different degree of energy trapping within the grating stopband in STW resonators may cause an asymmetry of the device frequency and group delay responses with regard to the resonant frequency which is rarely the case in well designed SAW resonators. Therefore, the 3-dB bandwidth of a STW resonator with an asymmetric response, cannot be used for correct calculation of the device Q. Also, as shown in Fig. 12 a), the peak value of the device group delay may not coincide with the frequency of minimum loss, but may be slightly displaced from it. That is why, in two-port STW resonators, the correct device loaded and unloaded Q should be determined at a desired point of the frequency- and/or group delay response curve. This is correct for a STW resonator-based fixed frequency oscillator because it is usually tuned in such manner that it operates at one frequency within the 3 dB device bandwidth. Alternatively, if the oscillator is a VCO, generally, it is tuned over a narrow bandwidth on both sides of that frequency. Typically, this is the frequency of minimum device loss or the frequency of maximum group delay which provides the highest phase slope and, therefore, the lowest oscillator phase noise.

In single-port STW resonators, the asymmetry of the acoustic response causes an admittance behavior that is not an ideal circle but a snail-shaped curve in the Smith chart, as shown in Fig. 12 b). Therefore, equivalent circuit parameter extraction from characteristic frequencies on this curve, may not provide sufficiently accurate results. As will be shown below, the better approach is to combine a Π-circuit measurement for determining the frequencies of series and parallel resonance with an impedance measurement for accurate extraction of the dynamic resistance R_d value.

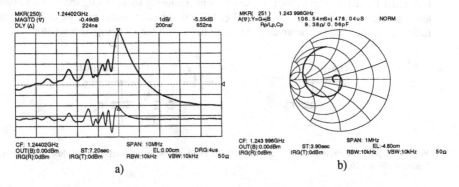

Fig. 12. Influence of the unsymmetry on the electrical characteristics of STW resonators: a) Frequency and group delay responses; b) Admittance plot of a 1.244 GHz device, (IEEE® 1997).

5.1. Electrical characterization of STW-based two-port resonators

The electrical performance of a two-port or single-port SAW or STW resonator is measured by means of a vector network analyzer which should be able to provide high-resolution data on the transmission (*S21*) and reflection characteristics (*S11* and *S22*) of the acoustic device. As evident from Fig. 13 which compares the narrowband frequency and phase responses *S21* of a SAW and a STW two-port resonator, there is basically no difference in their close to resonance behavior. Therefore, the equivalent circuit used for

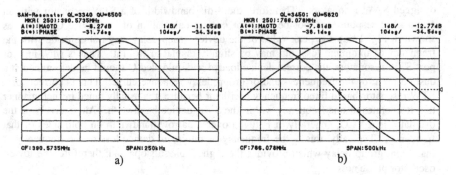

Fig. 13. Close to resonance frequency and phase responses of: a) a 390 MHz SAW and b) 766 MHz STW two-port resonator, (IEEE© 1993).

SAW two-port resonators, can also be used to describe the behavior of STW resonators. This circuit is shown in Fig. 14. It contains the static IDT capacitances C_t, the dynamic parameters R_d, L_d and C_d, which form the series resonant circuit, representing the resonance behavior, and an ideal transformer indicating the possibility of introducing an additional 180 deg. of phase shift at the resonant frequency by reversing the devices bond leads at one of the IDTs, as shown in Fig. 15. This phase reversal may be necessary to fulfill the phase condition in an oscillator stabilized with such a two-port resonator.

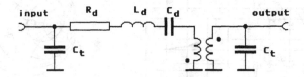

Fig. 14. Equivalent electrical circuit of a SAW/STW two-port resonator simulating the close to resonance behavior, (IEEE™ 1993).

Fig. 15. Altering the resonator phase by 180 deg. by reversing the bond lead contacting at one IDT.

The dynamic resonator parameters can be derived from the measured resonant frequency f_r, the insertion loss IL and the loaded group delay value τ_g at f_r using the following equations:[41]

$$Q_L = \pi f_r \tau_g \tag{2}$$

$$Q_U = Q_L / [1 - 10^{-(IL/20)}] \tag{3}$$

$$R_d = Q_L (R_g + R_l) / (Q_U - Q_L) \tag{4}$$

$$L_d = Q_U R_d / (2\pi f_r) \tag{5}$$

$$C_d = 1/[(2\pi f_r)^2 L_d] \tag{6}$$

where R_g and R_l (usually equal to 50 Ω), are the source and load impedances of the measurement system. The static IDT capacitances C_t can be measured directly with a capacitance meter or, if the number of IDT finger pairs N and the acoustic aperture W are known, the C_t values can be calculated as:

$$C_t = C_2 N W \tag{7}$$

where C_2 is the capacitance of one finger pair per unity length and has the value of 0.55pF/cm for rotated Y-cut quartz.

Figure 16 shows data from transmission ($S21$) measurements on a high-performance 1 GHz STW resonator for a microwave power oscillator application. With an insertion loss and group delay value of 3.64 dB and 864 ns respectively (see Fig. 16 b)), according to Eq. (2) and (3), the corresponding loaded and unloaded Q values are 2720 and 7945.

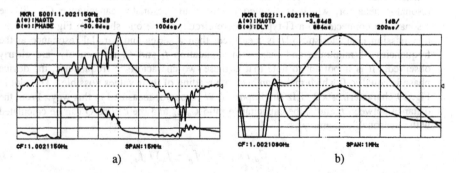

Fig. 16. Data from S21 measurements on a 1 GHz STW two-port resonator: a) broadband frequency and phase responses; b) narrowband frequency and group delay responses, (IEEE© 1994).

Using Eqs. (4), (5) and (6), the dynamic parameters are calculated as: R_d=52Ω, L_d=65,69µH and C_d=0.38fF.

The $S11$ and $S22$ plots in Fig. 17 a) and b) show the input and output impedances of the device from Fig. 16. The impedance plots are located in the capacitive part of the Smith chart which is due to the influence of the static IDT capacitances.

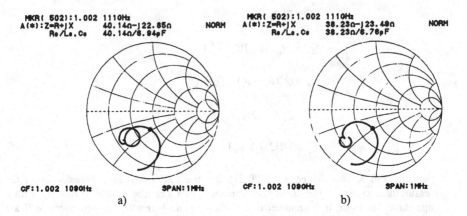

Fig. 17. Narrowband input impedance (a) and output impedance (b) of the device from Fig. 16, (IEEE© 1994).

5.2. Extraction of the equivalent electrical circuit parameters of single-port STW resonators

As shown in Section 3, single-port STW resonators use the Fabri-Perot principle and the geometry described in Fig. 3. In the vicinity of resonance they are modeled by the well known Butterworth-Van Dyke equivalent electrical circuit which is widely used to characterize the resonant behavior of BAW crystals.[42] As evident from Fig. 18, this circuit consists again of a series resonant branch (R_d, L_d and C_d), characterizing the motional resonator behavior, which is shunted by the static resonator capacitance C_o. If the resonator is connected to a Π-type resistive circuit, as the one shown in Fig. 18, and its transmission characteristic $S21$ is measured, then a series and parallel resonance at the frequencies of minimum and maximum attenuation f_s and f_p, respectively, are clearly observed. It is important to note that at these two characteristic frequencies, the device phase is equal to 0 which makes an accurate reading of their values possible. The plot in Fig. 19 characterizes a 2 GHz single-port STW resonator measured in this way. From the distance between f_s and f_p the capacitance ratio C_d/C_o and the device Q can be calculated with a reasonable accuracy as:

$$C_d / C_o = 2(f_p - f_s) / f_s \qquad (8)$$

$$Q = 1 / [4\pi C_o R_d (f_p - f_s)] = 1 / (2\pi f_s C_d R_d) \qquad (9)$$

From the data in Fig. 19 we find that the distance between series and parallel resonance is 860 KHz which yields a capacitance ratio C_o/C_d of 1167. Even in very high-frequency devices, the static capacitance C_o, including not only the IDT capacitance, but also possible package and mounting parasitics, can be measured with a capacitance meter. The C_o value of the device from Fig. 19 was measured as 1.63 pF, yielding a C_d value of 1.397 fF and a L_d value of 4.5 µH according to Eqs. (8) and (5) respectively. Now, the only unknown parameter is the motional resistance R_d. Unfortunately, due to the unsymmetry of the device characteristics, it cannot be extracted directly from an impedance or admittance measurement. However, since we are interested in its value only at one frequency which is f_s, a sufficiently accurate R_d measurement can be made by first tuning out C_o with a parallel inductor L_o and then performing the $S11$ measurement.[43] Since the bandwidth of the parallel L_o/C_o circuit is orders of magnitude higher than the bandwidth of the acoustic device, the L_o value does not need to be very precise. It is sufficient that it rotates the $S11$ plot in such manner that the latter crosses the real Smith chart axis at f_s. This is illustrated in Fig. 20 which is the admittance plot of the device from Fig. 19 with C_o tuned out with a parallel inductor. The reading of the R_d value on the real Smith chart axis is 8.35 Ω which, according to Eq. (9), yields a device Q of 6800.

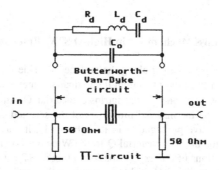

Fig. 18. The Butterworth-Van Dyke equivalent electrical circuit of a single-port acoustic wave resonator, embedded in a Π-type circuit for equivalent electrical parameter extraction, (IEEE© 1999).

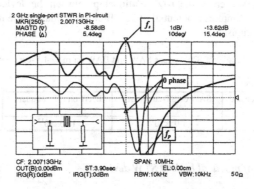

Fig. 19. Frequency and phase response data of a 2 GHz single-port STW resonator embedded in the Π-type circuit from Fig. 18. Note that the phase is 0 at the frequencies of series and parallel resonance, (IEEE© 1999).

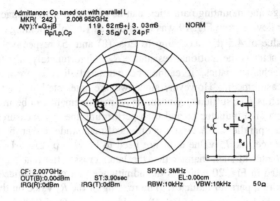

Fig. 20. Admittance plot of the 2 GHz single-port device from Fig. 19 with C_o tuned out with a parallel inductor, (IEEE© 1999).

6. Material Q and Loss Mechanisms in High-Q STW Resonators

Every acoustic wave, propagating along the surface or inside the bulk of a crystal suffers acoustic losses which limit the maximum theoretically achievable Q of a practical device using this acoustic wave mode. This intrinsic material Q-limit is defined as the $Q{\times}f$-product and depends on the material properties of the selected crystal cut orientation and the type of acoustic wave propagating in it.[20] Recently, it was shown theoretically and experimentally that the intrinsic material Q for STW devices on temperature compensated rotated Y-cut orientations of quartz, is superior to that of ST-cut quartz SAW devices and AT-cut quartz BAW devices.[44,45] Table 3 compares the material Q values of these three acoustic wave modes on quartz and shows that the intrinsic Q-limit for STW is in fact two to three times higher than that of the other two technologies. This is a great fundamental advantage of STW because the $1/Q^4$ relationship between the device Q and the close-to-carrier phase noise implies a potential of 15 dB close-in phase noise improvement in stable oscillator applications.[45]

Table 3. Intrinsic material Q of STW versus SAW and BAW on rotated Y-cut orientations of quartz, (IEEE© 1999).

Acoustic wave mode and cut orientation	SAW on ST-cut quartz	BAW on AT-cut quartz	STW on AT-cut quartz
$Q{\times}f$-product in Hz	$1.05{\times}10^{13}$	$1.35{\times}10^{13}$	$3.25{\times}10^{13}$

6.1. Examples of practical high-Q STW devices

In this section we will discuss data from practical devices that take advantage of the high intrinsic material Q of the STW mode.

The data in Fig. 21 were obtained experimentally from a 2 GHz in-line coupled resonator filter test structure. The coupling grating has been intentionally made 600 wavelengths long to maximize the cavity length. Since this long coupling grating has the same period as the IDTs and reflectors, it causes strong reflections along the cavity and a very abrupt cut-off behavior. Close to the cut-off frequency, the group delay value is extremely large: 1.932 μs. This corresponds to a loaded Q of 12200 and an unloaded Q of 12480 at the frequency of 2008 MHz, which is 81% higher than the material Q for AT-cut BAW. Please note that this type of CRF test device was intended to investigate the maximum achievable group delay for STW at 2 GHz. It cannot be used for practical applications since there is no well behaved resonance at the frequency of maximum group delay. If a STW device is to be used in an oscillator with maximized Q, the resonator geometry should be optimized in such manner that it supports one resonance with a magnitude by at least 5 dB higher than the adjacent resonances, to guarantee stable single-mode oscillator operation. Very high loaded Q values can be achieved with optimized multimode devices which have a long quasitransparent wave guide grating between the IDTs to minimize the device loss while keeping the cavity length as long as possible.[30] These devices use the multimode geometry from Fig. 5 c) and are designed according to the guide lines described in Sections 4 and 5. One such device is characterized in Fig. 22. With a group delay value of 978 ns and an insertion loss of 17.2 dB, this 2 GHz device has a loaded Q of 6170 and an unloaded Q of 7160, which is 37% higher than the material Q limit for ST-cut SAW at that frequency.

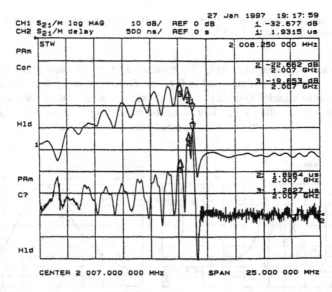

Fig. 21. Frequency and group delay responses of a CRF-type test structure with a very long coupling grating, (IEEE© 1999).

In the frequency range of 1.0 to 2.0 GHz, unloaded Q values as high as the material Q limit for AT-cut BAW can readily be achieved with practical short-cavity STW devices occupying a very small substrate area. As shown in Section 4.4., the Q of such devices is maximized by having two adjacent longitudinal modes coinciding at the lower stopband

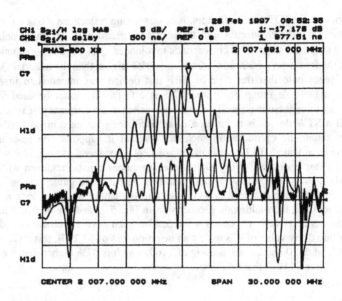

Fig. 22. Frequency and group delay responses of a 2 GHz multimode resonator for an oscillator application using maximized device loaded Q, (IEEE® 1999).

edge. In addition to high loaded Q, this approach also provides low device loss. One such short-cavity device is characterized in Fig. 23. The device group delay is 2250 ns which corresponds to a loaded Q of 8210. With an insertion loss value of 8.6 dB, the unloaded Q is found to be 13080 which is 97% of the material Q limit for AT-cut BAW at this frequency. Similar results also were obtained with 2.0 and 2.5 GHz short-cavity devices.[30]

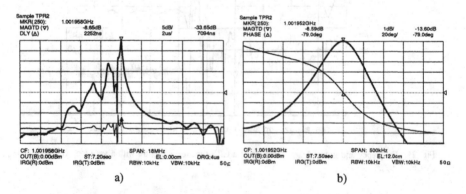

Fig. 23. Broadband frequency and group delay responses (a) and narrowband frequency and phase responses (b) of a 1 GHz short-cavity STW resonator, (IEEE® 1999).

The two main limiting factors to achieving Q values close to the material limit for STW in two-port resonators, are: 1) the ohmic loss of the IDT fingers and 2) the dissipation of wave energy in the bulk when the wave propagates between the IDTs. Both loss effects are substantially reduced in a single-port resonator configuration. Therefore, single-port STW resonators generally have higher device Q than two-port devices. This is evident from Fig. 24 showing data form a high-performance 1 GHz single-port resonator characterized according to the method described in Section 5. 2. As evident from Fig. 24 a), the distance between series and parallel resonance is 160 kHz, yielding a capacitance ratio of 3130. With a static IDT capacitance C_o value of 2.1 pF, this yields a dynamic capacitance C_d of 0.671 fF. The value of the dynamic resistance R_d is 7.7 Ω, as evident from the impedance plot in Fig. 24 b). The latter was obtained from a reflection measurement, after tuning out C_o with a parallel inductor. The R_d and C_d values obtained in this way, imply a device Q of 30750 which is very close to the theoretical material Q value of 32500 at the 1 GHz device frequency. This means that in fact, a single-port STW resonator features very small overall loss.

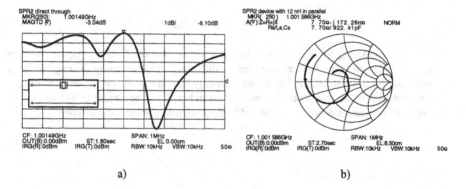

Fig. 24. Characteristic data from a 1 GHz high-performance single-port STW resonator: a) Frequency response obtained in a direct through measurement; b) Impedance plot with the static capacitance tuned out with a 12 nH parallel inductor, (IEEE© 1999).

6.2. Loss mechanisms in two-port STW resonators

As explained in the previous section, two-port STW resonant devices generally achieve lower Q-values than their single-port counterparts. Here, we will analyze the loss mechanisms in a two-port resonator and will try to identify the dominant source of loss in practical two-port devices.

Let us consider the 2 GHz multimode resonator from Fig. 22 which has been designed for maximum device Q. The characteristic device parameters were calculated from the data in Fig. 22 using the method described in Section 5. 1. They are listed in Table 4 containing also the peak value of the IDT conductance G_{max} which appears at the resonant frequency f_r and the ohmic resistance of the IDT fingers R_{ohm} which can easily be calculated from the metal film resistance and the finger geometry.[46] The G_{max} value was extracted from an admittance measurement on a test IDT which was fabricated separately on a free substrate area with exactly the same geometry and mass loading parameters as the one used in the resonator geometry. This was necessary because it is fairly difficult to extract precise data on the IDT admittance from experimental resonator data as the one shown in Figs. 16, 17 or 22. Once we know the values of the characteristic parameters listed in Table 4, we can apply the fairly comprehensive model used for analyzing the loss mechanisms of two-port SAW resonators to this particular STW device too.[47] According

Table 4. Characteristic parameters of the 2 GHz multimode device from Fig. 22, necessary for the description of the device losses according to Ref. /47/.

f_r / [MHz]	Q_u	R_d / [Ω]	L_d / [µH]	G_{max} / [mS]	R_{ohm} / [Ω]
2007.7	7160	622	352.94	2	8.1

to this model, the unloaded device Q can be described as a function of the effective cavity length m in acoustic wavelengths as follows:[47]

$$Q_u = 2\pi m / (\mu_{mat} + \mu_{ohm} + \mu_{other}), \qquad (10)$$

where μ_{mat} and μ_{ohm} are the material loss and ohmic loss coefficients, respectively, and μ_{other} includes all other loss mechanisms as diffraction, beam steering, leakage from the reflectors, air loading, scattering of wave energy into the bulk, etc. The parameter m represents the distance between the two mirrors in Fig. 3 and, according to Reference [48], it is related to the reflection coefficient $|\Gamma|$ along the resonant cavity as follows:[48]

$$m = 4 L_d f_r |\Gamma| G_{max}. \qquad (11)$$

Then, using the relationship:[48]

$$|\Gamma| = 1/(2R_d G_{max} + 1), \qquad (12)$$

the effective cavity length m is unambiguously calculated from the measured data in Table 4 as 1625 wavelengths. Then, according to Ref. /47/, the material loss coefficient is derived from the material Q value for STW, the device frequency and cavity length as:

$$Q_{mat} = (3.25 \times 10^7)/ f_r [MHz] = 2\pi m / \mu_{mat}. \qquad (13)$$

From Eqs. (10) and (13) it follows that:

$$\mu_{ohm} + \mu_{other} = \mu_{mat}[(Q_{mat}/Q_u) - 1] \qquad (14)$$

Finally, using the definitions in Ref. /47/, it is easy to show that there is a simple relationship between the ohmic loss coefficient and the actual ohmic resistance of IDT fingers, busbars and bond contacts which is:

$$\mu_{ohm} = 32 G_{max} R_{ohm} \qquad (15)$$

Using Eqs. (10) through (15) and the experimental data in Table 4, we calculated the values of the loss coefficients for the particular 2 GHz STW device from Fig. 22. The values are compared in Table 5. The material loss coefficient has the highest value which is to be expected, keeping in mind that this is a multimode device using an extremely long cavity. More indicative is the comparison between μ_{ohm} and μ_{other} which clearly indicates that the ohmic loss dominates all the other loss mechanisms, represented by μ_{other} in this device. This is typical for STW devices at frequencies above 1.5 GHz, therefore, special

Table 5. Comparison of the loss coefficients of the 2 GHz STW device from Fig. 22.

Loss coefficient	μ_{mat}	μ_{ohm}	μ_{other}
Value	0.63	0.52	0.28

design effort should be made to minimize ohmic losses in GHz range STW devices. Possible ways of achieving low ohmic losses in such devices include:

- minimizing the acoustic aperture;
- maximizing the number of active IDT finger pairs;
- maximizing the finger width;
- maximizing the film thickness or embedding the fingers if possible;
- minimizing the bus bar length or depositing bus bars with increased film thickness in a second metallisation step.

7. Power Handling Ability of STW Resonators

One of the most attractive features of STW resonators is their ability to handle very high levels of radio frequency (RF) power. As it will be shown later in this paper, this feature is of major importance to the design of high-power microwave oscillators with very low thermal noise floor since, in such oscillators, the STW device is run at incident power levels of up to 2 W. The maximum RF power level P_{max} that a STW resonator can dissipate, is calculated using the following equation:[23]

$$P_{max} = \left(\frac{|T_{12}|_p}{2050}\right)^2 \frac{mDW}{Q_u \lambda}, \tag{16}$$

where $|T_{12}|_p$ is the highest component of the shear stress that can occur in the STW device and here it is assumed to be as high as the degradation peak stress value of 2.8×10^8 N/m² for SAW resonators with Al-Cu metallisation, D is the effective penetration depth into the substrate which is typically 0.75λ and W is the acoustic aperture.[23] It should be noted that the factor J_m=2050 [Ns/m³]$^{1/2}$ in Eq. (16) is a constant for a given material and cut orientation.[49]

If we consider a practical STW device for a power VCO application, with an unloaded Q value of 2000, an effective cavity length m=2.2 mm and an aperture

$W=0.5$ mm, according to Eq. (16), we find that the maximum power at which degradation of the electrodes should occur in this device is 7.7 W. In fact, the author of this paper performed experiments in which such two-port devices were run in power oscillators at an incident continuous power level in the 1 to 2 W range for several months. No evidence of metal migration of the IDT fingers or degradation in overall device performance was observed. Only a slight up-shift of the device frequency by less than 100 ppm took place after the first few hours of operation.

For comparison, attempts were made to run SAW resonators of a similar geometry at high drive power levels. A complete damage in these devices occurred after a few seconds of operation at 50 mW of incident RF power. This degradation power level is 150 times lower than the damaging power level for a practical STW device.

8. Temperature Stability of STW Resonators

In the late 70-ies and early 80-ies, substantial efforts were dedicated to studying the temperature behavior of STW delay lines for oscillator applications.[4,12,13,25] Although the temperature stability of compensated cut orientations around the AT-cut of quartz have been available since 1977, designing a high-frequency STW device with a desired turn-over temperature is not trivial. This is because the mass loading parameters which are of major importance to the electrical device performance, strongly influence also its turn-over temperature due to the static stress caused by the Al film.[12,13] In most practical cases, the inexperienced designer must first optimize the STW device for desired electrical performance, and then a cut correction has to be made to precisely adjust the desired device or oscillator turn-over temperature. This is shown in Fig. 25 containing important design data obtained from practical 766 MHz high-Q STW devices with a film thickness $h/\lambda=1.6\%$. It is evident that, except for the turn-over point, the overall temperature behavior and instability remains unchanged with cut orientation.

The data plot in Fig. 26 is the typical temperature behavior of a microwave STW resonator fabricated on AT-cut quartz using thin Al metallisation. The typical turn-over temperature T_o of this cut is about 35 deg. C which has been intentionally selected to be 10 to 15 deg. higher than room temperature. The reason for that is that the active circuitry of the STW oscillator, stabilized with this device, shifts the turn-over point of the entire oscillator down so that finally, the packaged oscillator achieves its maximum stability at room temperature. It is interesting to note that the temperature behavior in Fig. 26 is not exactly parabolic which is consistent with the data from Ref. /23/. Its higher temperature slope is steeper than the lower temperature one which implies a cubic behavior similar to that of AT-cut BAW devices. In fact, the experimental data points in Fig. 26 are ideally fitted by a third order polynomial fit.

One disadvantage of AT-cut STW devices is that they generally have lower overall temperature stability than their ST cut SAW counterparts. As evident from Fig. 27, for small temperature ranges around the turn-over temperature, the difference is not significant. The STW device in Fig. 27 achieves its 50 ppm instability over a 55K temperature span on both sides of T_o while the SAW device features the same instability over a 80K span. Thus, the STW device has a 45% worse temperature instability than its SAW counterpart for the 50 ppm temperature induced frequency shift. Unfortunately, the degradation becomes worse, especially at fairly high temperatures above T_o, because the temperature slope becomes very steep at offsets from T_o which are higher than 50K. That

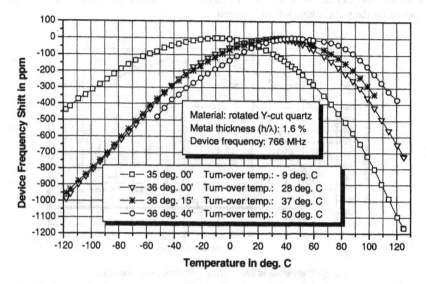

Fig. 25. Turn-over temperature versus cut orientation in practical AT-cut STW resonators, (IEEE© 1992).

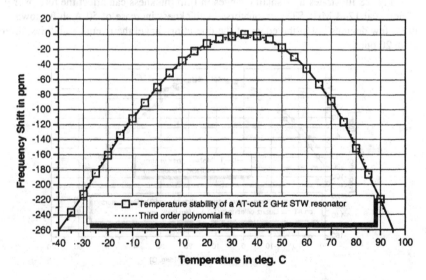

Fig. 26. Detailed temperature behavior of a 2.0 GHz STW resonator.

is why, precision STW oscillators should be ovenized when operated above 80 deg. C. An alternative approach would be to use the BT-cut of quartz for STW device applications in which the temperature stability is a major importance. BT-cut orientations around -50 deg. rotated Y-cut feature a much better temperature stability than ST-cut quartz for SAW devices, especially at high temperatures.[4] However, with a propagation velocity of about 3330 m/s the device frequency is only about 5.5% higher than that of ST-cut quartz devices for the same lithographic resolution.

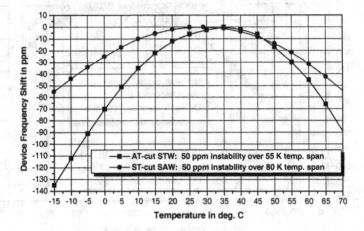

Fig. 27. Comparison of SAW versus STW device temperature stabilities.

Fig. 28 illustrates how small changes in film thickness can affect the turn-over point of practical 1244 MHz STW resonators. A thickness increase of 50 A shifts down T_o by just a few degrees but at the edges of the operating range, the overall stabilities differ by 15 to 20 ppm.

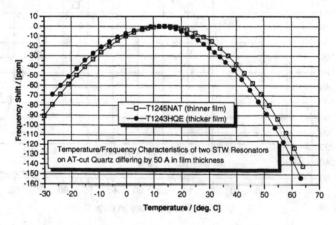

Fig. 28. Influence of film thickness variations on the temperature performance of practical STW resonators.

Finally, the effect of static film stress must be taken into account in a possible series production of STW resonators. Static stress occurs in the process of Al film deposition on the substrate surface and causes surface tensions which may affect the uniformity of the device temperature characteristics. In extreme cases, static stress may cause irreversible frequency jumps, especially in devices using thick metallization. That is why, such stresses must be released before the STW device is used. An efficient way to do this is to expose the devices to high-temperature bake-out for several hours before sealing, in the same way as this is done in SAW resonators to improve aging.[20] Another efficient way is to release the stress by means of a thermal cycling process. Figure 29 shows the temperature characteristics of a 2.1 GHz STW oscillator for three temperature cycles over the range (-80 ...+110) deg. C. It is evident that the first temperature run provided a fairly distorted temperature-frequency characteristic with a flat part in the 80 to 90 deg. C range. In the second run, the stress was almost completely released which results in a smooth curve. The latter was retained also during the third and subsequent temperature cycles. We found that two temperature cycles are enough to completely eliminate static stress in 2GHz STW resonators.

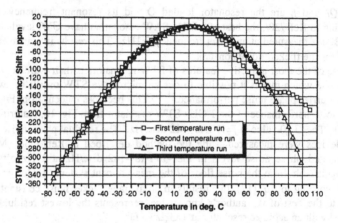

Fig. 29. Effect of thermal cycling on the temperature performance of 2.1 GHz STW resonators, (IEEE© 1995).

9. Vibration Sensitivity and Aging of STW resonators

First data on STW resonator vibration sensitivities indicate extremely low values in the 1 to 5×10^{-10}/g range.[50,51,76,77] These excellent results are enhanced by the fact that STW resonators can be made very small and selecting the optimum aspect ratio as well as applying header stiffening for minimum vibration sensitivity is not a problem.[52,53] The results in Refs. /50/ and /51/ are comparable to or better than the vibration sensitivity data obtained from the best SAW resonators built to date.

Although no data on aging of STW resonators has been published yet, first results indicate that all-quartz packaged 1 GHz devices feature aging rates comparable with the best SAW resonators built so far.[54] Typical aging rates of such devices are well below 1 ppm/year.

10. Residual 1/f Noise of STW Resonators

Residual $1/f$ phase noise in a SAW or STW resonator occurs when an unmodulated carrier is passed through the acoustic device. In this process, phase fluctuations which occur in the resonator cause a direct phase modulation (PM) of the carrier so that it appears with PM modulation noise side bands at the device output. The power spectral density $S'_\phi(f)$ of this PM noise has $1/f$ dependence and is commonly called flicker or $1/f$ noise.[20,47]

The $1/f$ noise behavior of SAW resonators is characterized by the flicker noise constant α_R which is calculated from the measured values of the resonator $1/f$ noise $\mathcal{S}'(f)$ as:[47]

$$\alpha_R = f\left\{\frac{1}{(2Q_L f_r)^2}\right\} 2 \times 10^{\left\{\frac{\mathcal{S}'(f)}{10}\right\}} \qquad (17)$$

where Q_L and f_r are the resonator loaded Q and its resonant frequency, respectively. According to Ref. /20/, the best 500 MHz SAW resonators have $\mathcal{S}'(f)$ levels in the -130 to -125 dBc/Hz measured at an offset frequency f=1Hz. This results in a an average value of 7×10^{-39} for the flicker noise constant.

One of the most attractive features of STW resonators is their extremely low residual phase noise which makes it possible to design microwave STW oscillators with very low close-to-carrier phase noise performance.[55, 51] This becomes evident from Fig. 30 which is the $1/f$ noise performance of the 1 GHz STW resonator from Fig. 16, featuring a loaded Q of 2720. The $\mathcal{S}'(f)$ value at f=1Hz is -142 dBc/Hz. The residual phase noise of several of these devices were measured using the method described in Ref. /55/. Most of them demonstrated phase noise levels in the -144 to -142 dBc/Hz at 1 Hz offset and were indistinguishable from the noise floor of the measurement system (the lower curve in Fig. 30). According to Eq. (17), this results in an average flicker noise constant of 3.4×10^{-40} which, to the best of the author's knowledge, represents the lowest residual phase noise achieved with an acoustic resonator at 1 GHz so far.

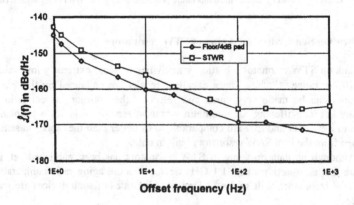

Fig. 30. Residual phase noise of the 1 GHz STW resonator from Fig. 16 (upper curve); noise floor of the measurement system (lower curve), (IEEE© 1994).

11. Frequency Trimming of STW Resonators

STW resonators on quartz are narrowband devices, the frequency of which is fairly sensitive to fabrication tolerances of the film thickness and finger-to-gap-ratio. For some low-cost applications a frequency tolerance of ±150 ppm is achievable with most of the fabrication equipment available to date. However, if the required frequency tolerance is in the ±10 to ±25 ppm range, then frequency trimming is necessary. The easiest way to do this is to fabricate the devices at a frequency which is about 1000 ppm higher than the desired one and then to trim it down to the required value. A very accurate method for frequency down trimming is bombarding the device surface with heavy xenon (Xe^+) ions in a vacuum chamber.[56] In this process the Xe^+ ions sputter off part of the quartz between the Al electrodes, while the latter remain protected by oxygen molecules which are much better adsorbed by metals than by quartz. In this way, shallow groves are formed between the metal strips along the resonant cavity. This results in a decrease in STW velocity and a linear device frequency down shift with trimming time, without degradation in device performance. Even a slight improvement of the device loaded Q and insertion loss is observed which results in an increase of the unloaded Q by 20%, as shown in Fig. 31. The plots in Fig. 32 are the frequency responses of a 772 MHz STW resonator during trimming. It is evident that the STW device can stand extreme frequency downshifts of more than 4 MHz (5200 ppm) without increase in device insertion loss. Serious degradation in overall performance is observed after 6.5 MHz down trimming.

The Xe^+ heavy ion bombardment method allows a very precise control of the ion current and the trimming rate. That is why, it can be used for frequency adjustment of STW resonators in an in-situ monitoring process within 10 ppm of accuracy.[55]

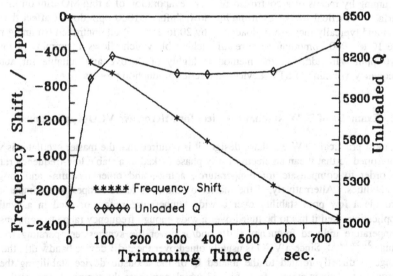

Fig. 31. Device frequency and unloaded Q versus trimming time, (IEEE© 1993).

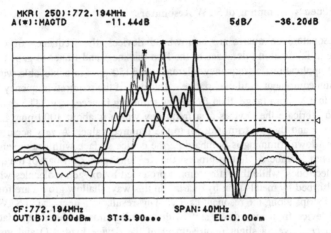

Fig. 32. Influence of Xe⁺ heavy ion bombardment on the overall performance of 772 MHz STW resonators. Right plot: untrimmed device; Medium plot: 4 MHz trimmed device; Left plot: 6.5 MHz trimmed device, (IEEE© 1994).

Since the STW velocity depends on the amount of mass loading on the substrate surface, any additional mass deposited on the device surface will shift its frequency down. If the deposited material is stable over temperature and time and if the deposition can be well controlled, then the mass loading phenomenon can also be used for device frequency trimming. Reference /58/ describes an easy method for STW resonator frequency trimming by means of a controlled resistive evaporation of a thin SiO film on the device surface. Practical frequency trimming downshifts of 1000 ppm do not affect the device loss and typically increase its loaded Q by 20 to 25%. Well controlled trimming rates in the 10 to 2000 ppm/minute are readily achievable which allows a trimming precision of ±10 ppm. In addition, the method is highly productive and suitable for automatic frequency trimming of STW devices in a series production process.

12. Examples of STW Resonant Devices for Microwave VCO Applications

In most practical STW oscillator designs it is required that the master oscillator is voltage controlled so that it can be electronically phase locked to a stable low-frequency reference in order to compensate for temperature-, aging-, and other medium- and long-term-instabilities. Alternatively, if the master oscillator is to be temperature compensated to provide a few ppm stability over a wide temperature range, or used in a synthesizer application, then it has to be tuned over a wide enough frequency range to compensate for temperature induced frequency shifts and/or cover the synthesizer operating frequency range.[57, 59, 60, 61] Since a VCO is usually tuned over the 3 dB device bandwidth, the tuning range is directly related to the loaded Q of the acoustic device stabilizing the VCO frequency. This is evident from Fig. 33 which compares the tuning characteristics of a 1 GHz VCO with two different device Q values of the STW resonators in the VCO loop, according to Ref. /55/. Oscillator A has a loaded Q of 3000 and provides a tuning range of about 100 kHz. Oscillators B and C use STW devices with a loaded Q of 1500. Therefore, it was possible to extend their tuning range to more than 600 kHz.

If in a SAW or STW based VCO the oscillator Q is too high, then the desired tuning range cannot be achieved. If it is too low, this will degrade the VCO phase noise performance since there is a $1/Q^4$ relationship between device Q and oscillator close-in phase noise. The comparison of the two VCO phase noise plots in Fig. 34 illustrates very well this relationship. The difference by a factor of two in device Q for oscillators A and B from Fig. 33 results in about 12 dB difference in phase noise performance over the offset range from 1 Hz to 10 KHz. That is why, an important requirement in the VCO design process is that the device loaded Q value is selected not lower than necessary to provide the best phase noise performance at the required tuning range.

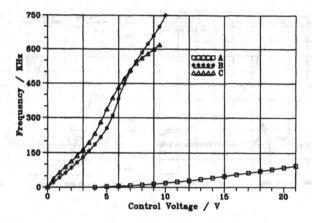

Fig. 33. Tuning characteristics of 1 GHz VCO with different values of the STW device loaded Q: 3000 for Oscillator A and 1500 for Oscillators B and C, (IEEE© 1996).

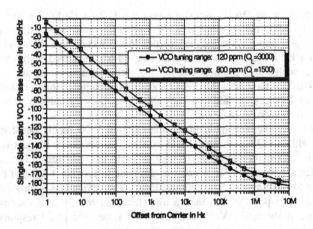

Fig. 34. Phase noise performance of Oscillators A and B from Fig. 33 using a STW device loaded Q of 3000 and 1500 respectively, (IEEE© 1992).

As shown in Sections 4.3. and 4.4., STWs propagating in resonant devices provide a very elegant way of adjusting the device loaded Q by exerting a control over the location of two adjacent longitudinal modes with respect to each other at the lower stopband edge. Thus, not only high-Q resonances but also low-Q resonator (LQR) or two-pole coupled resonator filter (CRF) responses can be obtained while keeping the device loss very low. Another advantage of this approach is that such devices do not need matching networks. Fig. 35 compares the frequency and group delay responses of two 2.48 GHz STW devices, the loaded Q of which differs by a factor of two while the unmatched insertion loss is about 10 dB for both devices.[30] As shown in Ref. /62/, STW two-port resonators can retain insertion loss values of about 10 dB at frequencies up to 3 GHz which greatly enhances oscillator design and improves its overall phase noise performance.

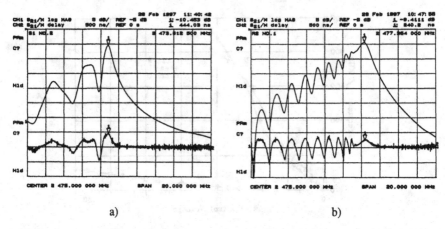

Fig. 35. Frequency and group delay responses of two 2.48 GHz low-loss STW resonators with two different values of the loaded and unloaded Q: a) Q_L=3450; Q_U=4930; b) Q_L=1870; Q_U=2830, (IEEE© 1997).

The plots in Fig. 36 characterize a very low-Q 2.1 GHz LQR which was designed to provide a 1.8 MHz tuning range in a VCO for a low-noise synthesizer application. The loaded Q is only about 1200 and the insertion loss is just about 5 dB which, to the best of the author's knowledge, is the lowest unmatched loss achieved with a two-port acoustic resonator at such a high frequency so far.[46] Another substantial advantage of low-Q STW resonators is that their loss, loaded Q and sidelobe suppression are fairly insensitive to fabrication tolerances. In addition, they allow frequency trimming over at least 1000 ppm without performance degradation.

In some synthesizer applications, the required VCO tuning range is too high to be achieved with a LQR stabilized VCO. In such cases STW based two-pole CRF can double the maximum tuning range that a LQR can provide. The data in Fig. 37 characterize a 1.95 GHz CRF which was designed to provide 4 MHz of tuning range in a VCO for a low-noise synthesizer application.[46] The device loss is only 10 dB and the dip between the two poles is less than 3 dB which can readily be compensated for by the excess gain of the active VCO circuitry. The symmetric phase response in Fig. 37 b) provides a smooth S-shaped VCO tuning characteristic.

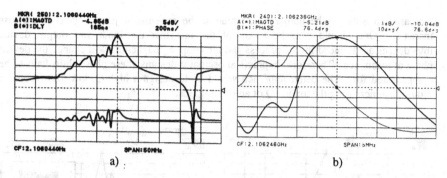

Fig. 36. Characteristics of an extremely low-loss 2.1 GHz STW low-Q resonator: a) broadband frequency and group delay responses; b) narrowband frequency and phase responses, (IEEE© 1995).

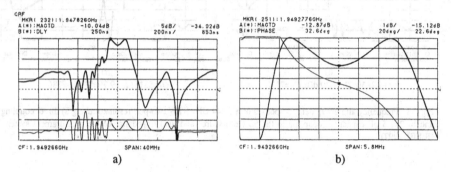

Fig. 37. Characteristics of a STW based 1.95 GHz two-pole CRF with 4 MHz of bandwidth: a) broadband frequency and group delay responses; b) narrowband frequency and phase responses, (IEEE© 1995).

12.1. *Increasing the loaded Q of an existing low-Q STW resonator*

As evident from Eqs. (3) and (4), the device loaded Q is related to its insertion loss and to the value of the source and load impedance. Therefore, altering the resonator coupling to the measurement system, which typically has a fairly low characteristic impedance of 50 or 75 Ohms, will change the loss and loaded Q of the acoustic device. This is valid for most microwave resonators, regardless of the physical principle they are based on.

In some applications, in which the close-to-carrier oscillator phase noise is of major importance, the design engineer may want to increase the loaded Q of an existing two-port STW resonator at the expense of an increase in its insertion loss. As shown in Fig. 38, this can be achieved by connecting coupling capacitors C_k of a properly selected value in series with the acoustic device which reduce the coupling to the load and source impedance of the measurement system.[63] The two 1.5 pF series capacitors almost double the device loaded Q at the expense of only 5.3 dB increase in insertion loss. Es evident from the data in Fig. 38, the device unloaded Q remains nearly unchanged in this process. Additional data in Ref. /63/ shows that with this practical method the device loaded Q can be increased to 80% of the unloaded Q value at the expense of just about 6 dB of loss increase. One should also be aware of the fact that the input and output device impedances are reduced in this process and the coupling capacitors will change the device phase at resonance. Therefore, in practical oscillator designs, these phase shifts have to be compensated for. The better approach would be to design a high-Q, low-loss resonator rather than increasing the loaded Q of an existing one.

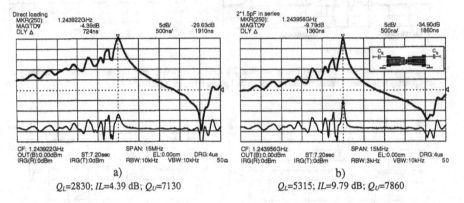

a) Q_L=2830; IL=4.39 dB; Q_U=7130
b) Q_L=5315; IL=9.79 dB; Q_U=7860

Fig. 38. Frequency and group delay responses of a 1.244 GHz STW two-port resonator using a) direct coupling and b) 2×1.5 pF capacitive coupling to the 50 Ohm source and load impedance of the measurement system, (IEEE© 1997).

13. High Performance Microwave Oscillators Using STW Resonant Devices

The unique properties of STW resonant devices in terms of high device Q at frequencies up to 3 GHz, extremely low $1/f$ noise, very high RF power handling capability, as well as low aging and low vibration sensitivity make these devices the best choice for designing high-performance microwave oscillators in the frequency range of 1.0 to 3.0 GHz, which represents a gap for low-noise microwave oscillators because other oscillator technologies do not provide satisfactory performance in this range. Below 1.0 GHz SAW oscillators work well. Above 3.0 GHz, DRO provide the best compromise between oscillator performance and cost. This section will present and discuss data from several examples of high-performance STW oscillators that successfully bridge this frequency gap.

13.1. *High-performance feedback loop STW oscillators*

The block diagram of a SAW or STW based feedback loop oscillator (FLO) is shown in Fig. 39. The oscillator loop consists of a loop amplifier (LA), a 3 dB power splitter (PS), a loop adjust line (LAL) and a two-port STW resonator (STWR). If the oscillator has to be tuned over the narrow 3 dB bandwidth of the STWR, as discussed in Section 12, then the loop must contain also an additional varactor tuned variable phase shifter (VPS) which provides a voltage controlled phase shift in the loop.[57,64] This type of FLO is called a voltage controlled oscillator (VCO). The well known amplitude and phase conditions of oscillation for the loop configuration in Fig. 39 are described by Eq. (18) and (19) respectively:

$$G_{LA} + L_{PS} + L_{LAL} + L_{STWR} + L_{VPS} \geq 0 \tag{18}$$

$$\varphi_{VPS} + \varphi_{LA} + \varphi_{PS} + \varphi_{LAL} + \varphi_{STWR} = 2n\pi \tag{19}$$

Here G_i and L_i are the gain and loss factors in dB of the 5 VCO building blocks, described by their indexes (i) according to Fig. 39, φ_i are their phase shifts measured in degrees accordingly, and n is an integer. The circuit in Fig. 40 is a simple practical strip line implementation of a 2 GHz FLO operating at a fixed frequency since no VPS is connected in the loop.[46] The LA consists of the transistors T1 and T2 which must have sufficient gain to compensate for all losses around the loop, satisfying the amplitude condition of oscillation according to Eq. (18). The strip line transformers provide matching between the different FLO building blocks to minimize reflections and maximize the loop power. The 3 dB Wilkinson type PS provides half of the loop power to the oscillator load and improves load pulling by isolating the loop from load changes at the oscillator output. Finally, if the phase condition for oscillation (Eq. (19)) is roughly fulfilled, the LAL is used for fine adjustment of the FLO frequency at the STWR resonant frequency where lowest phase noise is achieved.[20]

Although strip line FLO provide excellent performance at frequencies around and above 2 GHz, they occupy a fairly large area which is not acceptable for many applications. The FLO circuit in Fig. 41 has a very similar performance as the one from Fig. 40, however, it is entirely based on discrete SMD components and occupies an area of less than 2 cm^2. Instead of strip lines, discrete LC networks are used for matching around the loop and for the power splitter.[47] This approach allows the additional possibility of electronic tuning if some of the capacitors in the loop are replaced by varactors. One practical implementation of such a discrete design is shown in Fig. 42 which is the schematic of a high-performance, low-noise 2.48 GHz STW resonator based VCO.[64] The

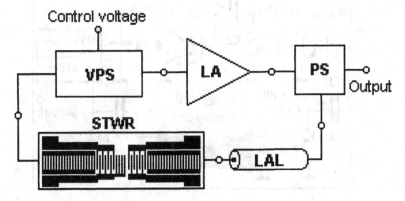

Fig. 39. Block diagram of a two-port STWR based FLO, (IEEE© 1998).

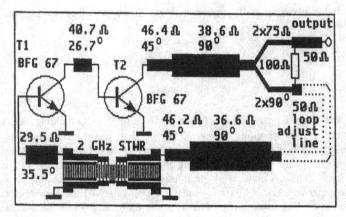

Fig. 40. A practical FLO implementation, (IEEE® 1995).

matching network between the STWR and the first transistor stage has been designed as a CLC-type varactor based VPS which provides smooth linear tuning of the VCO frequency over a frequency range of about 500 ppm within the 3 dB bandwidth of the low-Q STWR used in the circuit.

Explaining further details about the operation of STW based FLO, their electrical adjustment, phase noise performance optimization and construction would go beyond the scope of this paper. A variety of references provide sufficient design details and important guide lines on practical realization of high-performance SAW and STW based FLO.[20,46,47,50,51,55,57,64,65,66,67] Here, we will restrict ourselves just to presenting experimental data from a few characteristic examples.

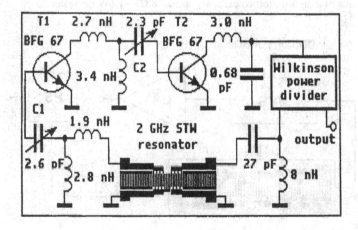

Fig. 41. STW based 2 GHz discrete component FLO design, (IEEE® 1995).

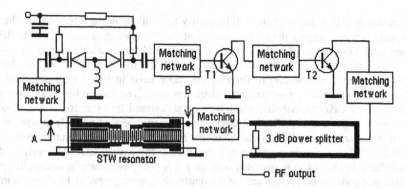

Fig. 42. Block diagram of a high-performance 2.48 GHz STW based VCO featuring very low phase noise, (IEEE© 1998).

The plot in Fig. 43 a) is the narrowband output spectrum of the 2.48 GHz VCO from Fig. 42, measured with a high-resolution spectrum analyzer over the frequency span of only 2 kHz and a 100 Hz resolution bandwidth. The curve is very smooth indicating a very low phase noise. This becomes evident from the data in Fig. 43 b) indicating a phase noise suppression of -15 dBc/Hz at 1 Hz offset and -105 dBc/Hz at 1 kHz offset from the

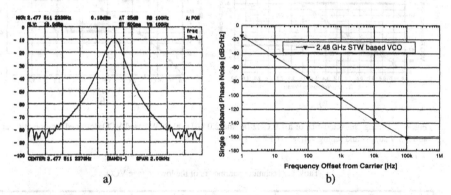

Fig. 43. Narrowband output spectrum (a) and phase noise performance (b) of the 2.48 GHz STW based VCO from Fig. 42, (IEEE© 1998).

carrier. To the best of the author's knowledge, this is the lowest close-to-carrier phase noise performance obtained with a STW oscillator at 2.5 GHz to date. Although impressive, these results are about 12 dB worse than the theoretical phase noise limit that should be expected from a 2 GHz STWR based oscillator, calculated in Ref. /46/, assuming a noiseless amplifier. As shown in Ref. /55/, this limit can be approached if specially selected silicon bipolar transistor amplifiers with very low 1/f noise levels are used in the loop. The best close-in phase noise performance reported to date was measured with 1 GHz STWR oscillators, indicating single side band phase noise suppression values in the -33 to -35 dBc/Hz range at 1 Hz offset from carrier.[51,55]

Recalculated in terms of short-term frequency fluctuations in the time domain, this value results in a one second short-term stability of 3×10^{-11}/s which is comparable to the short term stability of some of the best 10 MHz crystal oscillators available to date.

As shown in Section 12, the accurate control over the Q of the STWR stabilizing the oscillator frequency, allows minimizing the phase noise in VCOs with a desired tuning range. Figure 44 compares the tuning characteristics of two low-power VCOs operating at 1.0 and 2.0 GHz respectively, which have been designed for operation at low supply and tuning voltages and high RF/d.c. efficiency in battery operated portable equipment. The tuning linearity is excellent while the output power variation with tuning does not exceed ±1 dB. Table 6 compares additional parameters of these two VCO. Although the 1 GHz oscillator in Fig. 44 a) is operated only on a 3 V d.c. power supply voltage, it provides excellent phase noise performance at the remarkable output power of 12 dBm (16 mW) on a 50 Ohm load. With 26 mA of d.c. current consumption, this results in a RF/d.c. efficiency of 20%. In fact, because of its high efficiency, this VCO was used as a reference oscillator in a 1 GHz microwave data transmitter. The required output power of 0.5 W at 50% RF/d.c. efficiency, coupled directly to the transmitter antenna, was achieved by adding only one single-transistor B-class amplifier stage to the VCO output.

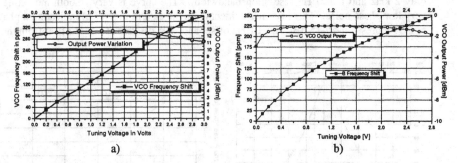

Fig. 44. Tuning characteristics of a 1.0 GHz (a) and 2.0 GHz (b) STW based VCO designed for operation at 3 and 2.8 V supply and tuning voltages, (IEEE© 1998).

Table 6. Technical parameters of the low-voltage VCO.

Parameter	1 GHz VCO	2 GHz VCO
VCO center frequency	1001.2 MHz	2010 MHz
VCO output power	12 dBm ± 1 dB	-1 dBm ± 1 dB
Tuning voltage	0 ... 3 V	0 ... 2.8 V
Tuning range	350 ppm (typical)	250 ppm (typical)
Phase noise suppression at:		
1 Hz offset	-12 dBc/Hz	-9 dBc/Hz
1 KHz offset	-105 dBc/Hz	-100 dBc/Hz
Thermal noise floor	-175 dBc/Hz	-162 dBc/Hz
Supply voltage	3 V	2.8 V
Supply current	26 mA	7 mA
Temperature stability (-30 ... +70)°C	within 150 ppm	within 130 ppm

13.2. Direct frequency modulation of STW based VCO. Applications to microwave data transmitters.

As evident from Fig. 13 b), a well designed two-port STW resonator has a linear phase response over its 3 dB bandwidth. This results in a very linear tuning characteristic of the VCO stabilized with such a device and the tuning data in Fig. 44 a) illustrates this behavior. If the VCO from Fig. 44 b) is biased at 1.5 V which is the center of its tuning characteristic and a modulating signal of a frequency f is superimposed with the bias voltage, then the VCO output becomes frequency modulated (FM). This means that it can transmit information at a FM rate f over a radio channel. The plots in Fig. 45 a) and b) are broadband FM spectra of a 1 GHz VCO, FM modulated with sinusoidal signals of 2 kHz and 70 kHz modulation frequency, respectively. The frequency deviation dF is ±300 kHz which results in a FM bandwidth of 600 kHz. The spectra are symmetric about the center (carrier) frequency which is an indication of low modulation distortion. VCOs with a tuning characteristic as the one shown in Fig. 44 a) provide typical FM distortion levels in the 0.1 to 1% range which makes them ideally suited for wireless transmission of analogue signals in a high fidelity (HiFi) quality.[68]

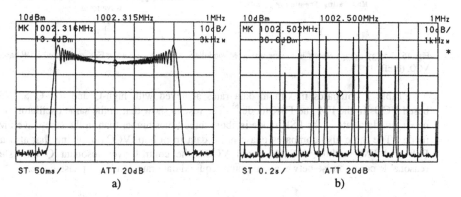

Fig. 45. Broadband FM spectra of a directly FM modulated 1 GHz STW based VCO with modulation frequency of a) 2 kHz and b) 70 kHz. The FM bandwidth is 600 kHz, (IEEE© 1992).

One important issue in a directly frequency modulated STW based VCO is the maximum data rate which it can transmit in the sinusoidal FM mode and in the frequency shift keying (FSK) mode in which the modulating signal is a square-wave data stream. Extensive studies on FM and FSK modulated FLO have shown that the maximum achievable data rates are limited by energy storage effects in the oscillator loop and will strongly depend on the loaded group delay of the STWR in the VCO.[57,65,69,70] This effect is shown in Fig. 46 a) and b) for sinusoidal and square wave modulation of a 1 GHz VCO stabilized with a low-Q STWR of 200 ns loaded group delay. As evident from Fig. 46 a), when the modulation frequency f becomes too high, the VCO frequency cannot follow the modulating signal and the FM deviation dF decreases from its low-frequency value dF_o. In this particular case, the 3 dB drop in frequency deviation occurs at a modulation frequency of about 100 kHz. This is the maximum usable data rate at which this VCO can transmit when sine-wave FM modulated.

In the FSK mode, the VCO behaves in a way very similar to a RC-integrator since the energy storage effect causes exponentially increasing and decreasing slopes on the detected wave form. This is evident from Fig. 46 b), comparing the modulating wave form with the one obtained after detection of the VCO output signal. In the FSK mode, the maximum usable data rate is limited by the oscillator Q and the amount of gain compression in the loop. As a rule of thumb, in a STWR stabilized VCO with a Q of 1000, the maximum FSK data rate is about $(20\tau_g)^{-1}$, where τ_g is the loaded group delay of the STWR in the loop.

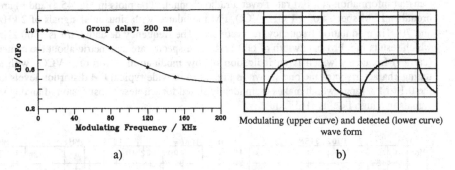

a)

b)

Fig. 46. Energy storage effects in a) sine-wave FM modulated and b) square wave FSK modulated STW based VCO, (IEEE© 1992).

Table 7 compares maximum data rates achieved with high-Q and low-Q STWR based VCOs in the FM and FSK mode.[65] The results show that in the same oscillator, the maximum data rate in the FSK mode is about three times higher than in the sine-wave FM mode. The table also compares phase noise data from both VCOs implying that also in FSK and FM modulated STW based VCOs, the control over resonant Q allows a reasonable compromise between maximum required data rate and VCO phase noise.

Table 7. Maximum data rates and phase noise performance of two STW based VCOs with different oscillator Q, (IEEE© 1992).

Parameter	767 MHz STWR based VCO	1 GHz CRF based STW VCO
Loaded group delay τ_g at f_r	1.16 µs	200 ns
Oscillator loaded Q ($Q_L = \pi \tau_g f_r$)	2800	630
Sine wave f at 3 dB deviation drop	9.2 kHz = $(94\tau_g)^{-1}$	100 kHz = $(51\tau_g)^{-1}$
Deviation rise and fall time	36 µs = $(31\tau_g)^{-1}$	3.6 µs = $(18\tau_g)^{-1}$
Phase noise suppression at: 100 Hz offset 1 kHz offset 10 kHz offset	83 dBc/Hz 111 dBc/Hz > 118 dBc/Hz	67 dBc/Hz 97 dBc/Hz >118 dBc/Hz

13.3. STW based high-power oscillators with extremely low thermal noise floor

As shown in Section 7, despite their small size, STW resonators can handle very high levels of radio frequency (RF) power without performance degradation. This unique feature is of major importance to the design of high-power microwave oscillators with very low thermal noise floor (TNF) since, in such oscillators, the STW device is run at incident power levels of up to 2 W.[71]

According to Ref. /20/, the single side band thermal noise floor of a FLO is calculated in dBc/Hz as:

$$\mathfrak{S}(f) = -174 + G + F - P_0 \qquad (20)$$

where G is the loop loss in dB, F is the noise figure of he loop amplifier in dB and P_0 is the FLO loop power in dBm. It is obvious that the TNF level can be improved if G and F are minimized and P_0 is maximized. Since STW resonators are very low-loss devices with typical insertion loss values in the 3 to 6 dB range and loop amplifiers typically have very low noise figure values in the 3 to 5 dB range, the only significant improvement of the TNF can be achieved if the loop power is increased to levels, practically limited to the RF power handling ability of the STW device. As shown in Ref. /71/, high loop power levels can be achieved by using highly efficient nonlinear AB-class amplifiers in the loop. One practical implementation of an AB-class amplifier based VCO, indicating also the loop power levels at different points of the loop, is shown in Fig. 47. It has the classical FLO architecture with the only difference being in the operation principle of the AB-class amplifier. As shown in Fig. 48, at small drive power levels, the AB-class amplifier operates as a linear A-class amplifier providing sufficient gain for safe FLO starting. After a few loops, the power at the amplifier input reaches a certain threshold level (8 dBm in Fig. 48), at which it starts operating in the switched nonlinear B-class mode. In this mode, the amplifier can provide several Watts of output power at an efficiency of up to 50%. The steady state is reached when the linear A-class amplifier in the circuit from Fig. 47 is driven into gain compression. Thus, it limits the maximum drive power at the input of the AB-class amplifier to a safe level, preventing it from overheating.

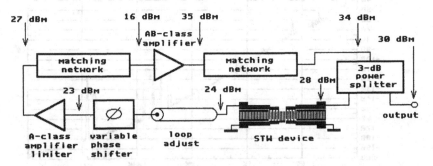

Fig. 47. Block and level diagram of a high-power STW based VCO using a highly efficient AB-class power amplifier in the loop, (IEEE© 1993).

As an example for the phase noise calculation, we will consider a power VCO which was practically built with a 1 GHz low-Q STWR.[71] With an overall G value of 9.6dB (5.2dB STWR loss plus 4.4 dB loss in the power splitter and phase shifter), a LA noise figure F of 5 dB and a loop power P_0 of 34.8 dBm, according to Eq. (20), the noise floor is calculated as -194.2 dBc/Hz. The phase noise plot in Fig. 49 confirms this result by presenting data from a practical phase noise measurement, using the delay line frequency discriminator method.[55,71] To the best of the author's knowledge, this result for the TNF is 10 dB better than the best 400 MHz SAW oscillator built to date.[47]

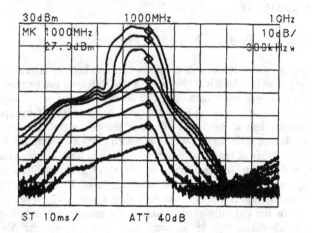

Fig. 48. AB-class amplifier frequency responses at drive power levels (top to bottom) of: 14, 11, 8, 4, -6, -16, -26 and -36 dBm, (IEEE© 1993).

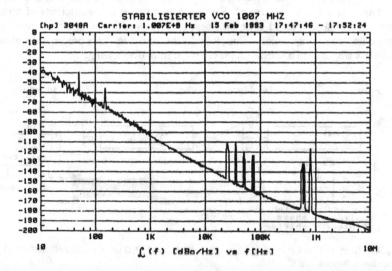

Fig. 49. Phase noise performance of a 1 GHz STW based power VCO, (IEEE© 1993).

Figure 50 shows the harmonic spectrum of the power VCO, discussed above. The output power level at the fundamental frequency of 1 GHz is 32 dBm (1.6 W), while the second and higher order harmonics are suppressed by more than 25 dBc without additional filtering. Another very attractive feature of this power VCO is its RF/d.c. efficiency η of

35% which makes it ideally suited for battery operated equipment.[55,71] In addition, as any other VCO, the STW power oscillator can be directly FM or FSK modulated as shown in Section 13.2, and operated as a highly efficient and stable low-noise data transmitter. Further details on the design, performance and applications of such and similar STW based data transmitters are discussed in Refs. /68/, /69/, /70/, /72/ and /73/.

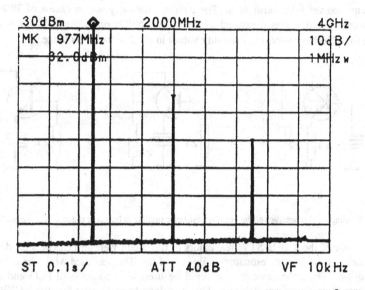

Fig. 50. Harmonic spectrum of the 1 GHz power VCO. P_{out} = 1.6W; η = 35%, (IEEE© 1993).

13.4. Negative resistance and feedback loop oscillators using single-port STW resonators

As implied in Section 6.1, single-port STWR can achieve much higher device Q values than their two-port counterparts and can approach the material Q-limit for STW on quartz. This and the high RF power handling ability make STW single-port devices very well suited for the design of simple microwave fixed frequency and voltage controlled oscillators with improved short-term stability, high output power and increased RF/d.c. efficiency. Among RF engineers the negative resistance oscillator (NRO) is very popular for this reason, especially at higher microwave frequencies. Single-port STWRs seem to be a good choice for stabilizing NROs, since they have very low R_d values.[43,74] On the one hand, this makes it easy to ground a transistor terminal through the STW device. On the other hand, the amount of negative resistance that has to be generated by the transistor to compensate for the R_d loss is accordingly low. Thus, oscillation starts well and the oscillator does not generate parasitic oscillations.

The first experiments performed by the author used the basic NRO circuits described in Fig. 51 a), b) and c).[43] In the circuits in Fig. 51 a) and b), the resonator operates at series resonance and grounds the base and emitter of the transistor, respectively, while the complex impedances Z1 and Z2 set the phase condition of oscillation. The circuit in Fig.

51 c) uses the popular grounded collector configuration which is known for its low phase noise in voltage controlled oscillator (VCO) applications.[75] With Z2 tunable, the circuit in Fig 51 c) was found to provide about 150 ppm of tuning range when stabilized with a 1GHz single-port STWR.

The circuit in Fig. 51 a) takes advantage of the high RF power handling ability of the STW device. It is used in a 1 GHz grounded base configuration and was run at 23dBm of output power for several days. The RF/d.c. efficiency was in excess of 30% while the short-term stability was measured as 7×10^{-10}/s. NROs operating at lower power levels were found to have short-term stability values in the 2 to 5×10^{-10}/s range.

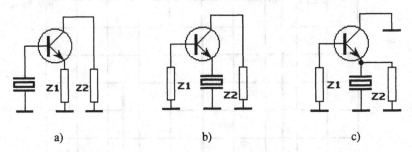

Fig. 51. Simplified schematics of the negative resistance oscillators built and tested in this study, (IEEE© 1999).

Most of the 1 GHz STW resonators in this study were in fact two-port devices. To use them as single-port resonators, only one of the IDT was used while the other one was left open. Thus, the same device could be used to evaluate both FLO and NRO, and compare their short term stability. This was done to make sure that the STWR $1/f$ phase noise contribution has the same source for both oscillator types. For the same transistor in the NRO and FLO circuit and the same oscillator output power, the NRO was found to provide a 2 to 4 times better short-term stability than the FLO using the same STW device in a two-port configuration. A possible explanation of this improvement could be that in a NRO, the oscillator power does not pass through all components in the FLO loop where each of them adds to the $1/f$ noise, but just bounces back and forth between the STWR and the active device. In this process, the $1/f$ noise phase modulation is significantly reduced.

A further way to improve the overall oscillator phase noise performance while keeping the $1/f$ noise contribution of the acoustic device at a minimum, is to use the resonator as a single-port device in a capacitive Π-circuit configuration which behaves as a two-port resonator. This is shown in Fig. 52 characterizing such a circuit. Since only one IDT is used, the $1/f$ noise contribution of a second IDT is eliminated. Increasing the shunt capacitance values improves the sidelobe suppression and increases the circuit Q at the expense of higher circuit loss. Thus, a trade-off between oscillator phase noise and tuning range or maximum FM or FSK data rate can be achieved if the circuit is used in a voltage controlled FLO as shown in Fig. 39. The lowest phase noise is achieved when the FLO is operated slightly above series resonance where the phase slope is maximum, (see phase response data in Fig. 19).

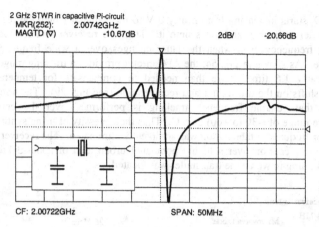

Fig. 52. Frequency response of a 2 GHz capacitive Π-circuit using a single-port 2 GHz STWR, (IEEE© 1999).

14. Other Applications of STW Resonators and Oscillators

STW based resonant devices and oscillators can greatly improve system performance in a variety of applications including microwave data transmitters and receivers,[68,71,72,73] linear low-noise detectors,[69,70] noise filters using injection locked STW oscillators[57] and in physical, chemical and biological sensors,[78,79,80,81] including wireless systems for transmission, reception and processing of sensor data.[82,83,84] Discussing all these applications would go beyond the scope of this paper, which is why we will very briefly present only two of them to create an impression on what one can do with STW.

14.1. *A high-performance STW based microwave transceiver for narrowband FM voice communications*

Reference /73/ describes a system application of low-noise STW based VCOs which greatly simplify the system concept and improve the performance of a battery operated 1.244 GHz transceiver for narrowband FM (NBFM) voice communications. The transmitter is a highly stable voice modulated 1.244 GHz VCO stabilized with a high-Q STWR as the one described in Sections 12 and 13. 1. Its output has been amplified to provide 250 mW transmitted power over at least 7 hours operation on one battery charge. The receiver uses the triple conversion superheterodyne method which allows the reception bandwidth to be reduced to 4 kHz, significantly improving receiver sensitivity and close-to-channel interference compared to previously described systems. The block diagram of the receiver circuit is shown in Fig. 53. The key to achieving high overall receiver performance is the first local oscillator which is a low-noise 1.27 GHz STW based VCO converting the incoming signal to a low first intermediate frequency of 26 MHz which is further processed by a highly sensitive, low-power consumption, low-cost integrated circuit. One important function of this circuit is that it generates a tuning proportional d. c. voltage used for automatic frequency control (AFC) and frequency tracking. When the receiver is turned on, the AFC voltage, controlling the 1.27 GHz STW based VCO, starts increasing from nearly 0 V to the value at which the receiver tunes to the transmitter frequency and locks onto it. Then the receiver remains locked onto the transmitter frequency even when the latter changes over a wide frequency range with temperature. As shown in Fig. 54, the AFC circuit provides a tracking range of 190 ppm. This is nearly 1.6 times more than needed to compensate for temperature induced frequency shifts on the transmitter and receiver side (see Fig. 29). The temperature tests confirmed that reliable frequency tracking is performed over the entire operating temperature range of (-30 to +60) deg. C. Thus the necessity of using synthesizer or PLL circuitry for frequency stabilization was completely eliminated. The receiver was found to provide good reception over a highly populated urban area of about 3 km^2 around the concrete building in which the transmitter was located.

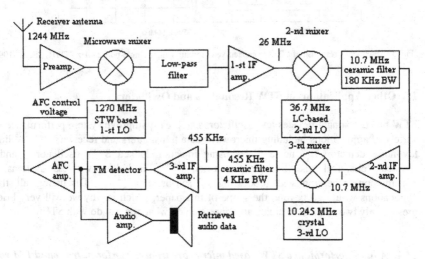

Fig. 53. Block diagram of the STW based 1.244 GHz microwave NBFM receiver.

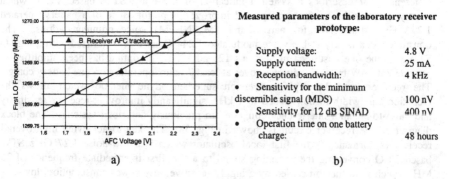

Measured parameters of the laboratory receiver prototype:

- Supply voltage: 4.8 V
- Supply current: 25 mA
- Reception bandwidth: 4 kHz
- Sensitivity for the minimum discernible signal (MDS): 100 nV
- Sensitivity for 12 dB SINAD: 400 nV
- Operation time on one battery charge: 48 hours

a)   b)

Fig. 54. Frequency tracking ability (a) and measured parameters (b) of the receiver from Fig. 53

14.2. A high-resolution humidity sensor using a polymer coated STW resonator

As discussed in Section 11, the STW propagation velocity is very sensitive to changes in the amount of mass, loading the surface of the acoustic device. Therefore, if a STW resonator is coated by a material which can change its mass by selectively absorbing certain gas or liquid phase chemical compounds, then the resonator frequency will change proportionally to the mass of the sorbed compound. This is the operation principle of STW-based sensors which can register mass changes of a few picograms if connected in an oscillator circuit with a few parts in 10^{-10}/s short term stability. Figure 55 shows a laboratory prototype of a high-resolution STW based humidity sensor in which the coating material is a hexamethyldisiloxane (HMDSO) polymer film deposited on the resonator surface in a glow discharge plasma process.[85] The data in Fig. 56 compares the relative sensitivities of HMDSO coated 767 MHz STW resonators at two different polymer film thicknesses, with the sensitivity of a BAW crystal, coated with the same polymer. With a relative sensitivity of 1.4 ppm/% change in relative humidity, the STW sensor is 6 times more sensitive than its BAW counterpart.

Extensive studies on polymer coated STW resonators for chemical and biological sensor applications are currently on the way[86].

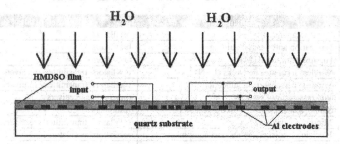

HMDSO + H2O => Increased mass loading => Reduced STW velocity => Sensor frequency downshift

Fig. 55. Operation principle of a HMDSO coated STW based humidity sensor, (IEEE© 1998).

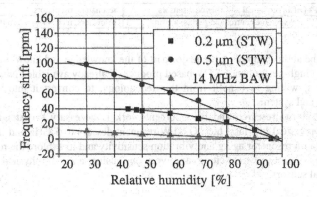

Fig. 56. STW versus BAW sensitivity of HMDSO coated humidity sensors (IEEE© 1998).

15. Summary and Conclusions

This paper has reviewed the historical development, current status, recent advances and application areas of STW based resonator and oscillator technology. Characteristic properties, peculiarities and limitations of STW propagation under periodic metal strip gratings and resonator structures on quartz have been highlighted and compared with widely used SAW technology. Important aspects for the practical design of low-loss STW resonant devices with a controlled Q have been discussed and illustrated with experimental data from laboratory prototypes. The advantages of STW resonator technology in terms of high intrinsic material Q, high-power handling capability and low residual 1/f noise have been confirmed with a variety of practical microwave resonators, oscillators and systems benefiting from these advantages. It has been shown that STW resonator technology can successfully bridge the frequency gap from 1.0 to 3.0 GHz for high-Q acoustic resonators and low-noise stable microwave oscillators, in which no other currently available technology can provide satisfactory performance at a reasonable cost. Table 8 lists current and future application areas of STW resonator technology.

Table 8. Areas of application of STW-based products

Application	STW-based product
Smart weapons	oscillators
Navigation and weapon guiding	oscillators
High-resolution RADAR	oscillators
Wireless local area networks (LAN)	oscillators, transmitters, receivers
Wireless personal communications	oscillators, transmitters, receivers
Wireless security systems	transmitters, receivers
Stable, low-noise microwave oscillators	resonators
Microwave data links	oscillators, transmitters, receivers
Precision measurement and test equipment	resonators, oscillators
Notch filters	resonators
Walkie-talkies	oscillators, transmitters, receivers
Radio tags	resonators, transmitters, receivers
Wireless home stereo and video appliances	oscillators, transmitters, receivers
Wireless microphones	oscillators, transmitters, receivers
Physical, biological and chemical sensors	resonators, oscillators
Wireless remote control and remote sensing	resonators, receivers, transmitters

In the author's opinion, the major part of the knowledge and know-how necessary for designing high-performance STW based products is already available. Now is the time for companies, willing to benefit from this technology, to convert it into market oriented products and systems.

Additional research and development work is necessary to extend the frequency range of practical low-loss, high-Q STW resonant devices to 5 GHz and higher, to obtain more data on resonator aging and vibration sensitivity and to explore the remarkable STW properties in the new, exciting and very promising area of physical, chemical and biological sensors.

Acknowledgments

The author wishes to gratefully acknowledge a variety of research institutions and companies for their invaluable technical and financial support, as well for the very fruitful and successful collaboration over more than a decade of time, without which, obtaining a great deal of the results, presented in this paper, would have been impossibile. These institutions and companies are listed in chronological order as follows:

- Institute of Solid Sate Physics, Sofia, Bulgaria — 1989-2000
- Institute of Semiconductor Physics, Novossibirsk, Russia — 1989-1994
- Technical University of Dresden, Germany — 1989-1990
- McMaster University, Hamilton, Ontario, Canada — 1989-1990
- Institute of Electronics, Sofia, Bulgaria — 1991-1998
- RF Monolithics, Dallas, Texas, USA — 1992-1993
- Technical University of Munich, Germany — 1993
- Advanced SAW Products, Neuchatel, Switzerland — 1993
- Raytheon Research Division, Lexington, Massachusetts, USA — 1994
- National Institute of Standards and Technology, Boulder, Colorado, USA — 1994
- National Research Fund of Bulgaria — 1994-1996
- Faculty of Physics, Sofia University — 1994-1998
- Research Institute of Posts and Telecommunications, Wuhan, P. R. China — 1995-1996
- Fujitsu Laboratories, Akashi, Japan — 1996-1997
- M-tron Industries, Yankton, South Dakota, USA — 1997-1998
- Vectron International, Norwalk, Connecticut, USA — 1998-2000
- Institute of Instrumental Analysis, Research Center Karlsruhe, Germany — 1999-2000
- LG Precision, Seoul, Korea — 1999-2000

References

1. A. E. Love, "Some Problems of Geodynamics", Cambridge, University Press, 1911.
2. J. L. Bleustein, Appl. Phys. Lett, 13, 412, (1968).
3. Yu. V. Gulyaev, Journal of Technical Physics Letters, 9, 63, (1969).
4. M. Lewis, Proc. IEEE 1977 Ultrasonics Symposium, pp. 744-752.
5. T. I. Browning and M. F. Lewis, Electronics Letters 13 (5), March 1977, pp. 128-130.
6. K. H. Yen, K. L. Wang, R. S. Kagiwada, Electronics Letters, 13 (1), 1977, pp. 37-38.
7. A. Renard, J. Henaff, B. A. Auld, Proc. IEEE 1981Ultrasonics. Symp., pp. 123-127.
8. K. F. Lau, K. H. Yen, R. S. Kagiwada, K. L. Wang, Proc. IEEE 1977 Ultrasonics Symposium, pp. 996-1001.
9. D. L. Lee, Proc. IEEE 1980 Ultrasonics Symposium, pp. 245-250.
10. K. F. Lau, K. H. Yen, A. M. Kong, K. V. Rousseau, Proc. IEEE 1983 Ultrasonics Symposium, pp. 263-266.
11. K. V. Rousseau, K. F. Lau, K. H. Yen, A. M. Kong, Proc. IEEE 1982 Ultrasonics Symposium, pp. 279-283.
12. K. F. Lau, K. H. Yen, R. S. Kagiwada, A. M. Kong, Proc. IEEE 1980 Ultrasonics Symposium, pp. 240-244.
13. J. Y. Duquesnoy, H. Gautier, Proc. IEEE 1982 Ultrasonics Symposium, pp. 57-62.
14. C. A. Flory, A. R. Baer, Proc. IEEE 1987 Ultrasonics Symposium, pp. 313-318.
15. C. A. Flory, A. R. Baer, Proc. IEEE 1988 Ultrasonics Symposium, pp. 53-56.
16. B. Fleischman and H. Skeie, Proc. IEEE 1989 Ultrasonics Symposium, pp. 235-239.
17. R. Huegli, Proc. IEEE 1987 Ultrasonics Symposium, pp. 183-187.
18. R. Huegli, Proc. IEEE 1990 Ultrasonics Symposium, pp. 165-168.

19. P. S. Cross, W. R. Shreve, Proc. IEEE 1979 Ultrasonics Symposium, pp. 824-829.
20. T. E. Parker and G. K. Montress, IEEE Trans. Ultrason., Ferroelect., Frequency Control, Vol. (UFFC-35), No. 3, May 1988, pp. 342-364.
21. W. J. Tanski, Proc. IEEE 1979 Ultrasonics Symposium, pp. 815-823.
22. A. Ronnekleiv, Proc. IEEE 1986 Ultrasonics Symposium, pp.257-260.
23. T. L. Bagwell and R. C. Bray, Proc. IEEE 1987 Ultrasonics Symposium, pp. 319-324.
24. D. F. Thompson and B. A. Auld, Proc. IEEE 1986 Ultrasonics Symposium, pp. 261-266.
25. B. A. Auld and D. F. Thompson, Proc. IEEE 1987 Ultrasonics Symposium, pp. 305-312.
26. I. D. Avramov, Electronics Letters, Vol. 5, No. 27, Feb. 1991, pp. 414-415.
27. I. D. Avramov, IEEE Trans. Ultrason., Ferroelect., Frequency Control, Vol. (UFFC-40), No. 5, Sept. 1993, pp. 459-468.
28. I. D. Avramov, Proc. 6-th Conference "Acoustoelectronics'93", Sept. 19-25, 1993, Varna, Bulgaria, pp. 193-198.
29. I. D. Avramov, Mei Suohai and Liu Wen, Proc. IEEE 1996 International Frequency Control Symposium, pp. 252-260.
30. I. D. Avramov, O. Ikata, T. Matsuda, T. Nishihara and Y. Satoh, Proc. IEEE 1997 International Frequency Control Symposium, pp. 807-815.
31. I. D. Avramov, IEEE Trans. Ultrason., Ferroelect., Frequency Control, Vol. (UFFC-37), No. 6, November 1990, pp. 530-534.
32. W. M. Bright and W. D. Hunt, Proc. IEEE 1990 Ultrasonics Symposium, pp. 173-178.
33. E. Bigler, B. A. Auld, E. Ritz and E. Sang, Proc. 45-th Annual Symposium on Frequency Control, 1991, pp. 222-229.
34. D. F. Thompson, A. R. Northam and B. A. Auld, 1989 Proc. 3-rd European Frequency and Time Forum.
35. B. Fleischmann and H. Skeie, Proc. IEEE 1989 Ultrasonics Symposium, pp. 235-239.
36. V. P. Plessky, Proc. IEEE 1995 Ultrasonics Symposium, pp. 195-200.
37. K. Hashimoto, M. Yamaguchi, Proc.IEEE 1995 International Symposium on Frequency Control, pp. 442-451.
38. B. Abbot and K. Hashimoto, Proc. IEEE 1995 International Symposium on Frequency Control, pp. 452-458.
39. B. I. Boyanov, K. D. Djordjev, V. L. Strashilov and I. D. Avramov, Trans. IEEE Ultrasonics, Ferroelectrics and Frequency Control, Vol. (UFFC-44), No. 3, 1997, pp. 652-657.
40. I. Boyanov, V. L. Strashilov, IEEE Trans. Ultrasonics, Ferroelectrics and Frequency Control Vol. (UFFC-39), No. 1, January 1992, pp. 119-121.
41. T. Morita, M. Tanaka, Y. Watanabe, K. Ono and Y. Nakazawa, TOYO's Tech. Bull., No. 37, 1987, pp. 1-11.
42. M. Schmid, E. Benes and R. Sedlaczek, Meas. Sci. Technol., No. 1, (1990), pp. 970-975 (UK).
43. I. D. Avramov, Proc. IEEE 1999 Joint Meeting of the European Time and Frequency Forum (EFTF) and the Int. Freq. Control Symp., pp. 863-866.
44. I. D. Avramov and Mei Suohai, IEEE Trans. Ultrason., Ferroelectrics and Freq. Control, Vol. (UFFC-43), No. 6, Nov. 1996, pp. 1133-1135.
45. J. A. Kosinski, R. Pastore, I. D. Avramov, IEEE Proc. 1999 Joint Meeting of the 13-th Eur. Time and Frequency Forum (EFTF) and the Int. Freq. Control Symp., pp. 867-870.
46. I. D. Avramov, V. S. Aliev, S. Denissenko and A. S. Kozlov, Proc. IEEE 1995 Int. Freq. Control Symp., pp. 459-468.
47. T. E. Parker and G. K. Montress, Tutorial on High-Stability SAW Oscillators, presented at the IEEE 1995 Int. Freq. Control Symposium.
48. J. Schoenwald, R. C. Rosenfeld and E. J. Staples, Proc. IEEE 1974 Ultrasonics Symposium, pp. 253-256.
49. W. R. Shreve, R. C. Bray, S. Elliott and Y. C. Chu, Proc. IEEE 1981 Ultrasonics Symposium, pp. 94-99.

50. R. C. Almar, M. S. Cavin, Microwave Journal, February, 1995, pp. 88-98.
51. M. S. Cavin and R. C. Almar, Proc. IEEE 1995 International Frequency Control Symposium, pp. 476-485.
52. J. R. Reid, V. M. Bright, J. T. Stewart and J. A. Kosinski, Proc. IEEE 1996 International Frequency Control Symposium, pp. 464-472.
53. J. A. Kosinski, J. G. Gualtieri, Trans. IEEE Ultrasonics, Ferroelectrics and Frequency Control, Vol. (UFFC-44), No. 6, Nov., 1997, pp. 1343-1347.
54. G. K. Montress, Raytheon Research Division, 1995, Private communication.
55. I. D. Avramov, F. L. Walls, T. E. Parker and G. K. Montress, IEEE Trans. on Ultrasonics, Ferroelectrics and Frequency Control, Vol. (UFFC-43), No. 1, Jan. 1996, pp. 20-29.
56. V. S. Aliev and I. D. Avramov, IEEE Trans. on Ultrasonics, Ferroelectrics and Frequency Control, Vol. (UFFC-41), No. 5, Sept. 1994, pp. 694-698.
57. I. D. Avramov, Proc. 46-th IEEE Freq. Control Symp., 1992, pp. 391-408.
58. M. A. Taslakov and I. D. Avramov, Proc. IEEE 1998 Int. Frequency Control Symp., pp. 497-501.
59. M. A. Taslakov, I. D. Avramov, Proc. 5th Conference with International Participation "Acoustoelectronics'91", Varna, Bulgaria, 10-13 September, 1991, pp. 89-98.
60. M. A. Taslakov and I. D. Avramov, Proc. IEEE 1991 Ultrason. Symp., pp. 275-278.
61. M. A. Taslakov, Trans. IEEE Ultrasonics, Ferroelectrics and Frequency Control, Vol. (UFFC-45), No. 1, January 1998, pp. 192-195.
62. S. Denissenko, E. Cavignet, S. Ballandras, E. Bigler and E. Cambril, Proc. IEEE 1995 International Symposium on Frequency Control, pp. 469-475.
63. I. D. Avramov, Proc. 1997 European Frequency and Time Forum, pp. 394-398.
64. I. D. Avramov, O. Ikata, T. Matsuda, and Y. Satoh, Proc. IEEE 1998 Int. Frequency Control Symposium, pp. 519-527.
65. I. D. Avramov, Proc. 45-th Annual. Freq. Control Symp., 1991, pp. 230-238.
66. I. D. Avramov, Proc. Seventh European Frequency and Time Forum, 1993, pp. 459-464.
67. I. D. Avramov, Tutorial on Design and Applications of STW Resonant Devices, presented at the 1996 IEEE Int. Frequency Control Symposium, 42 pages.
68. I. D. Avramov, Proc. IEEE 1992 Ultrasonics Symposium, pp. 249-252.
69. I. D. Avramov, P. J. Edmonson, P. M. Smith, Electronics Letters, 15th March, 1990, Vol. 26, No. 6, pp. 364-365.
70. I. D. Avramov, P. J. Edmonson, P. M. Smith, IEEE Transactions on Ultrasonics, Ferroelectrics and Frequency Control, Vol. (UFFC-38), No. 3, May 1991, pp. 194-198.
71. I. D. Avramov, Proc. IEEE 1993 Int. Frequency Control Symposium, pp. 728-732.
72. I. D. Avramov, Proc. Sixth International Conference on "Radio Receivers and Associated Systems", Bath, UK, 26-28 Sept. 1995, pp. 76-80.
73. I. D. Avramov, B. Arneson and R. Lutes, Proc. 1998 Intl. Symp. on Acoustoelectronics, Freq. Control and Signal Generation, 7 - 12 June, 1998, St. Petersburg, Russia, (in press).
74. D. F. Thompson and A. Northam, Proc. IEEE 1991 Ultrasonics Symposium, pp. 263-268.
75. R. W. Rhea, "Oscillator Design and Computer Simulation", P T R Prentice Hall, Englewood Cliffs, N. J. 07632, 1990.
76. J. R. Reid, V. M. Bright and J. A. Kosinski, Trans. IEEE Ultrasonics, Ferroelectrics and Frequency Control, Vol. (UFFC-45), No. 2, 1997, pp. 528-534.
77. E. Bigler, S. Ballandras, Proc. IEEE 1996 International Symposium on Frequency Control, pp. 422-429.
78. J. Andle, M. G. Schweyer and J. F. Vetelino, Proc. IEEE 1996 International Symposium on Frequency Control, pp. 532-540.
79. J. C. Andle, J. F. Vetelino, Sensors and Actuators A, Vol. 44-3, 1994, pp. 167-176.
80. M. G. Schweyer, J. T. Weaver, J. C. Andle, D. J. McAllister, L. French and J. Vetelino, Proc. IEEE 1997 International Symposium on Frequency Control, pp. 147-155.
81. R. L. Baer, C. A. Flory, M. Tom-Moy and D. S. Solomon, Proc. IEEE 1992 Ultrasonics Symposium, pp. 293-298.

82. G. Scholl, F. Schmidt, T. Ostertag, L. Reindl, H. Scherr and U. Wolff, Proc. IEEE 1998 International Symposium on Frequency Control, pp. 595- 607.
83. G. Fischerauer, F. L. Dickert and R. Sikorski, Proc. IEEE 1998 International Symposium on Frequency Control, pp. 608-614.
84. E. I. Radeva and I. D. Avramov, invited talk presented at the Workshop on Environmental Sensing, June 11-12, 1999, Tsukuba, Japan.
85. E. I. Radeva and I. D. Avramov, Proc. IEEE 1998 Ultrasonics Symposium.
86. I. D. Avramov, M. Rapp, A. Voigt, U. Stahl and M. Dirschka, "Comparative Studies on Polymer Coated SAW and STW Resonators for Chemical Gas Sensor Applications", to be presented at the IEEE 2000 Innt. Frequency Control Symposium, Kansas City, USA.

SAW ANTENNA DUPLEXERS FOR MOBILE COMMUNICATION

MITSUTAKA HIKITA
Central Research Laboratory, Hitachi Ldt.,
Kokubunji-shi, Tokyo 185-8601, Japan

Mobile communications systems such as cellular radios have recently become very widespread and important to both business and personal users. A key component in the radio transceiver is an antenna duplexer, which makes it possible to use a single antenna to transmit and receive RF signals simultaneously. In this chapter, block diagrams of radio transceivers are shown and the frequency characteristics required for duplexers are discussed with regard to the system requirements. Procedures for designing duplexers using SAW-resonator-coupled filters and experimental results relevant to several systems are presented. Non-linear characteristics of the duplexers are also discussed.

1. Introduction

There are three main kinds of cellular radio systems: those using analog cellular radios, such as Advanced-Mobile-Phone Service (AMPS)[1], and Total-Access-Communication System (TACS)[2], those using time-division-multiple-access (TDMA) cellular radios, such as Global System for Mobile communications (GSM)[3] and Personal Digital Cellular (PDC), and those using code-division-multiple-access (CDMA) cellular radios, such as Personal-Communication System (PCS) based on IS-95[4,5], and wide-band (W) CDMA systems.

Table 1. Frequency allocation for cellular-radio systems operating at 0.8-1.0 GHz.

Systems	f_T (MHz)	f_R (MHz)	Remarks
EAMPS (US)	824-849	869-894	Analog, CDMA
NADC (US)	824-849	869-894	TDMA
ETACS (EU)	872-905	917-950	Analog
NMT (EU)	890-915	935-960	Analog
EGSM (EU)	880-915	925-960	TDMA
NTACS (JPN)	898-901	843-846	Analog
	915-925	860-870	
PDC (JPN)	898-901	843-846	TDMA
	915-925	860-870	
	925-940	870-885	
	940-948	810-818	
J-CDMA (JPN)	887-901	832-870	CDMA
	915-925	860-870	

Table 2. Frequency allocation for cellular-radio systems operating at 1.5-2.2 GHz.

Systems	f_T (GHz)	f_R (GHz)	Remarks
PCS (US)	1.85-1.91	1.93-1.96	TDMA, CDMA
DCS (EU)	1.71-1.785	1.805-1.88	TDMA
PDC (JPN)	1.429-1.453	1.477-1.501	TDMA
IMT-200 (EU, JPN)	1.92-1.98	2.11-2.17	CDMA

The frequencies allocated to cellular radio systems operating at 0.8-1.0 GHz and at 1.5-2.2 GHz are listed in Tables 1 and 2[6,7]. The duplexers used in analog and CDMA cellular radio systems will be discussed taking as examples those used in Extended-AMPS (EAMPS) and Extended-TACS (ETACS). In these systems, cellular phones transmit and receive radio frequency (RF) signals simultaneously. This requires high attenuation levels for the transmitter and receiver SAW filters at the mutual frequency bands[8,9].

The duplexers used in TDMA cellular radio systems, which also include dual-band cellular phones, will be discussed taking as examples those used in Extended-GSM (EGSM) and Digital Communications System (DCS). Because the transmitter-burst signals and the receiver-burst signals in TDMA systems do not overlap, the duplexer in those systems can use an antenna switch and a low-pass filter instead of transmitter SAW filters. A new switch-type SAW duplexer for dual-band cellar phones is also discussed.

Electronic Industries Association (EIA) standards for the blocking characteristics and the spurious emissions of EAMPS are illustrated schematically in Figs. 1(b) and 1(c), the frequencies allocated to that system are shown in Fig. 1(a)[1]. In Figs. 1(b) and 1(c), the numerical values are defined within a 30-kHz bandwidth. The minimum receiver sensitivity (MRS) and the typical power transmitted from the antenna (P_{Ant}) given in the standards are respectively −116 dBm and 28 dBm. Though the standards specify that the spurious emissions at f_R must be smaller than −80 dBm, if the receiver

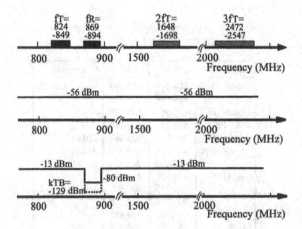

Fig. 1. System requirements for 800-MHz EAMPS. (a) Frequency allocation. (b) Blocking characteristics. (c) Spurious emissions.

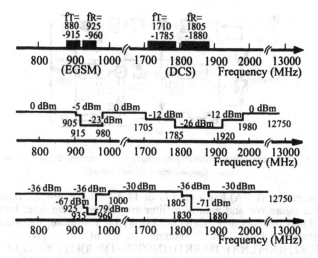

Fig. 2. System requirements for 800-MHz EGSM and 1.8-GHz DCS.
(a) Frequency allocation. (b) Blocking characteristics. (c) Spurious emissions.

sensitivity is not to be reduced, the spurious emissions must actually be smaller than the thermal noise $kTB = -174 + 10\log(30 \times 10^3)$ dBm, where k is the Boltzman constant $(1.384 \times 10^{-23}$ W·s/K), T is the absolute temperature, and B is the bandwidth (30 kHz). At 290 °K, the thermal noise is about −129 dBm, and this value is also shown by the broken line in Fig. 1(c).

The frequency allocations, blocking characteristics, and spurious emissions that GSM05.05 regulations[2] specify for EGSM and DCS are shown in Figs. 2. In Figs. 2(b) and 2(c), the values are defined within a 100-kHz bandwidth. The MRS and the $P_{Ant.}$ given in the regulations respectively are −102 dBm and 33 dBm.

When the cellular phones are receiving, they must protect themselves from the blocking-interference signals shown in Figs. 1(b) and 2(b). And when they are transmitting, the spurious emissions must be smaller than those shown in Figs. 1(c) and 2(c). The characteristics required of SAW duplexers are largely determined by the specifications illustrated in Figs. 1 and 2. This chaper shows block diagrams of transceivers, describes procedures for designing SAW-resonator-coupled filters and SAW antenna duplexers, shows examples of some types of SAW duplexers used in various systems, and discusses the non-linear distortion characteristics of SAW duplexers.

2. Block Diagrams and Frequency Characteristics

A block diagram of the RF-circuit configuration used in the kind of cellular phone used in analog and CDMA systems is shown in Fig. 3. In these systems a heterodyne-demodulation method is used in the receiver part of the phone and an orthogonal-modulation method is usually used in the transmitter part of the phone. As shown in the figure, the receiver part consists of a down-converter comprising a mixer, an intermediate-frequency (IF) filter, a de-modulator, etc. And the transmitter consists of a modulator, an up-converter comprising a mixer, a transmitter pre-filter (T2), etc.

As shown in Fig. 4(a), the receiver signal at f_R, the transmitter signal at f_T, the local signal at $f_{Lo} = f_R + f_{IF}$, the image signal at $f_{Im} = f_{Lo} - f_{IF}$, and the duplex-image signal at $f_{DIm} = 2f_T - f_R$ are very important in each section of the receiver parts. They determine the frequency characteristics of the duplexer from the antenna port to the receiver (Rx) port.

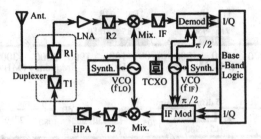

Fig. 3. RF-circuit configuration for conventional heterodyne-demodulation and orthogonal-modulation methods.

A low-noise amplifier (LNA) has a gain G of 14-17 dB and a noise figure NF of 1.2-2dB. The noise figure NF and gain of a filter are related to the filter loss Loss(R1) as follows: NF(R1)=Loss(R1) and G(R1)=1/Loss(R1). Therefore, the total noise figure of the receiver from the antenna terminal is given as follows: NF(total)=NF(R1)+(NF(LNA)–1)/G(R1)+(NF(R2)–1)/(G(R1) × G(LNA))+ · · · · =Loss(R1)×NF(LNA)+Loss(R1)×(Loss(R2)–1)/G(LNA)+· · · ·. We can see from this equation, that if G(LNA) is large enough, NF(total) - which has a large effect on the MRS of the receiver - is determined mainly by Loss(R1) and NF(LNA). An acceptable value for Loss(R1), i.e. the insertion loss at f_R for the duplexer from the antenna port to the Rx port, is 3.0-4.5 dB.

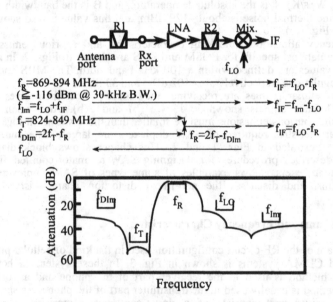

Fig. 4. Reciver part of EAMPS phones. (a) Signals at each section.
(b) Examples of frequency characteristics from the antenna port to the Rx port.

The large power at f_T from the transmitter to the receiver must be suppressed by 45-55 dB. Otherwise the noise figure is degraded because of the saturation characteristics of the LNA. Therefore, the attenuation level at f_T is also required to be larger than 45 dB for the duplexer from the antenna port to the Rx port. At the image frequency f_{Im} the attenuation level is required to be 35-40 dB. This is because the total attenuation of both the duplexer and the inter stage filter R2 must, as shown in Fig. 1(b), be larger than the difference between the blocking signal (–56 dBm) and the MRS (–116 dBm). As shown in Fig. 4(a), the spurious signal at the duplex-image frequency f_{DIm} is converted to the signal at $f_R=2f_T-f_{DIm}$ by mixing between the spurious signal and the large transmitter signal due to the 3rd-order non-linearity of the LNA. In general the sum of the attenuation at f_{DIm} and twice the attenuation at f_T is required to be larger than 110-120 dB.

According to the EIA standard[1], the leakage of the local signal from the antenna must be smaller than –47 dBm, whereas according to the ETACS standard[2] it must be smaller than –44 dBm. The attenuation at f_{LO}, however, is also related to both the attenuation level of R2 and the structure of the down-converter, e.g., whether it comprises a single-ended mixer or a double balanced mixer. In general, 15-20 dB attenuation level is required at this frequency for the duplexer. Examples of frequency characteristics required from the antenna port to the Rx port are shown in Fig. 4(b).

As shown in Fig. 5(a), the transmitter signals at f_T, the noise at f_R and the spurious signals at $f_{SP}=2f_T-f_R$ in each section of the transmitter parts are very important in determining the frequency characteristics of the duplexer from the transmitter (Tx) port to the antenna port. A general high-power amplifier (HPA) has a gain of 20-25 dB and a noise figure of 5-10 dB. The noise at f_R generated from the modulator, the up-converter, the driver amplifier, etc. is suppressed to almost thermal-noise level kTB by the transmitter pre-filter T2. In the HPA the noise is increased by G+NF, i.e. 30-35 dB. As shown in Fig. 1(c), this must again be suppressed to less than the thermal-noise level at the antenna terminal. Thus, the attenuation level required at f_R for the duplexer from the Tx port to the antenna port is 35-40 dB.

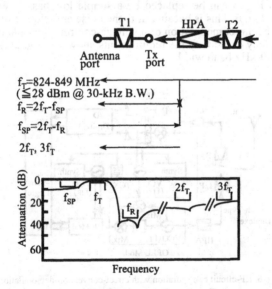

Fig. 5. Transmitter part of EAMPS phones. (a) Signals at each section.
(b) Examples of frequency characteristics from the Tx port to the antenna port.

Another attenuation at $f_{SP}=2f_T-f_R$ is necessary to suppress the spurious signal from the antenna, which as shown in Fig. 5(a) is converted to the signal at $f_R=2f_T-f_{SP}$ because of the 3rd-order non-linearity of the HPA. Thus, the attenuation level required at this frequency depends on the non-linear characteristics of the HPA. In general, the sum of the attenuation levels at f_R and f_{SP} is required to be 40-45 dB. Examples of the frequency characteristics required from the Tx port to the antenna port are shown in Fig. 5(b). The insertion loss at f_T must be small, i.e., 1.5-2.5 dB, if $P_{Ant.}$ is to be high. The attenuation levels at the 2nd- and 3rd-harmonic frequencies must be between 15 and 25 dB. This is because HPA's generally assure that the power of harmonic emissions is 30 dBc below the carrier power. This requires that the attenuation levels for duplexers to be more than 15 dB if the spurious-emission requirement shown in Fig. 1(c) (less than −13 dBm) is to be satisfied. For ETACS the requirement is less than -24 dBm^2.

A block diagram for the kinds of cellular phones used in the TDMA systems, (e.g., EGSM and DCS) is shown in Fig. 6. Many new circuit technologies for these systems have been investigated. The direct-conversion (DC) demodulation method is one of the new techniques making the receiver circuit configurations simpler than those using the conventional heterodyne method[10]. The offset-phase-lock-loop (OPLL) modulation method is another new technique that can make transmitter circuit configurations simpler than those using a conventional orthogonal modulation method[11]. Figure 6 shows the newest circuit configuration for EGSM or DCS, which uses the DC-demodulation and OPLL-modulation methods.

In the case of DC demodulation, R2 is not necessary because there is no image frequency. The receiver-top filter R1 must suppress the blocking-interference signals given in Fig. 2(b) in order to prevent saturation of the LNA and the mixer. Therefore, the required attenuation levels from the antenna port to the Rx port are mainly determined by the blocking-signal levels. The acceptable value for the insertion loss is also 3.0-4.5 dB.

In the case of OPLL modulation, the transmitter pre-filter T2 and the final-stage filter T1 shown in Fig. 3 can be replaced by a simple low-pass filter and an antenna switch as shown in Fig. 6. This is because at f_R the noise emissions from the antenna are small, because of the direct modulation of a voltage-controlled oscillator (VCO) with an OPLL circuit, and they can satisfy the requirements shown in Fig. 2(c): −79 dBm and −71 dBm within 100-kHz band width.

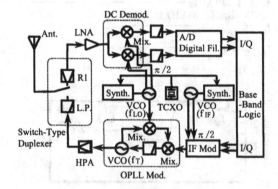

Fig. 6. RF-circuit configuration with direct-conversion-demodulation and offset-phase-lock-loop-modulation methods.

Examples of the required frequency characteristics from the antenna port to the Rx port for a switch-type duplexer used in EGSM or DCS are shown in Fig. 7. And corresponding examples of those from the Tx port to the antenna port are shown in Fig.

8. From the Tx port to the antenna port, the insertion loss must be 1.0-1.5 dB and the attenuation levels must be more than 30 dB at the harmonic frequencies if the spurious-emission requirements shown Fig. 2(c) are to be satisfied.

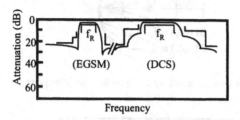

Fig. 7. Examples of frequency characteristics from the antenna port to the Rx port for EGSM and DCS.

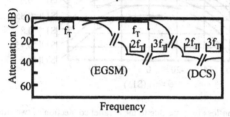

Fig. 8. Examples of frequency characteristics from the Tx port to the antenna port for EGSM and DCS.

3. Basic Configurations of SAW Antenna Duplexers

A duplexer configuration, that provides two functions, filtering and diverging, is shown in Fig. 9[8]. Two SAW filters - a receiver filter R1 and a transmitter filter T1 - are connected in parallel at the antenna port. Because of this connection, SAW filters require not only certain frequency characteristics but also a fixed antenna-side impedance. As shown in Fig. 9, $S11(Z_R)$ and $S11(Z_T)$ are defined as the S-parameters of R1 and T1 looking from the antenna side. And as shown in Fig. 10, $S11(Z_R)$ and $S11(Z_T)$ must be nearly 0, (i.e., $Z_R \doteq Z_T \doteq 50\,\Omega$) at the pass bands of the filters. Moreover, at the mutual frequency bands - that is at f_T for $S11(Z_R)$ and at f_R for $S11(Z_T)$ - they must become approximately 1+j0, (i.e., $Z_R \doteq Z_T \doteq \infty$). This is necessary if the insertion-loss increase caused by the mutual interaction between Z_R and Z_T when R1 and T1 are connected in parallel is to be made as small as possible.

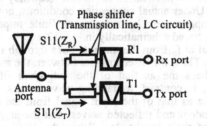

Fig. 9. Basic configuration of an antenna duplexer.

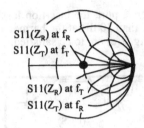

Fig. 10. Impedance looking from parallel-connection point.

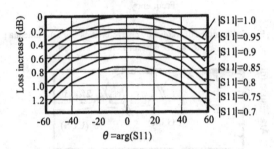

Fig. 11. Insertion-loss increase due to the parallel connection of two filters.

Increasing losses due to the parallel connection of the two filters are shown in Fig. 11. The increased loss values for one filter are shown, taking as parameter the magnitude of the S11 for the other filter, as a function of the angle of the S11 for the other filter. If the loss increase due to the parallel connection of the two filters is to be kept below 0.5 dB, the magnitude of the S11 of the filters must be larger than 0.8 - 0.95 and the angle of the filters must be less than $\pm 20°$ - $\pm 60°$ respectively, at mutual frequencies. Therefore, not only frequency characteristics but also impedance characteristics must be designed when SAW filters are used in the duplexer. In general, SAW filters are designed to have the S11's that have large magnitudes at mutual frequencies. And phase shifters (i.e., transmission lines or LC circuits) are arranged between the filters and the connection point (i.e. the antenna port) to optimize the angles of the S11's.

The attenuation at f_T from the antenna port to the Rx port becomes larger for the duplexer than for individual filter R1. At f_R this is also true for the attenuation from the Tx port to the antenna port. These phenomena are also caused by the mutual interaction between Z_R and Z_T. In an ideal-connected condition (i.e., Z_R and Z_T are infinite at f_T and f_R) a 6-dB improvement of the attenuation in mutual frequency bands can be expected after parallel connection. Under actual connection conditions, however, 4.5-5.5 dB is a reasonable attenuation improvement because of the finite impedance of Z_R and Z_T. These phenomena are illustrated schematically in Fig. 12.

In Fig. 12(a) the signal at f_R from the antenna goes through to R1 via a connection point without attenuation. Thus, the insertion loss between the antenna port and the Rx port is fundamentally the same as that of the individual filter R1. Between the connection point and T1, in contrast, a standing wave occurs. The amplitude of the standing wave is the same as that of the input signal from the antenna. The standing wave is composed of incident and reflected waves, and the amplitude of the standing wave is given by sum of their amplitudes. When the incident wave is completely reflected from T1, the amplitude of the signal incident to T1 is equal to half that of the signal from the antenna. Therefore, at f_R the attenuation between the antenna port and the Tx port is 6 dB greater than that of the individual filter T1.

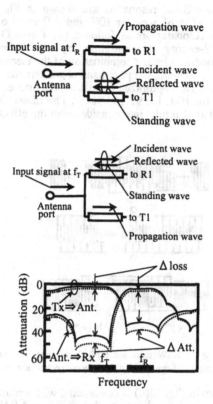

Fig. 12. Comparison between an individual filter and parallel-connected filters.
(a) Signal behaviors when the signal from the antenna port is at f_R.
(b) Signal behaviors when the signal from the antenna port is at f_T.
(c) Frequency-characteristic changes due to parallel connections.

When the input signal from the antenna is at f_T, it goes through to T1 via a connection point without attenuation and, as shown in Fig. 12(b), a standing wave occurs between a connection point and R1. Therefore, at f_T the attenuation between the antenna port and the Rx port is 6 dB greater than that of the individual filter R1.

The frequency characteristics of the duplexer are illustrated schematically in Fig. 12(c), where they are compared with those of the individual filters R1 and T1. Insertion losses at the pass bands are also increased a little because of the parallel connection between R1 and T1, but the attenuation in the mutual bands is improved by 4.5-5.5 dB.

4. SAW-Resonator-Coupled Filter

Filters constructed with SAW resonators (e.g., ladder-type filters) are usually used in SAW duplexers. SAW-resonator configurations and filter circuits using SAW resonators are described in this section.

4.1 SAW Resonators

Examples of well-known SAW resonators are shown in Fig. 13[12-15]. The resonator shown in Fig. 13(a) consists of a center IDT and reflectors on both sides of it. The features of this type of resonator are as follows: (1) a high Q value can be obtained because of the SAW-energy concentration by the reflectors; (2) impedance characteristics can be modified by the combination of the electrode pitches between the IDT and the reflectors. In general, the pass-band characteristics of the filter constructed with this type of resonators are optimized by making the pitches of the reflectors slifgtly different from those of the IDT, i.e. 97-103 %[16,17]. The filters should be designed using computer simulations that also take into consideration the effect of the thickness of the Al electrode.

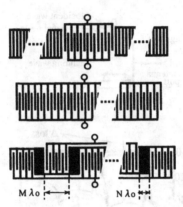

Fig. 13. SAW resonators. (a) Reflector-type resonator.
(b) Multi-finger IDT resonator. (c) Serially connected multi-IDT resonator.

The resonator shown in Fig. 13(b) is constructed with a multi-finger IDT. Backward SAW's with almost the same amplitude as that of forward SAW's can exist within an IDT with many finger pairs. This is because in a periodic structure, forward- and backward-travelling SAW's interact through internal reflections. Despite the rather simple configuration of this resonator, it provides wide band resonant-impedance characteristics as well as high-power-handling capabilities[12].

The resonator shown in Fig. 13(c) is constructed with serially connected multi-IDT's. The features of this type of resonator are as follows: (1) the aperture of the resonator can be widened by a factor of 4 without changing the impedance, which enables it to handle more power than can be handled by the resonator shown in Fig. 13(b)[13], and (2) the capacitance ratio defined by γ =fr/(2(fa-fr)) can be changed by choosing M and N in Fig. 13(c) properly. The new γ' is given approximately by the relation $1/\gamma$'=$(1/\gamma)\cdot(M/(N+M))$, where γ is the capacitance ratio of the basic configuration shown in Fig. 13(b). The latter feature is very convenient for achieving sharp-cutoff frequency characteristics, which require resonators with rather large γ's, (i.e., smaller bandwidths between the resonant frequency fr and the anti-resonant frequency fa)[18].

The SAW resonators differ from conventional resonators like the crystal resonator, the dielectric electromagnetic resonator, etc. The most remarkable difference is the bulk-wave excitation from the IDT into the substrate, which produces resistive components in the impedance characteristics. Examples of the results of fundamental experiments and computer simulations confirming bulk-wave radiation phenomena are shown in Fig. 14 with Smith charts. Using Fig. 14(a)'s multi-finger-IDT resonator, the impedance characteristics for 36° YX-LiTaO$_3$ [19], 64° YX-LiNbO$_3$ and 41° YX-LiNbO$_3$[20, 21] substrates are shown in Fig. 14(b), (c) and (d), respectively[22]. The resistive components produced by bulk-wave radiation may degrade the magnitudes of S11's for the filters, which as shown in Fig. 11 will increase the insertion losses of the parallel-connected filters because of the mutual interaction. The influences of the bulk-wave radiation should therefore be avoided by choosing fr and fa appropriately. This is because fr and fa strongly depend on the thickness of the Al-electrode fingers, but fb - from which frequency the bulk-wave radiation occurs - depends on the substrate intrinsically and is independent of the thickness of the Al-electrode fingers.

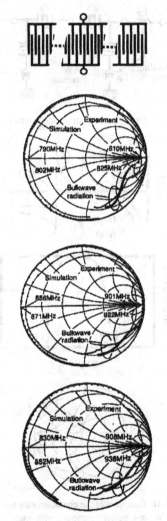

Fig. 14. Impedance characteristics of SAW resonators.
(a) Resonator pattern used in the experiment.
(b) 36° YX-LiTaO$_3$. (c) 64° YX-LiNbO$_3$. (d) 41° YX-LiNbO$_3$.

4.2 Ladder-Type SAW Filters

Ladder-type filters[23-26] constructed with SAW resonators are usually used as the R1 and T1 filters in SAW duplexers. An example of the equivalent circuit for a simple ladder-type filter configuration is shown in Fig. 15. As the influences of bonding-wire inductors, capacitors against earth, floating capacitors between electrode patterns, etc. must be considered, only the bonding-wire inductors are included in the equivalent circuit.

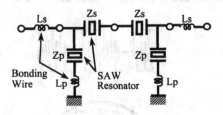

Fig. 15. Example of a simple ladder-type configuration.

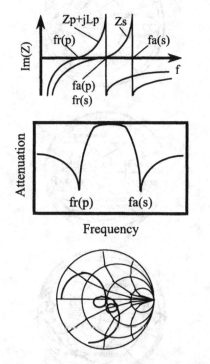

Fig. 16. Examples of the characteristics and impedance of ladder-type configurations with internal matching.
(a) Impedance characteristics for each resonator.
(b) Frequency characteristics. (c) Input impedance.

As shown in Fig. 16(a), if an impedance for the parallel-arm element (i.e., $Z_p + j\omega L_p$) has an anti-resonant frequency $fa(p)$ nearly equal to a resonant frequency $fr(s)$ of an impedance for the series-arm element (i.e., Z_s), frequency characteristics with a pass band near $fa(p) \doteqdot fr(s)$ can be obtained. The stop bands are at frequencies lower than $fr(p)$ and higher than $fa(s)$. The frequency characteristics shown in Fig. 16(b) are obtained without any outer matching circuits[16, 17, 24]. An example of the typical internal matching is shown in Fig. 16(c) with a Smith chart.

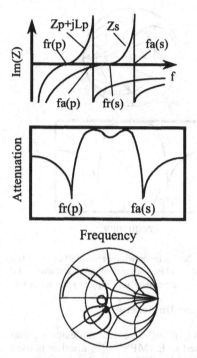

Fig. 17. Examples of the characteristics and impedance of ladder-type configuration using resonators with a smaller band width.
(a) Impedance characteristics for each resonator.
(b) Frequency characteristics. (c) Input impedance.

To achieve frequency characteristics with sharper cutoffs, resonators with a smaller bandwidth between fr and fa - for example, the one shown in Fig.13(c) - should be used[18]. As shown in Fig. 17(a), $fa(p)$ is lower than $fr(s)$ in this case. The frequency characteristics shown in Fig. 17(b) have within the pass band large ripples caused, as can be well understood from the Smith chart in Fig. 17(c) by the impedance mismatching. If external matching circuits are arranged at the input and output terminals shown in Fig. 18(a), 50-Ω impedance matching as well as very sharp-cutoff frequency characteristics can be achieved. This is shown in Figs. 18(b) and 18(c).

In general, different frequency characteristics are required for SAW filters depending on the kinds of cellular-radio systems, such as EAMPS, ETACS, EGSM, DCS and so on. Designers therefore must first determine what kinds of resonators are fit for their purposes, and whether or not matching circuits should be used.

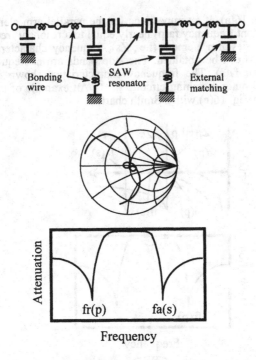

Fig. 18. Configuration with sharp-cutoff characteristics.
(a) External matching for the configuration shown in Fig. 17.
(b) Input impedance. (c) Frequency characteristcs.

5. Examples of SAW Antenna Duplexers

Some examples of duplexers that will be useful in designing duplexers for other systems are given here[27-40]. One is used in EAMPS[34, 37-39], another is used in ETACS[28, 32], and the last is used in EGSM/DCS dual-band phones[40]. As shown in Table 1, EAMPS has a 25-MHz bandwidth for each of the transmitter and receiver frequency bands and has a 20-MHz spacing between the bands. An example of the miniature duplexer, which consists of ladder-type filters for this system will be described. ETACS, on the other hand, has a 33-MHz bandwidth for each of the transmitter and receiver bands and only a 12-MHz spacing between them. The technologies used to obtain frequency characteristics with very sharp cutoffs for this system will be described. As shown in Tables 1 and 2, EGSM and DCS are serviced in 800 MHz and 1.8 GHz, respectively. Therefore, a new dual-band duplexer which consists of not only SAW filters but also switching diodes will also be described.

5.1 Antenna Duplexer for EAMPS Cellular Phones

Examples of equivalent circuits of the typical ladder-type SAW filters used as R1 and T1 in the EAMPS duplexer are shown in Fig. 19(a) and (b), respectively[37-39]. SAW resonators with the reflectors on both sides, as shown in Fig. 13(a), are used in these filters. Bonding-wire inductors (e.g., 0-3 nH) and the equivalent inductors produced by the package-ground lines (e.g., 0-0.5 nH) are also included in these figures[34]. The bonding-wire inductor provides about 1 nH per 1 millimeter wire length.

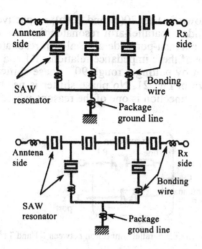

Fig. 19. Examples of SAW filters for EAMPS duplexer[37, 38].
(a) R1 filter. (b) T1 filter.

The number of SAW resonators is determined by the frequency characteristics of the required duplexer (i.e., insertion losses, attenuation levels, etc). Many forms of ladder-type filters can be used, but the configurations shown in Fig. 19 provide two remarkable features. One is that the influences of the bonding wires and the package-ground lines are positively used in designing the frequency characteristics[34, 37, 38]. Wide band flat-attenuation levels can be achieved by optimizing these parasitic inductors. Moreover, the isolation characteristics between the Rx port and the Tx port in a duplexer can also be improved by making small changes in the inductors caused by the Rx- and Tx-common package-ground lines[37, 38].

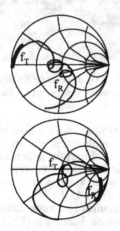

Fig. 20. Input impedance of Fig. 19's filters looking from the antenna side.
(a) R1 filter. (b) T1 filter.

The other feature is the impedance characteristics of the R1 and T1 filters. An impedance of R1 looking from the antenna port becomes low at f_T and that of T1 becomes high at f_R, as shown in Fig. 20(a) and 20(b), respectively. This is because R1 has on the antenna-port side a parallel-arm resonator that has the resonant frequency near f_T and T1 has on the antenna-port side a serial-arm resonator that has the resonant frequency near f_R. Because of these impedance relations, R1 and T1 can be connected in parallel at the antenna port by putting a rough 90° phase-shifter only between R1 and the antenna port, as shown in Fig. 21. No phase shifter is needed between T1 and the antenna port. This simple connection is one of the reasons why the duplexer module can be small[37, 38].

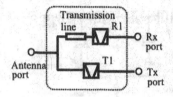

Fig. 21. Simple parallel connection between R1 and T1[37, 38].

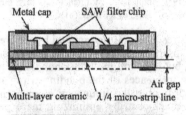

Fig. 22. SAW duplexer for EAMPS[37, 38].
(a) Schematic view of ceramic package.
(b) Photograph of the duplexer ($9.5 \times 7.5 \times 2$ mm^3).

As shown in Fig. 22(a), a duplexer with a size of $9.5 \times 7.5 \times 2.0$ mm^3 has beeen achieved. A photograph of the duplexer is shown in Fig. 22(b) and its frequency characteristics are shown in Fig. 23. From the antenna port to the Rx port, the insertion loss at f_R (869-894 MHz) is about 3.6 dB and the rejection level at f_T (824-849 MHz) is about 53 dB. From the Tx port to the antenna port, the insertion loss at f_T is about 2.0 dB and the rejection level at f_R is about 45 dB[37, 38].

This type of SAW duplexer requires two more factors to be taken into consideration. One is protection against electrostatic discharges (ESD's). ESD's with potentials of 10-20 kV may be applied to an antenna terminal. As shown in Fig. 21, the antenna port of the duplexer is electrically floating, which leaves SAW filters to be destroyed by ESD's. Therefore, external circuits including a serial-arm DC-blocking capacitor and a parallel-arm inductor to earth are necessary between the antenna terminal and the antenna port of the duplexer.

The other factor is the suppression at the 2nd- and 3rd-harmonic frequencies. In general, ladder-type SAW filters can not achieve high-rejection levels at the off bands. The general HPA's have spurious emissions at the 2nd- and 3rd-harmonic frequencies that are about 30 dB lower than the carrier power. Therefore, according to the requirements shown in Fig. 1(c), attenuation of 15 dB or more at the 2nd- and 3rd-harmonic frequencies is required. An external low-pass filter consisting of serial-arm inductors and parallel-arm capacitors is also put between the transmitter and the Tx port of the duplexer.

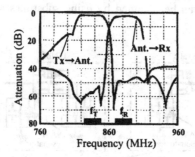

Fig. 23. Frequency characteristics of the EAMPS duplexer[37, 38].

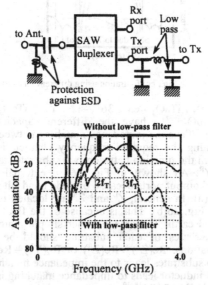

Fig. 24. SAW duplexer together with circuits for ESD protection and harmonic suppression[37, 38]. (a) Equivalent circuit. (b) Off-band frequency characteristics from the Tx port to the antenna port.

An example of an equivalent circuit including an ESD-protection circuit as well as a low-pass filter is shown in Fig. 24. Examples of the wide band frequency characteristics from the transmitter to the antenna terminal are also shown in Fig. 24 together with the frequency characteristics obtained without a low-pass filter[37, 38].

5.2 Antenna Duplexer for ETACS Cellular Phones

Both wide band characteristics and frequency characteristics with very sharp cutoffs are required in ETACS because the bandwidths for the transmitter and receiver are each 33 MHz and they are only 12 MHz apart. These characteristics can be achieved by a combination of ladder-type filters formed on different kinds of substrates. This is because, as shown in Fig. 25, sharp-cutoff characteristics can be obtained by using substrates such as $LiTaO_3$ with not so large k^2's but good TCD's. And wide band suppression characteristics can be obtained by using substrates such as $LiNbO_3$, with large k^2's[18, 30].

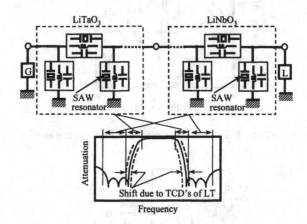

Fig. 25. Ladder-type filter configuration using different kinds of substrates.

As shown in Fig. 26(a), $LiTaO_3$ - e.g., 36° -42° YX-$LiTaO_3$[19, 41, 42] - and $LiNbO_3$ - e.g., 64° /41° YX-$LiNbO_3$[20-22] - have the different capacitance Co between the electrodes and different k^2. Thus the impedance matching between the two substrates is very important in obtaining the required frequency characteristics. The impedances looking left and right from the junction between two substrates are respectively defined by $Z^L=1/Y^L$ and $Z^R=1/Y^R$. As shown in Fig. 26(a), external matching circuits are necessary at the input and output ports because of wide band filter configuration, which is also shown in Fig. 18(a). In general, Z^L and Z^R are located under the X-axis of the Smith chart as shown in Fig. 26(b). This is because of the floating capacitors between the resonator patterns and earth. To achieve rather simple matching between the two substrates, filter characteristics of each substrate should be designed having the following impedance relations: $Re(Z^L) \doteqdot Re(Z^R)$ or $Re(Y^L) \doteqdot Re(Y^R)$. A simple serial inductor between the two substrates leads to the impedance matching in the former case, whereas a simple parallel inductor leads to impedance matching in the latter case. These inductors are also shown in Fig. 26(b)[18, 30].

Examples of equivalent circuits for SAW filters used as R1 and T1 in the ETACS duplexer are shown in Fig. 27(a) and (b), respectively. Each filter is constructed with 36° YX-$LiTaO_3$ and 64° YX-$LiNbO_3$ substrates. R1 is constructed with a ladder-type configuration similar to that used in the EAMPS duplexer. Sharp-cutoff frequency characteristics near the pass band are obtained by forming the type of resonators shown in Fig. 13(c) on a 36° YX-$LiTaO_3$ substrate. Resonators of the type shown in Fig. 13(b) are formed on a 64° YX-$LiNbO_3$ substrate in order to obtain wide band suppression characteristics. The relation $Re(Z^L) \doteqdot Re(Z^R)$ can be also achieved, which leads to the impedance matching by introducing a simple serial inductor between the two substrates as shown in Fig. 27(a)[30].

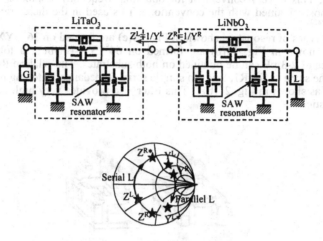

Fig. 26. Impedance matching between two substrates.
(a) Impedance looking from the junction point.
(b) Impedance relations for the simple serial- and parallel-inductor matching.

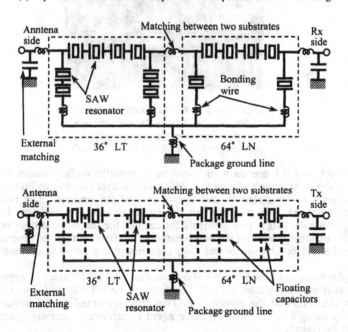

Fig. 27. Examples of SAW filters for the ETACS duplexer. (a) R1 filter. (b) T1 filter.

The filter T1, in contrast, has a new ladder-type configuration using as the parallel-arm elements capacitors between the resonator patterns and earth. This configuration makes it possible for the pass band of the filters to be near the resonant frequencies of the resonators, and for the stop bands to be near the anti-resonant frequencies of the resonators. This is very convenient for obtaining frequency responses that are almost same as those obtained with the conventional T1's used in the dielectric-resonator-type duplexers[28, 30].

As the types of resonators shown in Fig. 13(c) are formed on 36° YX-LiTaO$_3$ while the types shown in Fig. 13(b) are formed on 64° YX-LiNbO$_3$, a total parallel-arm capacitance of 2-3 pF can be obtained on both substrates. The relation $Re(Z^L) \doteqdot Re(Z^R)$, which is the same as for R1, leads to simple serial-impedance matching between the two substrates as shown in Fig. 27(b)[30]. This filter also provides very high-power durability characteristics.

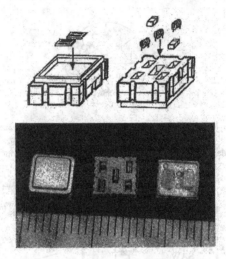

Fig. 28. ETACS SAW filter mounted on a package.
(a) Schematic view of a multi-layer ceramic package
(left: upper side, right: reverse side).
(b) Photograph of the filter (6.3 × 6.3 × 2 mm^3).

Filters R1 and T1 are each mounted on a specific surface-mount package. The package is shown schematically in Fig. 28(a). Multi-layer ceramic technology is used to form five pocket-type holes on the reverse side of the package. The two substrates, LiTaO$_3$ and LiNbO$_3$, are mounted on the upper side of the package. Five external matching elements, that is two elements for each of the input and output ports and one element for matching between the two substrates, are fixed within the reverse-side holes. Miniature helical or chip coils and chip condensers can be used as matching-circuit elements.

The photographs of the filter are shown in Fig. 28(b). The external appearance after hermetic sealing is shown on the left. The size is 6.3 × 6.3 × 2 mm^3. The reverse side is shown in the center of the photograph. Two chips mounted on the upper side of the package by using a conductive adhesive agent to achieve capacitors against earth are shown on the right[18].

Figure 29 shows frequency characteristics of the ETACS duplexer formed directly on the mother board of the radio transceiver using R1 and T1 filters. The insertion loss from the antenna port to the Rx port is about 4.5 dB at f_R (917-950 MHz) and the rejection levels at f_T (872-905 MHz) are higher than 49 dB. The insertion loss from the Tx port to the antenna port is about 2.8 dB at fT and the rejection levels at f_R are higher than 36 dB. These characteristics are shown in Fig. 29(a)[30].

As shown in Fig. 27(b), T1 has at the antenna port a parallel-arm inductor that provides protection against ESD's without the need for external protection circuits. Moreover, a serial-inductor between the two substrates, also shown in Fig. 27(b), make the attenuation at the 2nd- and 3rd-harmonic frequencies large without need for an external low-pass filter between the transmitter and the Tx port of the duplexer. These characteristics, which provide levels higher than 30 dB at harmonic frequencies, are shown in Fig. 29(b).

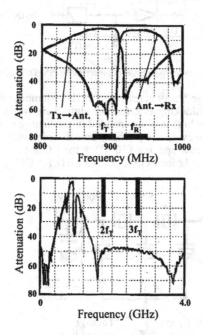

Fig. 29. Frequency characteristics of the ETACS duplexer. (a) Pass-band characteristics. (b) Off-band characteristics from the Tx port to the antenna port.

5.3 Switch-Type Antenna Duplexer for EGSM/DCS Dual-Band Cellular Phones

A block diagram of the switch-type duplexer for 800-MHz EGSM and 1.8-GHz DCS dual-band cellular phones[3] is shown in Fig. 30. And as shown in Fig. 6, the direct-conversion (DC) demodulation method[10] for the receiver and the offset-phase-lock (OPLL) modulation method[11] for the transmitter are assumed. The dulpexer not only switches the transmitter and receiver signals but also separates 800-MHz and 1.8-GHz signals. As shown in Fig. 30, two SAW filters are used in the EGSM and DCS receiver parts. Two diode switches are also used in the EGSM and DCS transmitter parts.

Recently, RF circuits and intermediate frequency (IF) circuits are largely integrated, which results in one or two IC's. Therefore, the effects of common-mode noise caused by the IC substrate, the source lines, ect. are reduced by driving almost all circuit elements within the IC's differentially. Therefore, Rx ports of the duplexer are also required to differently operate. Passive balun circuits are therefore put between the R1 filters and the Rx ports in the duplexer (Fig. 30).

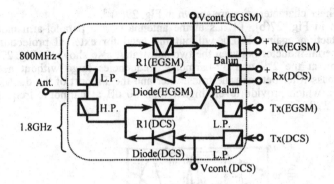

Fig. 30. Block diagram of a switch-type SAW duplexer for EGSM/DCS dual-band phones.

Because there are no image frequencies when the DC-demodulation method is used, R1's must suppress only the blocking-interference signals shown in Fig. 2(b). Thus, the rejection levels need not be so high as those needed in the analog cellular radios, such as EAMPS, ETACS, etc. However, very sharp-cutoff frequency characteristics are necessary because of the narrow spacing (10 MHz for EGSM and 20 MHz for DCS) and the large bandwidth (35 MHz for EGSM and 75 MHz for DCS).

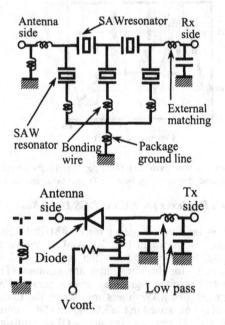

Fig. 31. EGSM parts of the dual-band duplexer shown in Fig. 30.
(a) R1 filter. (b) Switching circuit with a low-pass filter.

An example of the equivalent circuit for the R1 filter used in the EGSM parts of the dual-band duplexer is shown in Fig. 31(a). A diode-switching circuit with the low-pass filter providing attenuation at the 2nd- and 3rd-harmonic frequencies is shown in Fig. 31(b). Sharp-cutoff frequency characteristics are obtained by using the type of resonator shown in Fig. 13(c), which is the same as that used in ETACS. The configurations used in the DCS part are also almost the same as those used in ETACS[40].

One of the most significant features of the duplexer shown in Fig. 30 is that a SAW filter and a switching circuit are directly connected in parallel, which makes it possible to reduce the number of switching diodes and to construct a simple and small duplexer module. An insertion-loss increase due to the mutual interaction between the SAW filter and the switching circuit is the most serious problem for the direct-parallel connection. Therefore, similar design procedures are necessary to make the impedances of the SAW filters and the switching circuits as large as possible at mutual frequencies[30].

		EGSM		DCS	
		Rx	Tx	Rx	Tx
Vc (EGSM)	0V	on	off	—	—
	2.5V (5mA)	off	on	—	—
Vc (DCS)	0V	—	—	on	off
	2.5V (5mA)	—	—	off	on

Fig. 32. Control-voltage combinations.

Combinations of the switching-control voltages for the dual-band duplexer are summed up in Fig. 32, where the voltage values have no meaning but the current values of about 4-5 mA are important. These current values ensure the complete ON-state of the diode.

Frequency characteristics from the antenna port to the EGSM Rx port are shown in Fig. 33(a) and 33(b). The curves are obtained in the condition that the antenna port is loaded by 50-Ω single-ended input impedance and the Rx port is loaded by 100-Ω balanced out impedance. The insertion loss is about 3.0 dB, and the frequency characteristics show sharp cutoffs. Frequency characteristics from the EGSM Tx port to the antenna port are shown in Fig. 34(a) and 34(b). These curves are obtained when both the antenna port and the Tx port are loaded by 50-Ω single-ended impedances. The insertion loss is as small as 1.0 dB, and the suppression levels at the 2nd- and 3rd-harmonic frequencies are over 30 dB.

Frequency characteristics from the antenna port to the DCS Rx port are shown in Fig. 35 (a) and 35(b). Here again the antenna port has 50-Ω single-ended input impedance and the Rx port has 100-Ω balanced output impedance. The insertion loss is about 3.3 dB, and suppression levels sufficient for the blocking-interference signals given in Fig. 2(b) have been obtained. Frequency characteristics from the DCS Tx port to the antenna port are shown in Fig. 36(a) and 36(b). These are the curves obtained when both the antenna port and the Tx port have 50-Ω single-ended impedances. The insertion loss is as small as 1.2 dB, and the suppression levels at the 2nd- and 3rd-harmonic frequencies are also over 30 dB.

A photograph of the switch-type SAW duplexer for dual-band cellular phones is shown in Fig. 37. The duplexer measures $10 \times 8 \times 2$ mm^3, and the Rx output ports for both the EGSM and DCS parts are of the balanced type, which makes it possible to connect this duplexer directly to an IC using the DC-demodulation method.

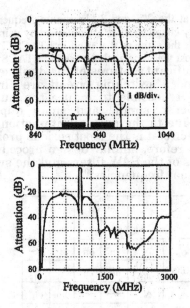

Fig. 33. Frequency characteristics from the antenna port to the EGSM Rx port (differential output). (a) Pass-band characteristics. (b) Off-band characteristics.

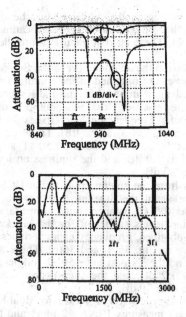

Fig. 34. Frequency characteristics from the EGSM Tx port to the antenna port. (a) Pass-band characteristics. (b) Off-band characteristics.

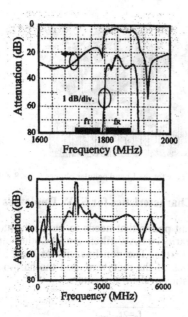

Fig. 35. Frequency characteristics from the antenna port to the DCS Rx port (differential output). (a) Pass-band characteristics. (b) Off-band characteristics.

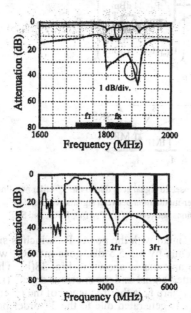

Fig. 36. Frequency characteristics from the DCS Tx port to the antenna port. (a) Pass-band characteristics. (b) Off-band characteristics.

Fig. 37. SAW duplexer for dual-band cellular phones ($10 \times 8 \times 2$ mm^3).

6. Non-Linear Distortion Characteristics for SAW Duplexers

One of the other very important points that needs to be taken into account when using SAW duplexers is the non-linearity of SAW filters and the other circuit elements such as the pin diodes within the duplexers. The non-linearity of these elements degrades the spurious-response characteristics of the receivers, and the harmonic emissions and the power-compression characteristics of the transmitters.

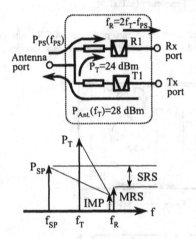

Fig. 38. Generation of pseudo signal at f_R due to the 3rd-order non-linearity of R1.
(a) Inter-modulation between signals at f_T and f_{SP}.
(b) Relations between $P_T(f_T)$, $P_{SP}(f_{SP})$, MRS, and SRS.

As shown in Fig. 38(a), a signal with the power P_T, which is smaller than the power $P_{Ant.}$ transmitted from the antenna by 4.5-5.5 dB, goes into R1, where it is mixed with the spurious signals from the antenna as shown in Fig. 38(b). The difference between the frequencies of the transmitted signal and the spurious signal, (i.e., f_T and f_{SP}) is given by the relation of $f_T-f_{SP}=f_R-f_T$, where f_R is the frequency of the received signal. Therefore, the pseudo signal at $f_R=2f_T-f_{SP}$ is generated due to the 3rd-order non-linearity of the R1 filter, which is one of the inter-modulations between the SAW's, a very similar effect to that observed in amplifiers and mixers. This pseudo signal at f_R degrades the sensitivity of the receivers.

The 3rd-order non-linear characteristics of a conventional IIDT-type filter constructed with image-impedance connected IDT's[28] and of a ladder-type filter are compared in Fig. 39(a) and 39(b)[18]. Within the former filter, the input electrical energy is immediately converted to SAW's as schematically illustrated in Fig. 40(a). Within the latter filter, however, only a small fraction of the input electrical energy is converted to SAW's as shown in Fig. 40(b). This is because the excitation direction of the SAW's is perpendicular to the electrical energy flow within the filter.

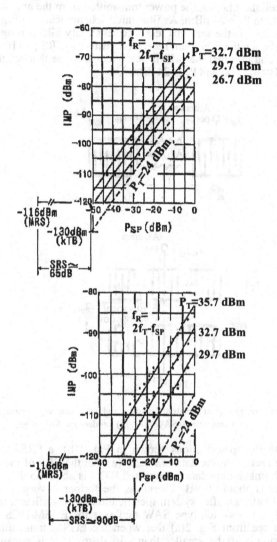

Fig. 39. Experimental results evaluating IMP due to the 3rd-order non-linearity.
(a) Conventional IIDT-type filter with image-impedance connected IDT's.
(b) R1 with a ladder-type configuration.

The magnitude of the inter-modulation product IMP, shown in Fig. 39 is roughly proportional to the product of P_T squared, the spurious-signal power P_{SP} and the interaction length between the input and output IDTs. In the conventional filter shown in Fig. 40(a), the interaction length is given by L. Thus, IMP=$C \cdot P_T^2 \cdot P_{SP} \cdot L$ (C: appropriate constant). In the ladder-type filter shown in Fig. 40(b), the electrical energy coupled to SAW's is very small. Moreover, there are no output IDTs in one-port resonators, so the interaction length L becomes ideally small.

In Fig. 39(a) and 39(b), the IMP's are plotted against the P_{SP} with P_T as parameter. In the EAMPS cellular phones, the power transmitted from the antenna $P_{Ant.}$ is about 28 dBm, which leads to $P_T \doteqdot 24$ dBm. As illustrated schematically in Fig. 38(b) and shown quantitatively in Fig. 39, the spurious response sensitivity SRS is roughly determined by the difference between the minimum receiver sensitivity MRS (–116 dBm for EAMPS) and the P_{SP}, on condition that the IMP is nearly equal to the thermal noise kTB (about -130 dBm when B=30 kHz) and P_T is 24 dBm[18].

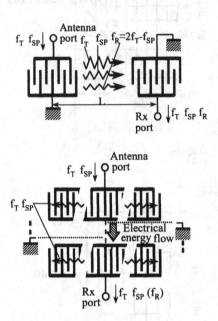

Fig. 40. Propagation of signals at f_T and f_{SP}, and generation of pseudo signal at f_R.
(a) Conventional SAW filter. (b) Ladder-type SAW filter.

According to the specifications shown in Fig. 1(b), a SRS≥60 dB is generally required. We can see from the data in Fig. 39(a) that the SRS of the conventional filter constructed with image-impedance connected IDT's is about 65 dB, whereas that of the ladder-type filter is about 90 dB. Therefore, the former is very marginal for cellular-radio use but the latter satisfies system specifications with sufficient margins[18].

With regard to the switch-type SAW duplexer for EGSM/DCS dual-band cellular phones, we can see from Fig. 2(a) that when the EGSM transmitter is ON, the 2nd-harmonic emission must be smaller than –36 dBm. This is because the twice f_T is partially overlapped with the receiver band of DCS. This emission is caused by the 2nd-order non-linear characteristics of the switching diodes and the SAW filters due to 35-dBm EGSM maximum transmitter power. The 2nd- and 3rd-harmonic emissions at frequencies other than the f_R of DCS must be smaller than –30 dBm on condition that the transmitter powers are 35 dBm and 32 dBm for EGSM and DCS, respectively.

Examples of the 2nd- and 3rd-harmonic generation characteristics of the EGSM parts in the dual-band duplexer are shown in Fig. 41(a) and 41(b), respectively. The figures show that values small enough have been obtained. Almost the same performances have been also obtained for DCS.

The output power compression also occurs because of non-linear or saturation characteristics of the switching diodes. Examples of power-compression characteristics of the EGSM parts in the dual-band duplexer are shown in Fig. 42. A 1-dB-compression point of 39.6 dBm - that is, about 10W - has been obtained. This value is sufficient for cellular phone use.

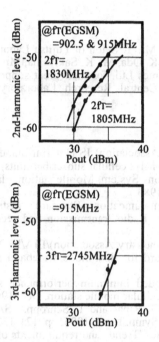

Fig. 41. Non-linear characteristics of the EGSM Tx part in the EGSM/DCS dual-band duplexer (a) 2nd-harmonic spurious emission. (b) 3rd-harmonic spurious emission.

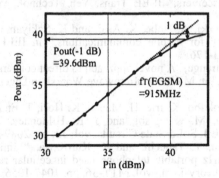

Fig. 42. Power-compression characteristics of the EGSM Tx part for the dual-band duplexer.

7. Conclusions

The frequency characteristics required for antenna duplexers used in the cellular phones in various systems (e.g., EAMPS, ETACS, EGSM, and DCS) have been derived and this derivation was based on the system specifications. SAW resonators, SAW-resonator-coupled filters (i.e., ladder-type filters) and procedures for designing antenna duplexers have been discussed. Published examples of duplexers, which have achieved miniature-module sizes and high-performance frequency characteristics have also been presented.

Acknowledgements

The author would like to thank Mr. Y. Satoh of Fujitsu Laboratory Ltd. for his useful comments and is grateful to Mr. K. Oda, Mr. K. Sakiyama, Mr. S. Wakamori, and Mr. T. Shiba of Hitachi Media Electronics Ltd. for their great support in this work. He also thanks Mr. N. Shibagaki of Central Research Laboratory, Hitachi Ltd., for his collaboration.

References

1. EIA (Electronic Industries Association) IS (Interim Standard) -19, Recommended minimum standards for 800-MHz cellular subscriber units, Feb. 1986.
2. Total Access Communications System, Mobile station - land station compatibility specification, Issue 3, Oct. 1984.
3. ETSI (European Telecommunication Standard Institute), Digital cellular telecommunications system: Radio transmission and reception (GSM 05.05 ver. 6.0.0), Jan. 1998.
4. TIA (Telecommunications Industry Association)/EIA/IS-95, Mobile station – base station compatibility standard for dual-mode wideband spread spectrum cellular system, May, 1995.
5. TIA/EIA/IS-98, Recommended minimum performance standards for dual-mode wideband spread spectrum cellular mobile station, Dec. 1994.
6. J. Machui, J. Bauregger, G. Riha, and I. Schropp, "SAW device and cordless phones," in IEEE Ultrason. Symp. Proc., 1995, pp. 121-130.
7. J. D. Kim and D. W. Weiche, "Trends and requirements of SAW filters for cellular system applications," in IEEE Ultrason. Symp. Proc., 1997, pp. 293-302.
8. M. Hikita, T. Tabuchi, and A. Sumioka, "Miniaturized SAW devices for radio communication transceivers," IEEE Trans. Veh. Technol., vol. VT-38, pp. 2-8, 1989.
9. M. Hikita, N. Shibagaki, A. Isobe, K. Asai, and K. Sakiyama, "Recent and future RF SAW technology for mobile communications," in IEEE Microwave Symp. Digest, 1997, pp. 173-176.
10. S. Atkinson and J. Strange, "A novel approach to direct conversion RF receivers for TDMA applications," in MWE'99 Microwave Workshop Digest, Yokohama, Japan, Dec. 1999, pp. 53-57.
11. T. Yamawaki, M. Kokubo, K. Irie, H. Matsui, K. Hori, T. Endou, H. Hagisawa, T. Furuya, Y. Shimizu, M. Katagishi, and J. R. Hildersley, "A 2.7-V GSM RF transceiver IC," IEEE J. Solid-State Circuits, vol. 32, pp. 2089-2096, 1997.
12. M. Hikita, Y. Ishida, T. Tabuchi, and K. Kurosawa, "Miniature SAW antenna duplexer for 800-MHz portable telephone used in cellular radio systems," IEEE Trans. Microwave Theory Tech., vol. MTT-36, pp. 1047-1056, 1988.
13. M. Hikita, N. Shibagaki, T. Akagi, and K. Sakiyama, "Design methodology and synthesis techniques for ladder-type SAW resonator coupled filters," in IEEE Ultrason. Symp. Proc., 1993, pp. 15-24.

14. T. S. Hickernel, "The dependencies of SAW-transducer equivalent-circuit-model parameters on transducer geometry," in IEEE Ultrason. Symp. Proc., 1997, pp. 127-130.
15. C. K. Campbell, "One-port leaky-SAW resonators: building blocks for low-loss RF front-end systems," FR design, Oct. 1999, pp. 22-28.
16. T. Matsuda, H. Uchishiba, O. Ikata, T. Nishihara, and Y. Satoh, "L and S band low-loss filters using SAW resonators," in IEEE Ultrason. Symp. Proc., 1994, pp. 163-167.
17. M. Ueda, O. Kawachi, K. Hashimoto, O. Ikata, and Y. Satoh, "Low loss ladder type filter in the range of 300 to 400 MHz," in IEEE Ultrason. Symp. Proc., 1994, pp. 143-146.
18. M. Hikita, N. Shibagaki, K. Sakiyama, and K. Hasegawa, "Design methodology and experimental results for new ladder-type SAW resonator coupled filters," IEEE Trans. Ultrason. Ferroelec. Freq. Contr., vol. UFFC-42, pp. 495-508, 1995.
19. N. Nakamura, M. Kazumi, and H. Shimizu, "SH-type and Rayleigh-type surface wave on rotated Y-cut LiTaO3," in IEEE Ultrason. Symp. Proc., 1977, pp. 819-822.
20. K. Yananouchi and K. Shibayama, "Propagation and amplification of Rayleigh waves and piezoelectric leaky waves in LiNbO3," J. Appl. Phys., vol. 43, pp. 856-862, 1972.
21. K. Yamanouchi and M. Takeuchi, "Applications for piezoelectric leaky surface waves," in IEEE Ultrason. Symp. Proc., 1990, pp. 11-18.
22. M. Hikita, A. Isobe, A. Sumioka, N. Matsuura, and K. Okazaki, "Rigorous treatment of leaky SAW's and new equivalent circuit representation for interdigital transducers," IEEE Trans. Ultrason. Ferroelec. Freq. Contr., vol. UFFC-43, pp. 482-490, 1996.
23. M. Hikita, T. Tabuchi, N. Shibagaki, and T. Akagi, "SAW filters applicable to high-power - Comparisons between IIDT-type and SAW-resonator-coupled filters -," in Proc. Int. Symp. SAW Devices for Mobile Comm., pp. 105-112, 1992.
24. Y. Satoh, O. Ikata, T. Matsuda, T. Nishihara, and T. Miyashita, "Resonator-type low-loss filters," in Proc. Int. Symp. SAW Devices for Mobile Comm., pp. 179-185, 1992.
25. K. Hashimoto and M. Yamaguchi, "General-purpose simulator for surface acoustic wave devices based on coupled-of-mode theory," in IEEE Ultrason. Symp. Proc., 1996, pp. 117-122.
26. O. H. Huor, N. Inose, and N. Sakamoto, "Improvement of ladder-type SAW filter characteristics by reduction of inter-stage mismatching loss," in IEEE Ultrason. Symp. Proc., 1998, pp. 97-102.
27. K. Anemogiannis, C. Beck, A. Roth, P. Russer, and R. Weigel, "A 900 MHz SAW microstrip antenna-duplexer for mobile radio," in IEEE Ultrason. Symp. Proc., 1990, pp. 729-732.
28. M. Hikita, T. Tabuchi, N. Shibagaki, T. Akagi, and Y.Ishida, "New high-performance and low-loss SAW filters used in ultra-wideband cellular radio systems," in IEEE Ultrason. Symp. Proc., 1991, pp. 225-230.
29. O. Ikata, Y. Satoh, H. Uchishiba, H. Taniguchi, N. Hirasawa, K. Hashimoto, and H. Ohmori, "Development of SAW small antenna duplexer using SAW filters for handheld phones," in IEEE Ultrason. Symp. Proc., 1993, pp. 111-114.
30. N. Shibagaki, T. Akagi, K. Hasegawa, K. Sakiyama, and M. Hikita, "New design procedures and experimental results of SAW filters for duplexer considering wide temperature range," in IEEE Ultrason. Symp. Proc., 1994, pp. 129-134.
31. M. Hikita, N. Shibagaki, K. Asai, K. Sakiyama, and A. Sumioka, "New miniature SAW antenna duplexer used in GHz-band digital mobile cellular radios," in IEEE Ultrason. Symp. Proc., 1995, pp. 33-38.
32. T. Makkonen, V. P. Plessky, S. Kondratiev, and M. M. Salomaa, "Electromagnetic modeling of package parasitics in SAW-duplexer," in IEEE Ultrason. Symp. Proc.,

1996, pp. 29-32.
33. N. Shibagaki, K. Sakiyama, and M. Hikita, "Precise design technique for a SAW-resonator-coupled filter on 36° YX-LiTaO3 for use in a GSM SAW duplexer module for satisfying all GSM system specifications," in IEEE Ultrason. Symp. Proc., 1996, pp. 19-24.
34. S. Mineyoshi, O. Kawachi, M. Ueda, Y. Fujiwara, H. Furusato, and O. Ikata, "Analysis and optimal SAW ladder filter design including bonding wire and package impedance," in IEEE Ultrason. Symp. Proc., 1997, pp. 175-178.
35. N. Shibagaki, K. Sakiyama, and M. Hikita, "Miniature SAW antenna duplexer module for 1.9 GHz PCN system using SAW-resonator-coupled filters," in IEEE Ultrason. Symp. Proc., 1998, pp. 499-502.
36. N. Kamogawa, S. Dokai, N. Shibagaki, M. Hikita, T. Shiba, S. Ogawa, S. Wakamori, K. Sakiyama, T. Ide, and N. Hosaka, "Miniature SAW duplexers with high power capability," in IEEE Ultrason. Symp. Proc., 1998, pp. 119-122.
37. O. Ikata, T. Nishihara, Y. Satoh, H. Fukushima, and N. Hirasawa, "A design of antenna duplexer using ladder type SAW filters," in IEEE Ultrason. Symp. Proc., 1998, pp. 1-4.
38. H. Fukushima, N. Hirasawa, M. Ueda, H. Ohmori, O. Ikata, and Y. Satoh, "A study of SAW antenna duplexer for mobile application," in IEEE Ultrason. Symp. Proc., 1998, pp. 8-12.
39. Y. Satoh, T. Nishihara, O. Ikata, M. Ueda, and H. Ohomori, "SAW duplexer metallizations for high power durability," in IEEE Ultrason. Symp. Proc., 1998, pp. 17-26.
40. M. Hikita, N. Matsuura, N. Shibagaki, and K. Sakiyama, "New SAW antenna duplexers for single- and dual-band handy phones used in 800-MHz and 1.8-GHz cellular-radio systems," to be published in IEEE Ultrason. Symp. Proc., 1999.
41. O. Kawachi, G. Endoh, M. Ueda, O. Ikata, K. Hashimoto, and M. Yamaguchi, "Optimum cut of LiTaO3 for high performance leaky surface acoustic wave filters," in IEEE Ultrason. Symp. Proc., 1996, pp. 71-76.
42. K. Hashimoto, M. Yamaguchi, S. Mineyoshi, O. Kawachi, M. Ueda, G. Endoh, and O. Ikata, "Optimum leaky-SAW cut of LiTaO3 for minimised insertion loss devices," in IEEE Ultrason. Symp. Proc., 1997, pp. 245-254.

LADDER TYPE SAW FILTER AND ITS APPLICATION TO HIGH POWER SAW DEVICES

YOSHIO SATOH and OSAMU IKATA

Fujitsu Laboratories Ltd., Peripheral Systems Laboratories
64 Nishiwaki, Ohkubo-cho, Akashi 674-8555, Japan

SAW filters with one-port SAW resonators being connected in a ladder structure, which is called ladder type SAW filter, have many features such as low insertion-loss, wide passband and simple filter design and so on. The design methodology, features of characteristics, and developments of higher frequency filter are described for an application in cellular phones. Furthermore, in order to apply to high power SAW devices such as the antenna duplexer of cellular phones, the material of inter-digital transducers (IDT) was improved to have high power durability. The optimum design in a ladder type SAW filter, the improvement in materials for IDT, and the optimum design of the package, all contributed the success of realizing a small SAW antenna duplexer.

1. Ladder Type SAW Filters

1.1. Introduction

The architecture of ladder type filter has been well known in the design of quartz filters and ceramic filters. As the equivalent circuit of component resonators of these mechanical filters is similar to that of one-port SAW resonator, the idea of the ladder type filter can be simply applied to SAW filter design. S.C.C. Tseng first proposed the ladder type SAW filter in 1974.[1] In that paper, however, the series arm in a ladder structure was only made of one-port SAW resonators while the parallel arm consisted of a capacitance from an inter digital electrode. Therefore, the characteristics of this ladder type SAW filter with center frequency of around 33MHz didn't have a perfect pass-band shape. The reason why a SAW resonator wasn't used in the parallel arm was that the precise adjustment of the resonant frequency of SAW resonator was very difficult in those days. The resonant frequency difference between the series arm resonator and the parallel arm resonator must be precisely controlled to get the optimum band-pass characteristics. It was very hard to control the period of SAW resonators precisely by using the photolithography technique in the 1970's. The main interest of SAW researchers seemed to focus on acoustically coupled resonator filters such as the multi-mode SAW resonator filter [2,3] to get a narrow band-width and a higher out-of band rejection. After more than 10 years after Tseng's paper, the ladder type SAW filter with a capacitance in its parallel arm reappeared in 1988 in order to realize low loss filters for the antenna duplexer of cellular phones.[4] This filter had a low insertion loss of 1~2 dB. However, it had band elimination characteristics and a weak out-of-band rejection of 20 dB. The regular ladder type SAW filter which has SAW resonators both in the series arms and the parallel arms and hence perfect band-pass characteristics was first reported by the authors [5,6] and by M. Hikita et.al.[7,8] at the same time in 1992. In the development of Ref. 6, the precise difference of the period of inter-digital transducers between the parallel arm resonators and the series arm resonators, which is 0.18 μ m for 900 MHz-band filters, was realized by using a combination process of a g-line stepper or an i-line stepper with dry etching technique, as described later in this section. This advanced lithography technique, widely used in semiconductor technology, enabled us to realize a ladder type SAW filter composed of all SAW resonators. This high-frequency ladder type SAW filter were found to be low-loss and wide-band without the need for a matching circuit, and was confirmed to be suitable for cellular phone application [5-8]. Since these reports, many papers describing on this type of filter have been published, and these filters including the lattice configurations have together been called SAW impedance element filters (IEF).[9] In this section, the structure and principle of ladder type SAW filters, the simulation methods, the designing methods for cellular phone applications, the fabrication process for high frequency filters, and the features of characteristics are described in detail.

1.2. Structure and principle

The nominal circuit structure of the ladder type SAW filter is shown in Fig.1. In this figure, the notations Rs and Rp represent the series-arm resonator and the parallel-arm resonator, respectively. Both Rs and Rp are one-port SAW resonators and their resonant frequencies are slightly different from each other. In order to consider the principle, the basic section of a ladder type filter is shown in Fig.2 where the LC equivalent circuits without lossy elements represent one-port SAW resonators for convenience. In the basic section, the impedance of the series-arm resonator (Z_s) and the admittance of the parallel-arm resonator (Y_p) are given by the following equations,

$$Z_s = jX_z. \qquad (1)$$
$$Y_p = jB_p. \qquad (2)$$

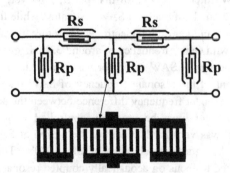

Fig.1. The nominal structure of the ladder type SAW filter.

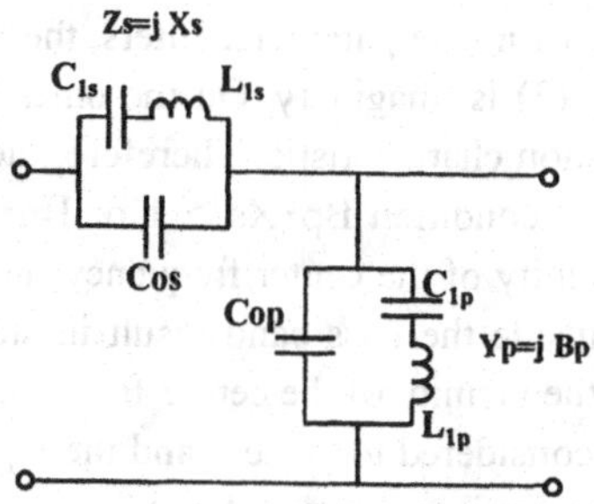

Fig.2. A basic section of ladder type filter represented by the LC equivalent circuit.

The variation of the reactance Xs and the susceptance Bp as a function of frequency are plotted in Fig.3. The one-port SAW resonator is well known to have a double resonant characteristic like a ceramic resonator. That is, it has a resonant frequency (fr) and an anti-resonant frequency (fa). Here, the anti-resonant angular frequency of the parallel-arm resonator (ω_{ap}) is designed to nearly equal the resonant angular frequency of the series-arm resonator (ω_{rs}) as shown in Fig.3. It becomes easy for today's fabrication technique to design and fabricate such a fine frequency adjustment as described in the introduction. In this situation, the image propagation constant (θ) of the basic filter section of Fig.2 is investigated. It is generally expressed in terms of Xs and Bp in the following equation

$$\tanh(\theta) = \sqrt{Bp \cdot Xs / (Bp \cdot Xs - 1)}. \qquad (3)$$

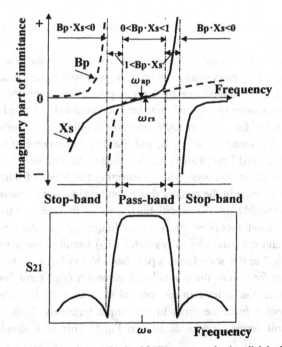

Fig.3. Formation of band-pass filter by using two kinds of SAW resonators having slightly different period.

According to the theory of image parameter filters, the basic section shows a pass-band characteristic when Eq. (3) is imaginary. On the other hand, when it is real, the basic section shows an attenuation characteristic. Therefore, the condition $0 < Bp \cdot Xs < 1$ gives the pass-band, while the condition $Bp \cdot Xs > 1$ or $Bp \cdot Xs < 0$ gives the stop-band as shown in Fig.3. The vicinity of the center frequency ω_o ($=\omega_{ap}=\omega_{rs}$) results in a pass-band and both sides outside the pass-band result in stop-bands. Since Xs and Bp are nearly equal to zero in the vicinity of the center frequency, the insertion loss of the basic section is theoretically considered to be zero and the input/output impedance is equal to the line impedance whatever its value may be.

For a more intuitive interpretation, we can use the concept of a one-port SAW resonator being a frequency switch as shown in Fig.4. As described above, the one-port

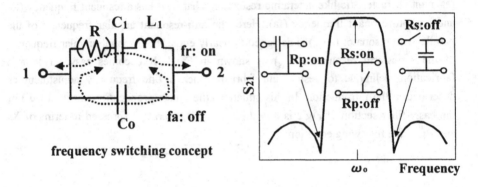

Fig.4. Frequency switching concept and an interpretation of the principle of a ladder type SAW filter.

SAW resonator has a double resonant characteristic. At the resonance frequency (fr), which is caused by a resonance between the motional capacitance C_1 and the motional inductance L_1, the impedance of the one-port SAW resonator is ideally zero. This situation corresponds to the "on" state between terminals of 1 and 2. On the other hand, the impedance of the one-port SAW resonator is infinity at the anti-resonance frequency (fa), which is determined by C_1, L_1 and the IDT capacitance of Co and slightly higher than fr as described later. This is the "off" state. The one-port SAW resonator acts as a frequency switch in this way. When connecting two kinds of frequency switches into a ladder configuration in the manner that the anti-resonance frequency of the parallel-arm resonator is roughly coincident with the resonance frequency of the series arm resonator, they act as a band-pass filter. At the center frequency ω_o, the series-arm resonator (Rs) turns "on" and the parallel-arm resonator (Rp) simultaneously turns "off". The ladder circuit of Fig.2 in this state forms a pass-band like in Fig.4. At the lower frequency side of the center frequency, the parallel-arm resonator (Rp) turns "on". And at the higher frequency side, the series-arm resonator (Rs) turns "off". In either case, the RF power flow is rejected from the input to the output terminals. Under these conditions, the ladder circuit forms stop-bands like in Fig.4. This is a simple explanation of the principle.

1.3. Simulations

This section briefly describes several kinds of simulation methods for ladder type SAW filters. The first step of the simulation is the calculation of the one-port SAW resonator. There are mainly three models for this calculation.

    a) LC circuit model [10]
    b) Smith's equivalent circuit model [6,11,12]
    c) Coupling-of-modes (COM) model [13,14]

The LC circuit model uses the structure of Fig.2 and is the simplest method. This calculation method roughly agrees with the experimental data around the center frequency. In this model, the resonance frequency (fr) and anti-resonance frequency (fa) of the one-port SAW resonator is expressed by the next equations using the parameters shown in Fig.4.

$$fr = 1 / 2\pi\sqrt{L_1 \cdot C_1} \quad (4)$$

$$fa = fr \cdot \sqrt{(1+C_1/Co)} \quad (5)$$

The resonance frequency is determined by experimental data or the value of acoustic velocity under the IDT divided by the period of the IDT. From Eq.(4), we can get the product of $L_1 \cdot C_1$ using the fr value. The capacitance ratio Co/C_1 ($=\gamma$) of Eq.(5) is roughly determined by the substrate used, and is about 15 for 36Ycut-X $LiTaO_3$. Then, fa is determined by γ using Eq.(5). The value of fa can be also determined by experimental data and we can confirm the γ value by Eq. (5) vice versa. The value of Co is the electrostatic capacitance of the IDT and is expressed by the next equation assuming 50% metallization of the IDT.

$$Co = N \cdot W \cdot \varepsilon_0 \sqrt{\varepsilon_{11} \cdot \varepsilon_{33}} \quad (6)$$

Here, N is the number of finger pairs of the IDT and W is the aperture length in MKS. ε_0 is the dielectric constant of air and ε_{11}, ε_{33} are specific inductive capacities of the substrate. From the given values of Co, γ, and fr, the values C_1, L_1 are derived using the above relations. If necessary, we can add a resistance (R) term to the motional arm of Fig.4. In that case, the resistance means a loss term such as an electrical resistance of the IDT and the bulk wave radiation and is empirically determined. After determining all the LCR parameters of the series-arm resonator and the parallel-arm resonator, the fundamental matrix (F) of the basic section of the ladder filter shown in Fig.2 is expressed by the following formula using Zs and Yp

$$F = \begin{bmatrix} A & B \\ C & D \end{bmatrix} = \begin{bmatrix} 1 & Zs \\ 0 & 1 \end{bmatrix} \cdot \begin{bmatrix} 1 & 0 \\ Yp & 1 \end{bmatrix}. \quad (7)$$

The Smith's equivalent circuit model gives a more accurate calculation method. By using Smith's equivalent circuit model, we can simulate the effect of actual design conditions of a SAW resonator such as the numbers of finger pairs of the IDT and the reflectors, the relative positions between the IDT and the reflectors, the aperture length and so on. Fig.5 shows an example of Smith's equivalent circuit expressed by a 4-port fundamental matrix with two acoustic ports and two electric ports [5]. Smith's equivalent

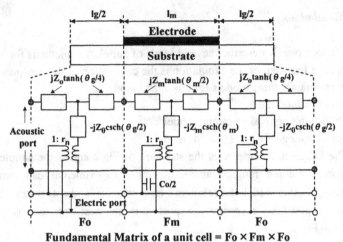

Fig.5. Smith's equivalent circuit of a unit cell by 4port matrix.

circuit is originally and generally expressed by a 3-port admittance matrix [11,12]. The total fundamental matrix of the one-port SAW resonator is fully obtained by cascading the unit cells of Fig.5 by the necessary number to construct the IDT and the reflectors. The admittance of the one-port SAW resonator is easily derived from the total fundamental matrix by setting the acoustic impedance of the free surface at the ends of the acoustic ports. Finally, we can obtain the fundamental matrix of a ladder circuit by substituting the impedance or admittance of the one-port SAW resonator for Eq.(7). Fig.6 shows an example of a comparison between the simulation by Smith's equivalent circuit model and experimental values. Fig.7 is the tested circuit condition. The effect of the bonding wires is taken into account in this investigation. The substrate used is 36Ycut-X LiTaO$_3$ and the film thickness of the electrode is 300nm. The series-arm resonator has 150 finger pairs, 60 μm aperture, and 4.12 μm period for the IDT and 80-shorted stripes for the reflectors, while the parallel-arm resonator has 40 finger pairs, 160 μm aperture and 4.30 μm period for the IDT, and 80-shorted stripes for the reflectors.

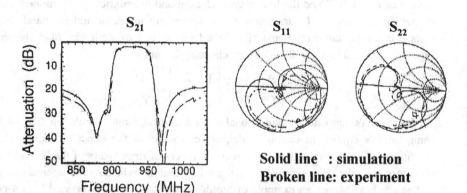

Fig.6. The comparison between simulation and experiment for S parameters.

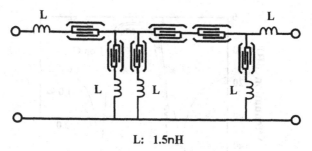

Fig.7. The tested circuit condition. Ls are inductance elements of the bonding wires.

The coupling-of-mode model is also useful for obtaining precise results. The coupling-of-mode theory is said to give shorter calculation speed and more precise result than the Smith's circuit model when treating periodic IDT structures, especially for analysis of unidirectional transducers. However, the determination of the COM parameters such as the self-coupling factor and the mode-coupling factor is more complicated than the determination of the circuit parameters of Smith's model, such as the SAW velocity, the acoustic impedance and the electro-mechanical coupling factor. The results obtained for the analysis of a one-port SAW resonator are said to be almost the same for the COM model and the Smith's equivalent circuit model. For details on the COM analysis for a one-port SAW resonator, see Ref.14.

1.4. Application to RF filters of Cellular Phones

1.4.1. Design

The method of designing a ladder type SAW filter is basically the same as for ceramic or quartz filters. However, the variation in design of a ladder type SAW filter is rather limited compared to quartz or ceramic filters because the capacitance ratio γ is almost fixed to the value of 36Ycut-X $LiTaO_3$ substrate, which is selected for application to RF filters of cellular phones due to its low temperature coefficient and high electro-mechanical coupling factor. Then, the main parameters for designing the RF filter for cellular phones restricted to the IDT's static capacitance of the series-arm resonator (Cos) and the parallel-arm resonator (Cop), and their periods which determine the resonance frequencies. The periods of the resonators are simply decided by considering the specification of the center frequency of the desired filters. In this section, the effect of Cos and Cop, the effect of the IDT resistance, and the method of increasing the band-width for a given γ are described for designing a ladder type SAW filter.

The capacitance ratio between the parallel-arm and the series-arm, Cop/Cos, mainly affects the out-of-band rejection as shown in Fig.8. This figure shows the simulation results for the filter with three stages of the basic section shown in Fig.7 and a center-frequency of 933 MHz. As Cop/Cos increases, the out-of-band is improved. However, the insertion-loss and the band-width deteriorate. This trade-off relationship in Cop/Cos

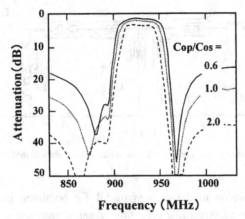

Fig.8. S_{21} dependence on static capacitance ratio Cop/Cos.

is similar to one in the case of the stage number of cascading basic sections.

The product of Cop by Cos affects impedance of the filter in the vicinity of the center frequency. We can demonstrate this relation by using the simple LC circuit model shown in Fig.2. The impedance of the series-arm resonator Zs and that of the parallel-arm resonator Zp(=1/Yp) are given by the following equations.

$$Zs = (\omega^2 - \omega_{rs}^2) / j\omega Cos(\omega^2 - \omega_{as}^2) \qquad (8)$$

$$Zp = (\omega^2 - \omega_{rp}^2) / j\omega Cos(\omega^2 - \omega_{ap}^2) \qquad (9)$$

Here, ω_{rs}, and ω_{as} are the resonant angular frequency and anti-resonant angular frequency of the series-arm resonator, respectively, while ω_{rp} and ω_{ap} are the resonant angular frequency and anti-resonant angular frequency of the parallel-arm resonator, respectively. These resonant angular frequencies are determined from Eqs. (4) and (5) for each resonator. For the matching condition of the constant K-type filter, the following relationship must be satisfied in the case of a line impedance R(Ω).

$$Zs \times Zp = R^2 \qquad (10)$$

As shown in Fig.3, the following relationships are assumed,

$$\omega_{ap} \approx \omega_{rs} \approx \omega_0,$$
$$\omega_0 - \omega_{rp} \approx \omega_{as} - \omega_0,$$
$$2\omega_0 \gg \Delta\omega, \quad \Delta\omega \equiv (\omega_{as} - \omega_{rp})/2 \qquad (11)$$

where ω_0 is the center angular frequency of the filter. From Eq. (10) and Eq. (11), the next relationship is obtained

$$R^2 \approx 1/(\omega_0^2 Cos\, Cop) \qquad (12)$$

The relationship of Eq. (12) for 50 Ω matching is plotted as the solid line for the ladder type SAW filter with a center frequency of 933MHz in Fig.9. On the other hand, this relationship is also obtained by using Smith's equivalent circuit model. The shaded area in Fig.9 is obtained by the condition that the standing wave ratio at the input port of the filter shown in Fig.7 is less than 2.0 against the 50 Ω line. This area is

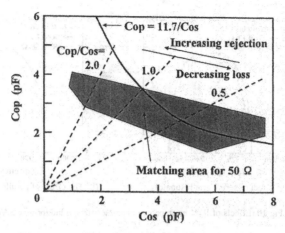

Fig.9. Impedance matching condition for the 933 MHz filter.

practically the matching condition of the 933 MHz filter for 50 Ω. The mismatch between the solid line of the LC circuit model and the shaded area from Smith's circuit model at high Cop/Cos values seems to be due to the effects of the inductance of the bonding wires and the disagreement between ω_{ap} and ω_{rs}. Note that optimum values of Cop and Cos for matching conditions exist in almost all practical Cop/Cos values. For the higher frequency filters, it is easily understood from Eq. (12) that the values of both Cop and Cos must be designed to be smaller with inverse proportionality to ω_o.

After determination of the Cop and Cos values, the next design step is how to apportion the aperture length and finger pairs' number of the IDT according to Eq. (6). This is directly related to how to design the electrical resistance of the IDT. The effect of the IDT resistance on the filter characteristics is described in this paragraph. Table 1 shows the experimental conditions for investigating the effect of the resistance. The tested circuit is the same as that shown in Fig.7. The capacitance values of the parallel-arm resonator and the series-arm resonator are kept to 2.6pF and 3.6pF, respectively in this investigation. Under these limitations, the aperture length and the number of finger pairs are varied while keeping their product constant in order to change the resistance of the IDT only. Table 1(a) shows the resistance variation of the series-arm resonator while keeping the parallel-arm resonator constant at condition number 6 of table 1(b). The table 1(b) shows the resistance variation of the conditions of parallel-arm resonator while keeping the series-arm resonator constant at condition number 1 of table 1(a). The

Table 1. The conditions of SAW resonator for investigation of IDT resistance.

(a) The condition of series-arm resonator				(b) The condition of parallel-arm resonator			
No.	aperture length	finger pair number	resistance	No.	aperture length	finger pair number	resistance
1	60 μm	150	0.06 Ω	4	53 μm	120	0.06 Ω
2	90 μm	100	0.13 Ω	5	80 μm	80	0.14 Ω
3	180 μm	50	0.50 Ω	6	160 μm	40	0.56 Ω

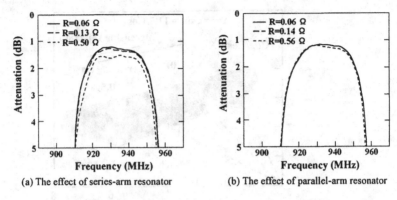

Fig.10. Effects of IDT resistance on characteristics of ladder type SAW filters.

obtained results are shown in Fig.10. From these figures, it is obvious that the insertion-loss of the filter is more sensitive to the resistance of the series-arm resonator than to the resistance of the parallel-arm resonator. Therefore, it is better to design the series-arm resonator such that its aperture length is as short as possible and the number of finger pairs is as large as possible in order to decrease the resistance of the IDT without increasing loss by interference of the SAW. This design is also suitable for a high power durability of the filter.

The last paragraph describes methods for increasing the band-width for a given γ. There are two methods for this. One is to use an inductance and the other is to enlarge the frequency difference between the resonance frequency (frs) of the series-arm resonator and the anti-resonance frequency (fap) of the parallel-arm resonator.

The former method is further classified into two methods, the insertion of L into the parallel-arm or into the series-arm. Fig.11(a) and Fig.11(b) show the effects of the inductance L in the parallel-arm and in the series-arm, respectively, obtained by Smith's model for 933 MHz ladder type SAW filters. Note that even very small inductances affect the characteristics of the filter. In the case of the parallel-arm insertion of Fig.11(a), it is observed that an increased band-width and an improved out-of-band rejection are simultaneously realized. This is a remarkable effect for SAW filter characteristics because an improvement of the out-of-band rejection is usually accompanied by a reduced band-width as shown in Fig.8. On the contrary, an increasing band-width is realized while the out-of-band rejection deteriorates in the case of the series-arm insertion as shown in Fig.11(b). Hence, the insertion of L into the parallel-arm is preferable. The reason why the insertion of L into the parallel-arm results in both increased band-width and improvement of the rejection band simultaneously is due to the shift of the resonance frequency and the increase in the admittance of the parallel-arm in the rejection band. A small inductance value of around 1~3 nH can be realized by the bonding wire and/or by the parasitic inductance of the lead patterns in a ceramic package. For this, it is important to avoid the common earth on the piezoelectric

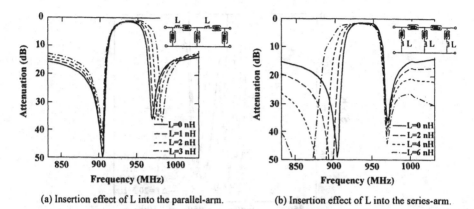

(a) Insertion effect of L into the parallel-arm. (b) Insertion effect of L into the series-arm.

Fig.11. Effects of inductance L for a ladder type SAW filter.

substrate but instead place it in the bottom layer of the multi-layered ceramic package.

The second method of increasing the band-width is to enlarge the difference between the resonance frequency of the series-arm resonator (frs) and the anti-resonance frequency of the parallel-arm resonator (fap). Important in this design is to know the upper limit of Δf (=frs-fap) where deterioration of insertion-loss or ripple in pass-band sets in. Fig.12 shows the dependence of the S_{21} on Δf. Fig.13 shows the dependence of the standing wave ratio (SWR) on Δf at the input port of the filter. The filter structure tested is the same as that in Fig.11 without inductance. It is found from these figures that a medium value of Δf is better for the insertion-loss and SWR than $\Delta f=0$. The limit of Δf in the case of a 933 MHz filter is 16MHz by requiring SWR<2. This Δf limit depends on the value of Cop/Cos and the cascading number(n) of basic sections. Increasing Cop/Cos or n reduces the limit of Δf. For instance, when Cop/Cos is 0.7 and n is 3, the ratio of $\Delta f/fo$ becomes up to 2% under the condition of SWR<2.

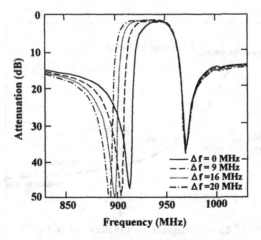

Fig.12. The dependence of Δf on S21 characteristics.

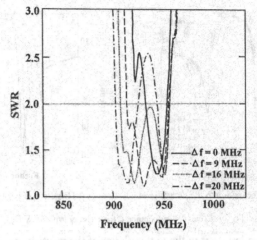

Fig.13. The dependence of Δf on SWR.

1.4.2. Fabrication process for high frequency filter

In this section, the fabrication process for high-frequency filters greater than 1GHz is briefly described. Fig.14 shows the relationship between the minimum pattern width and the center frequency of the filter for a fixed acoustic velocity under the IDT (V_{idt}=4000m/s) on 36Y-X LiTaO$_3$. This data is based on the following simple relationship assuming that the pattern width of a finger electrode is a quarter of the IDT period (λ)

$$W = V_{idt} / 4fo \qquad (13)$$

Here, W is the pattern width and fo is the center frequency of the filter. For example, when using 36Ycut-x LiTaO$_3$, which is suitable for cellular phone RF filters, the pattern

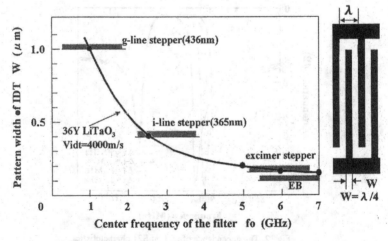

Fig.14. The pattern width vs. the center frequency for 36Y-LiTaO$_3$.

width becomes 0.4 μm at 2.5 GHz. Recent high resolution photolithography techniques in semiconductors make GHz SAW filters possible. If we use an i-line stepper, which uses a 365nm light source, we can realize filters of up to 2.5GHz. When using an excimer stepper having a resolution of 0.13~0.18 μm, filters in the range of 5.5~7.7 GHz can be produced. Electron beam lithography can produce 10 GHz SAW filter as demonstrated by Yamanouchi [15]. The specification and market size of the demanded GHz filter, through put, yield, and the price of equipment must be considered in deciding which apparatus is suitable. From this point of view, the i-line stepper is suitable for mass-production of 1.5 GHz~2.5 GHz SAW filters. This reduction-type stepper has another merit for high resolution because it can use the 5 times or 10 times magnified photo-masks delineated by the electron beam.

The etching method is another key technique in the fabrication of GHz filters. The validity of dry etching methods, especially the reactive ion etching (RIE), was successfully demonstrated by Yuhara et. al.[16] for high frequency SAW devices to replace the conventional wet etching methods. As the damage on the piezoelectric substrate by the reactive ion etching is well suppressed compared with other dry etching method such as ion-milling, no deterioration in filter characteristics and good reliability are observed. Using BCl_3 gas, we can cleanly etch Al or slightly doped Al film without corrosion or residuum as shown later in this chapter. This RIE method has already been put to practical use. Besides the etching method, the lift-off method is also widely used for SAW devices [15]. The main merit of the lift-off method is no need for any expensive dry-etching apparatus and no surface damage. But the drawback is not to be able to use sputtered films, which are favorable for adhesion and power durability, because lift-off needs the deposition of films at normal incidence to the substrate under low temperatures.

Combining these two techniques, exposure by i-line stepper and reactive ion etching, we fabricated an 0.4 μm-rule pattern and produced 2.5GHz filters. Fig.15 shows a SEM photograph of an 0.4 μm-rule IDT made by the two techniques. Fig.16 shows the S_{21} characteristic of a 2.45 GHz filter designed by a ladder type structure for wireless LAN

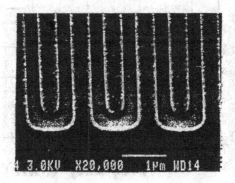

Fig.15. SEM photograph of 0.4 μm ruled IDT.

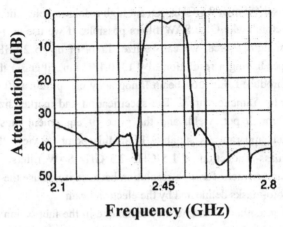

Fig.16. 2.45GHz filter for wireless LAN or Bluetooth systems.

or the Bluetooth systems. This filter has a 100 MHz band-width with a 4dB maximum insertion-loss, an out-of-band rejection higher than 30 dB and a high shape factor as shown in Fig.16.

1.4.3. Basic characteristics and features

The nominal S_{21} characteristic of a ladder type SAW filter designed for the AMPS (Advanced Mobile Phone System) specification is shown in Fig.17. It is compared with a conventional interdigitated interdigital transducer (IIDT) type filter previously obtained by authors[17]. These filters have a center frequency of 836 MHz and their design conditions are listed in Table 2. From Fig.17, it is confirmed that the insertion-loss, the out-of-band rejection, and the shape factor are all improved and an external matching circuit is not needed for ladder type SAW filters. The comparison of the characteristics are listed in Table 3 including other characteristics. From Table 3, the features of a ladder type SAW filter compared with IIDT are summarized as follows.

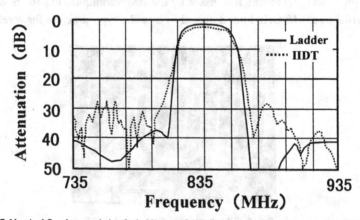

Fig.17. Nominal S_{21} characteristic of a ladder type SAW filter compared with a conventional IIDT filter.

Table 2. Design conditions for comparison between a ladder type filter and a IIDT type filter.

a) Ladder type SAW filter

series-arm resonator: aperture= 120 μm, finger pairs number= 100, Cos=4.8pF
parallel-arm resonator: aperture= 150 μm, finger pairs number= 48, Cop=2.9pF
the number of basic section: 4 stages
Al film thickness: 330nm
matching circuit: nothing

b) IIDT type SAW filter (See Ref.17)

input IDT: aperture=80 μm, finger pairs number=18~22(number's weighting), seven IDTs
output IDT: aperture=80 μm, finger pairs number=26~30(number's weighting), six IDTs
Al film thickness: 160nm
matching circuit: exist(LC circuit)

Table 3. The comparison of characteristics between ladder and IIDT.

Items	Ladder	IIDT
Max. insertion-loss in pass-band	2.4 dB without matching	3.6 dB with matching
Min. out-of-band rejection	38 dB	28 dB
Group delay ripple in pass-band	12 ns	20 ns
Max. SWR in pass-band	1.9	1.6
Reflection coefficient S_{11} in rejection band	0.8~0.9	0.7~0.8
Intercept point	61 dBm	37 dBm

a) Low insertion-loss.
b) High rejection level.
c) Low group delay ripple in pass-band.
d) High reflection coefficient in rejection-band.
e) High intercept point.

The above features except for item (c) are all suitable for the realization of SAW antenna duplexer as described in the next section.

Another design method for low-loss RF filter is the Double Mode SAW (DMS) filter reported by Morita et al. [18] almost at the same time as the ladder type SAW filter. This device was first reported using 64Y-X LiNbO$_3$ which has a large electromechanical coupling factor (11%), but a large frequency/temperature coefficient (70ppm/°C). However, a DMS filter made of 36~42Y-X LiTaO$_3$, which has a medium electromechanical coupling factor (5%) and a medium frequency/temperature coefficient (35ppm/°C), was recently reported [41]. Thanks to this development, the total performance of DMS is improved. Fig.18 shows a comparison of S_{21} between a ladder type filter and a DMS filter. The substrate used for both filters is 42Y-X LiTaO$_3$, which is rotated Y cut further than 36 degrees and is found to have lower loss for thick Al films as reported by Kawachi et al. [19]. The advantages of the ladder type SAW filter when compared with the DMS filter are,
a) Wider band-width.
b) Higher power durability.

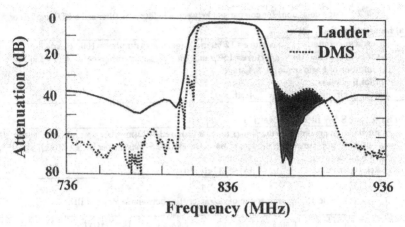

Fig.18. The comparison between a ladder type SAW filter and a DMS filter.

c) Steeper and higher out-of-band rejection of the high frequency side near pass-band (the shaded area in Fig.18).

The item of b) results from the large number of finger pairs in the series-arm resonator of the ladder type SAW filter and also from having a high intercept point. On the contrary, the advantages of the DMS filters are,

a) Extremely high level of out-of-band rejection especially in the far out-of-band region.
b) Steep skirt characteristic at the lower band edge.
c) Possibility for realizing a balanced filter.

Both design methods are widely used for RF filters in cellular phones and share the market. Ladder type SAW filters are mainly used when high rejection in the upper side band, high power durability such as for antenna duplexer, or extremely wide band are required. On the other hand, DMS filters are applied when high rejection on lower side band or balanced configuration are needed.

These RF filters are packed and hermetically sealed in a small SMT package of LCC (lead-less chip carrier) type as shown in Fig.19. This miniaturization trend started with the first report by the authors in 1990 [17]. The SMT package size was then $5 \times 5 \times$

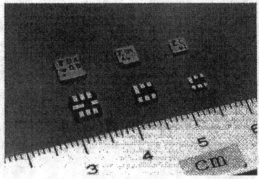

Fig.19. SMT package of SAW filters. The left is 5x5x1.6mm, the middle is 3.8x3.8x1.6mm, and the right is 3x3x1.2mm.

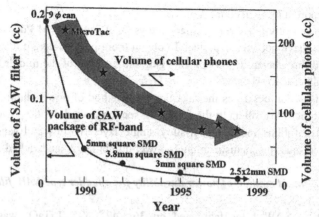

Fig.20. The size reduction trend of SAW filter and cellular phone.

1.6 mm. Since then, the SMT package size of RF SAW filters have been gradually reduced along with the size reduction of the cellular phones as shown in Fig.20. Today, a size of 2.5×2.0×0.9 mm is achieved using the flip-chip techniques [20].

2. Improvement on Power Durability of Ladder Type SAW Filters

2.1. Introduction

High-power durability is a very important property to enlarge the application domain of SAW devices. The high-power durability of SAW devices is initially needed for SAW resonators for high-power and high frequency oscillators. Recently, it is also essential for SAW antenna duplexers, which are one of the key devices for the miniaturization of cellular phones. The process of stress-migration in the IDT electrodes limits the power durability of SAW devices. So far, much effort has been spent on improving the resistance to stress-migration by finding new materials to replace the current material Al. Cu or Si doped Al films were initially introduced into SAW device technology for high-power SAW resonators [21,22] on the strength of being invulnerable to electro-migration in IC wiring technology. After that, Ti doped films were reported for the development of SAW antenna duplexers[23]. Other doping materials for binary alloys such as Al-Zn[24], Al-Ge[24], Al-Ta[25], and ternary alloys such as Al-Cu-Ni[26], Al-Cu-Mg[27] have so far been tested. However, it seems that these materials have not drastically improved power-durability. The effect of the doping metal is generally interpreted to make the grain size smaller and to block the movement of Al by the Al-alloy being segregated in the grain boundary. Contrary to this effect, the idea of decreasing the grain boundaries as much as possible, that is to say, to use single crystal Al, was proposed by Ieki et al.[28]. The growth of Al single crystal film, however, is sensitive to the substrate material. Al single crystal films have not been successfully grown directly on the rotated Y-X LiTaO$_3$ substrate, which is widely used for RF filters in cellular phones. Similar to the concept of Al-single crystal, the use of a highly textured Al film oriented [111] direction on a Cu thin film was also reported by Matsukura et al.[29].

We have recently tried some binary alloys of Al such as Al-Pd, Al-Y and Al-Sc, which so far have not been reported. To obtain more power-tolerant materials, we have investigated three-layered films by changing the material of the middle layer, which is sandwiched between Al films.

The next section describes the measurement method of power durability inherent to the ladder type SAW filter. In the following sections, we describe power-durability of many kinds of films including Al-alloy single layered films and sandwiched films. Finally, we discuss mechanisms of high-power durability of sandwiched films.

2.2. Measurement method of power durability for a ladder type SAW filter

A ladder-type SAW filter fabricated on $36 \sim 42°$ Y-X $LiTaO_3$ was used for the measurement of power durability. All filters were designed to have a center frequency of either 800 MHz or 1.9GHz and to have a fractional band-width of $3 \sim 4$ %[6,30]. The experimental data shown in this section are those obtained by using conventional Al-2%Cu films.

By applying a constant high RF power to the SAW filter, it was reported that the metal migration of the inter-digital transducer (IDT) occurs due to RF stress and hence it results in burn-out of the IDT[21]. In a ladder type SAW filter, the same phenomenon is observed using the simple measuring system shown in Fig.21 (a). Applying high power at a fixed frequency within the pass-band, a typical output power curve like Fig.21(b) is observed using the measuring system of Fig.21 (a). After applying RF power for a given time, the output power gradually deteriorates and finally drops off rapidly. Fig.22(a) shows the SEM photograph of IDT associated with deterioration of insertion loss and Fig.22(b) shows the effect of burn-out, corresponding to the regions shown in Fig.21(b). It is noted that Al migration causes the degradation of the insertion loss and it leads to burn-out of electrodes by increasing the resistance.

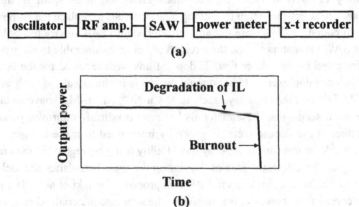

Fig.21. (a) The simple measurement system of power durability.
(b) Schematic characteristics of output power under high-power.

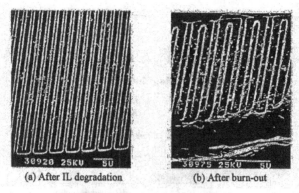

(a) After IL degradation   (b) After burn-out

Fig.22. SEM photograph of damaged IDT.

The specific measurement conditions for a ladder type SAW filter are described next. We found that the power durability of a ladder type SAW filter has a large dependence on frequency. Fig. 23 shows the frequency dependence of the lifetime for a transmitting filter at the accelerated power condition indicated in the figure. The lifetime was determined at 0.5 dB degradation of IL using the measurement system of Fig.21. We see that there is a convex frequency dependence of the lifetime and the degradation was most rapid at the frequency corresponding to the highest pass-band edge.

In an antenna duplexer, high power is also applied to the receiving filter. Therefore, we also have to evaluate the power durability of such filters. There are two possible frequency ranges for the power application to a receiving filter corresponding to the transmitting frequency band. One is the upper frequency side of the center frequency, and the other is the lower side. The former corresponds to Japanese cellular system, and

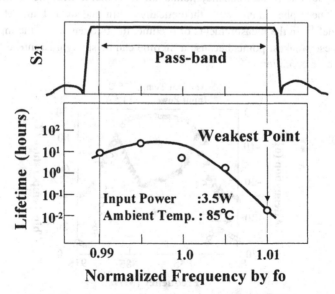

Fig.23. Frequency dependence of life time for a transmitting SAW.

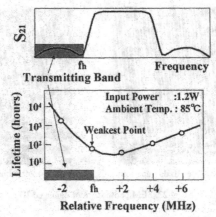

Fig.24. The frequency dependence of the lifetime for a receiving filter.

the latter corresponds to GSM, AMPS and PCS systems. As will be described later, the latter case had very weak power durability compared with the former case. We investigated the frequency dependence of the lifetime for the latter case. Fig.24 shows the frequency dependence of the lifetime for a receiving filter that has the transmitting band in the lower side of the center frequency. In Fig. 24, fh indicates the highest frequency of the transmitting band. As in the case of transmitting filter, the weakest frequency point is found to be the highest frequency of the rejection band, that is, the transmitting band for a receiving filter. These weakest points relate to the maximum temperature rise of the filter chip as shown in Fig. 25. The maximum temperature rise occurs at two frequency points, which are both transition bands of the filter. Nishihara et al. pointed out that these frequency points almost coincide with the weakest frequency points [31]. These phenomena were theoretically confirmed by J. Laine et al. using the COM model [32]. In the measurements of lifetime, the frequencies of the input power are fixed at these weakest points because a lifetime can be safely guaranteed by evaluating the worst case conditions.

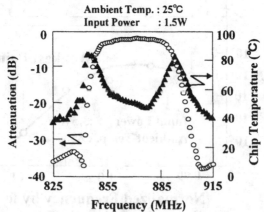

Fig.25. Frequency dependence of the chip temperature and transmission characteristics under high power conditions.

Next, the error mode should be considered for reproducibility of the lifetime measurements. When applying a high power signal at the weakest frequencies described above, the error mode schematically indicated in Fig.26 occurs in the filter characteristics. For a transmitting filter, the nearest series-arm resonator to the input terminal is most vulnerable for Al-migration. As a result, the upper side of the pass-band deteriorates as shown in the lower left of Fig. 26. On the contrary, the nearest parallel-arm resonator is destroyed in a receiving filter and the lower side of pass-band is distorted as shown in the lower right of Fig.26. Considering these error modes, we defined the lifetime limit from whether the insertion loss increases 0.5 dB or not, or whether the bandwidth at 1 dB reduction is decreased by 2 MHz or not. These precise measurements are carried out by an automatic evaluation system described later.

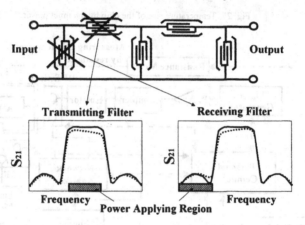

Fig.26. Nominal error mode of a ladder type SAW filter.

Before describing the automatic evaluation system, we briefly discuss the model of accelerated lifetime testing for SAW filters. The lifetime to failure is assumed to follow the Eyring's model generalized from Arrhenius's model:

$$\ln L_t = A + B/T - n \ln S \quad (14)$$

Here, L_t is the lifetime, A, B and n are constants, T is the absolute temperature, and S is the stress which is proportional to the input power in our case. If the input power, that is, S is kept constant, the lifetime is inversely proportional to temperature as reported by Yamada et al.[33]. It was also confirmed by Ikata et al.[34] that this relationship is kept in the case of ladder type SAW filters and that the lifetime is doubled if the temperature drops 10°C. When we investigate the relationship between the lifetime and the input power according to Eq. (14), we have to keep the temperature constant. However, the chip temperature rises as the input power increases as shown in Fig.27. Then, we controlled the ambient temperature so as to keep the chip temperature constant by measuring the chip temperature during testing. The measurement of the chip temperature was carried out by a resistive pattern fabricated on the filter chip. Considering the above-discussed conditions, we built the automatic evaluation system shown by Fig.28. It is composed of two systems, which are changed by RF switches.

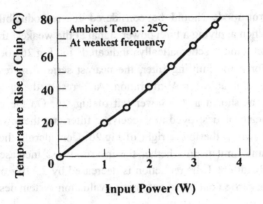

Fig. 27. Temperature rise of the chip by the input power.

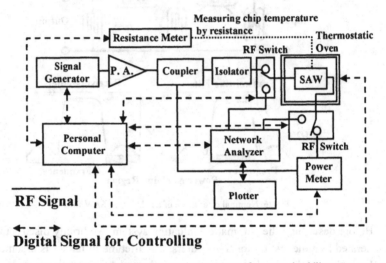

Fig. 28. Block diagram of automatic evaluation system for power durability test.

One is a path from a signal generator, through a power amplifier, a coupler, an isolator, a first RF switch, a SAW filter, and a second RF switch, to a power meter as shown in Fig.28. In this situation, the RF power with a programmed magnitude and frequency is applied to a SAW filter for a programmed period. Input power and output power during loading are monitored by a power meter and stored in a personal computer memory. The second is a loop made of a network analyzer and a SAW filter when turning the two RF switches to the opposite sides. In a programmed time, the network analyzer automatically measures filter characteristics and judges success or failure according to the rule of lifetime limit described before. A sequence of these measuring processes are all programmed and controlled by a personal computer. Using this evaluation system, we can precisely and easily measure the lifetime and the power durability of both transmitting filters and receiving filters. This system is especially useful for measuring receiving filters because we cannot get an output signal when applying a power to the out-of-band of filter by using the simple equipment as indicated in Fig.21.

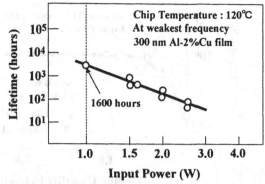

Fig. 29. Lifetime dependence on input power for transmitting filters.

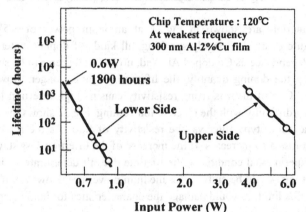

Fig. 30. Lifetime dependence on input power for receiving filters.

Fig.29 and Fig.30, respectively, show the relationship between the lifetime and the input power for transmitting filters and receiving filters measured by the present system. These plots show the linear correlation between lifetime and input power that is predicted by Eq.(14) for a constant chip temperature. From Fig.29, we also notice that the current Al-2%Cu film is insufficient for a transmitting filter of the antenna duplexer because its lifetime at 1W is only 1600 hours. In the receiving filter of Fig.30, two dependencies are plotted. One is for the case of applying power at the weakest frequency in the lower rejection-band near pass-band, and the other is for the upper side as described above. It is found that the former has very weak power durability and has insufficient lifetime as a receiving filter of an antenna duplexer with an applied power of 0.6W.

2.3. Single layered film

We investigated a few kinds of Al-alloy single layered film that have previously not been tested for SAW. We tested Al-0.3%Y, Al-0.9%Y, Al-0.6%Pd and Al-0.2%Sc, where the numbers indicate the weight % of the sputtering target used. The lifetime of

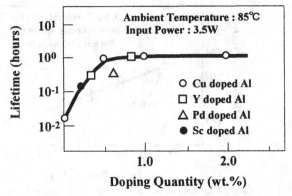

Fig. 31. Lifetime dependence on doping quantity to Al for single layered films.

these materials are shown in Fig.31 at an input power of 3.5W and an ambient temperature of 85°C. Roughly speaking, all kinds of doping materials seem to have a similar dependence as Cu doped Al. And, no material seems to be superior to Al-Cu. By increasing the doping quantity, the lifetime becomes longer. However, we have two problems. One of them is rising resistivity causing high insertion loss and the other is increasing difficulty with the reactive ion etching. Fig.32 shows the minimum insertion-loss of a ladder type filter vs. the resistivity of film. As the resistivity increases, the insertion-loss also increases. If the increase of the insertion-loss stays at 0.2~0.3 dB as in our experimental conditions, the lifetime doesn't deteriorate by increasing resistivity and the films would be suitable for medium power use. However, if the doping quantity exceeds this limit, it would damage the characteristics for both insertion-loss and power durability.

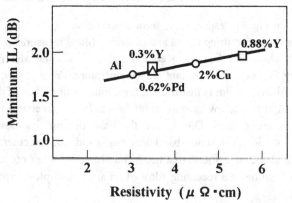

Fig. 32. Minimum insertion-loss vs. resistivity for slightly metal doped Al films.

Regarding the problems with RIE, Table 4 lists the situation for IDTs after etching by BCl_3, chlorides, and boiling temperature for some slightly doped Al films. For residue and corrosion, we also refer to the SEM photograph in Fig.33. In Table 4, we note that we cannot successfully make a reactive ion etching for highly doped Al films with chlorides of a high boiling temperature such as Al-4%Cu and Al-0.9%Y due to the presence of residues and corrosion.

Table 4. Possibility of RIE for slightly doped Al films.

Material	Residue	Corrosion	Chloride	Boiling Temp.(°C)
Al-4%Cu	Exist	Exist	CuCl	1490
Al-2%Cu	No	No		
Al-0.9%Y	Exist	Exist	YCl₃	1507
Al-0.3%Y	No	No		
Al-0.6%Pd	No	No	PdCl₂	680
Al	No	No	AlCl₃	182.7

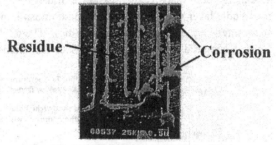

Fig. 33. SEM photograph of IDT after RIE.

2.4. Laminated film

We could not find a more power tolerant material than sputtered Al-2%Cu for use as single layered film. Next we tried to change the film structure itself. In IC wiring, there is also the concept of laminated films for the electro-migration durability. J.K.Howard et al. [35] reported sandwiched type films in which the middle layer is made of an inter-metallic compound of aluminum and transition metals, and sandwiched between pure aluminum or Al-Cu as shown in Fig. 34. As the middle layer in this structure is made of a hard inter-metallic compound formed through high temperature annealing, it can prevent voids from penetrating the film as schematically shown in Fig.34. Materials of

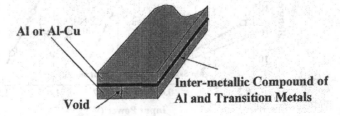

Fig. 34. The structure of sandwiched film used in IC wiring technology.

Table 5. Conditions for sandwiched films tested

Name	Filter type	Top and bottom layer	Middle layer
Cr sandwiched	Receiving	Al-2%Cu,145nm	Cr, 30nm
Ta sandwiched	Transmitting	Al-2%Cu,120nm	Ta, 20nm
Mo sandwiched	Receiving	Al-2%Cu,128nm	Mo, 30nm
Cu sandwiched	Transmitting	Al-2%Cu,120nm	Cu, 20nm
	Receiving	Al-2%Cu,125nm	Cu, 30nm

the middle layer which had the best lifetime were reported to be Cr or Ta [35]. We tested several kinds of sandwiched film including Cr and Ta in order to investigate useful materials for SAW devices. Table 5 lists experimental conditions for evaluating sandwiched films. The thickness of the middle layer was kept at less than approximately 10% of total thickness not to increase the resistance of the films. In this experiment, the thickness of the middle layer was not optimized. Both transmitting and receiving filters with a 800 MHz center frequency were used for testing. The total mass loading of each film was adjusted to keep the center frequency at 800MHz. Some of these metals were

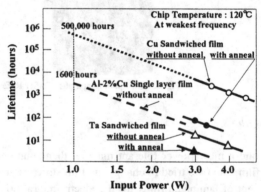

Fig. 35. Lifetime vs. input power for a transmitting filter.

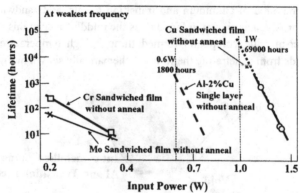

Fig. 36. Lifetime vs. input power for a receiving filter.

annealed at 400°C for one hour in vacuum. The results are indicated in Fig.35 and Fig.36, for transmitting and receiving filters, respectively. From Figs. 35 and 36, we notice that the Cu sandwiched film without annealing shows the longest lifetime over the other materials for both transmitting and receiving filters. Annealing at high temperatures deteriorated. Ta, Cr, and Mo, which are reported to be optimum materials for IC wiring, do not show superior lifetime over the conventional Al-Cu single layered film. On the other hand, Cu sandwiched film shows 200 times lifetime of the Al-Cu single film. The extrapolated values of lifetime at 1W, which are 500,000 hours for a transmitting filter and 69,000 hours for a receiving filter, are sufficient for practical duplexer applications.

We then focused on Cu sandwiched film to investigate practical properties in detail. First, we optimized the thickness of the middle Cu layer for a constant mass loading. Fig. 37 shows the maximum input power vs. the thickness ratio of Cu and the resisitivity of the sandwiched film vs. the thickness ratio of Cu. The maximum input power was measured by increasing the input power step by step, sweeping the frequency from the lower rejection band to the pass band. The optimum thickness is obtained to be around 10%. When increasing the Cu thickness over 10%, the power durability was reduced probably due to the increased Joule heating through the increased resistivity of the sandwiched film.

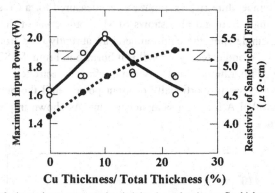

Fig. 37. Maximum input power and resistivity dependencies on Cu thickness ratio.

Table 6. Experimental condition for laminated film.

Laminating Number	Bottom layer ⟵				⟶ Top layer	
	Cu	Al-2%Cu	Cu	Al-2%Cu	Cu	Al-2%Cu
1	106 nm	—	—	—	—	—
	—	305 nm	—	—	—	—
2	40 nm	215 nm	—	—	—	—
	—	230 nm	30 nm	—	—	—
3	—	140 nm	30 nm	140 nm	—	—
4	—	100 nm	30 nm	100 nm	30 nm	—
5	—	67 nm	30 nm	67 nm	30 nm	67 nm

We next investigated the relation between the laminate number and the power durability. Thickness conditions for each laminated film are shown in Table 6. A single layered Cu film was also tested as a reference. Two types of double layered films were tested, one with a Cu bottom film and one with a Al-Cu bottom film. The film thickness of all laminated films was controlled to keep the mass loading constant in order to adjust the center frequency to the same value around 800 MHz. All samples were fabricated combining two types of dry etching processes, RIE and ion milling. For the laminated film with a bottom layer of Al-2%Cu, we used ion milling until the Cu film was etched away, and finally used RIE not to damage the substrate. For the Cu film and the double layer with a bottom layer of Cu, we used ion milling only. The results of lifetime and resistivity for the various laminate numbers are shown in Fig.38. The lifetime of this figure is the extrapolated value for 1W at chip temperature of 120 °C. It is noticed that the three layered film is the best structure for high power durability and that the resistivity of the laminated film abruptly increases above three layers. It is also noticed that the Cu film has almost the same lifetime as the Al-2%Cu film and that the double layer with Cu lowest, symbolized as Cu/Al-Cu in Fig.38, is slightly superior to the one with Al-Cu lowest, which agrees with the Al texture effect reported by Matsukura [29].

Finally, we confirmed that the Cu sandwiched film burned out without Al-migration. Fig.39 shows temperature rise vs. input power measured for a Cu sandwiched film and an Al-2%Cu single film and also shows SEM photographs of the IDTs after burnout. For the Al-Cu single film, the chip temperature abruptly rises from 1.1W of the input power, and at 1.5W indicated by symbol B it burns out. From the SEM photograph of B, we notice that Al migration occurs between the finger electrodes. On the other hand, the chip temperature rises proportionally to input power until the burnout point at 2.0 W indicated by symbol A for a Cu sandwiched film. As shown in photograph A, little Al migration occurs.

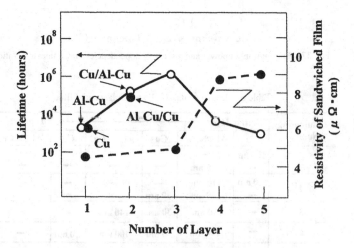

Fig.38. The dependencies of lifetime and resistivity on the number of layers.

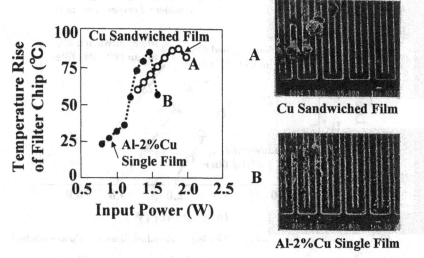

Fig.39. Temperature rise of chip vs. input power and SEM photographs of IDT after burnout.

For higher frequency applications such as 1.9 GHz band filters, the pattern width and film thickness must be reduced. In addition to this disadvantage, the SAW stress also increases proportionally to frequency [21]. As a result, the power durability becomes weaker and Cu sandwiched films cannot be used any more. For this application, we had to develop a more tolerable material. As described in the investigation of the Cu sandwiched film, transition metals are not suitable for high power durability of SAW devices. The drawback of transition metals is the need for high annealing temperatures to form a mechanically hard layer of Al and transition metal compounds. We tried a Mg sandwiched film expecting it to harden the Al film because Mg is known as a component material of duralumin, without knowing whether Mg is suitable for low annealing processes or not. Table 7 shows the experimental conditions for the Mg sandwiched film compared to the Cu sandwiched film. All processes were carried out without thermal exposure higher than 200°C. The lifetime tests were done at the weakest frequencies of the transmitting band of a 800 MHz filter and a 1.9 GHz filter compared to the 800 MHz filter made of the Cu sandwiched film. The results of lifetime tests depending on the input power are shown in Fig.40. In a comparison between the 800 MHz filters, the Mg sandwiched film has a lifetime approximately 100 times longer than that of the Cu sandwiched film. It is also shown that the Mg sandwiched film designed for a 1.9 GHz filter of the PCS system has sufficient power durability for a 1W power application of CDMA continuous wave.

Table 7. The experimental condition for a Mg sandwiched film.

Name	Frequency band	Top and bottom layer	Middle layer
Cu sandwiched	800 MHz	Al-2%Cu, 120 nm	Cu, 30 nm
Mg sandwiched	800 MHz	Al, 175 nm	Mg, 30 nm
	1.9 GHz	Al, 86 nm	Mg, 14 nm

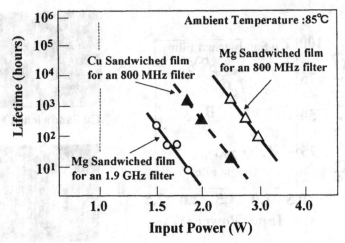

Fig.40. The lifetime vs. the input power for Mg sandwiched films and Cu sandwiched films.

Other advantages were investigated for the Mg sandwiched film. Fig. 41 shows the dependencies of maximum input power and resistivity on Mg thickness ratio of a sandwiched film, which has the same dependencies as Fig. 37. The highest power durability is obtained at a Mg thickness ratio of 4.6% where resistivity is $4.8\,\mu\Omega\cdot cm$. The thickness ratio is lower and the resistivity is almost the same compared with the Cu sandwiched film depicted in Fig.37. This means that the resistance of IDTs made from Mg sandwiched films are lower than those of Cu sandwiched films, because the thickness of the Al portion of the former can be designed to be higher than that of the latter, since the density of Mg is five times less than for Cu. Actually, the effect of low resistance is beneficial for the insertion loss of 1.9 GHz filters in which the IDT fingers become finer. The minimum insertion loss was improved by 0.5 dB compared to Cu sandwiched film when using Mg sandwiched film.

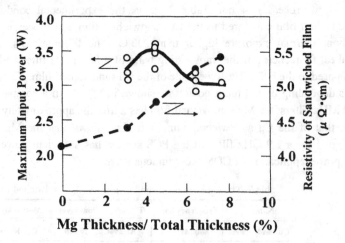

Fig.41. Maximum input power resistivity dependencies on Mg thickness ratio.

2.5. Mechanism for high power durability

Though sandwiched films with a middle layer of transition metal are known to be resistant to electro-migration in IC wiring, these metals are not suitable for SAW devices in our experiment. We notice from our experiments that it is important for the metal of the middle layer to provide some mechanically hardening effect to the Al films without high temperature annealing. In order to further investigate why the Cu sandwiched film and the Mg sandwiched film are better, we recorded cross-sectional TEM views of various sandwiched films as shown in Fig.42. These films are all annealed below 200°C. All the sandwiched films have smaller grain size than the Al-Cu

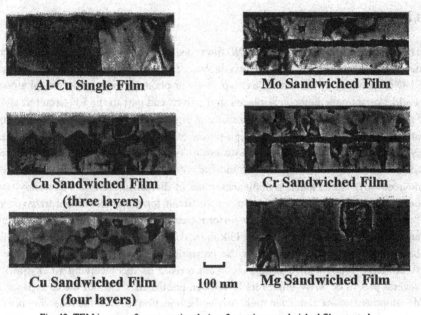

Fig. 42. TEM images of cross-sectional view for various sandwiched films tested.

single film. We find that the grains of the Cu sandwiched film form a piling block structure and the middle block layer is confirmed to be $CuAl_2$ by EDX (Energy Dispersive X-ray) analysis and X-ray diffract-meter. The piling block structure including the middle layer of $CuAl_2$ seems to make the film harden. The thickness of the middle layer should keep an optimum balance between the hardening effect and the electrical resistance. The TEM photograph of the Cu sandwiched film with four layers in Fig. 42 suggests that the excess of multi-layers results in increased electrical resistance due to an increased number of $CuAl_2$ grains in spite of the hardening as shown in Fig. 38. Contrary to the Cu sandwiched film, Mo and Cr sandwiched films, which are transition metals, don't show any Al-compound formation because the annealing temperature of 200°C is too low for these metals. From EDX analysis, the middle layers of these films are analyzed to be still Mo and Cr. Therefore, these transition metal sandwiched films don't have much mechanical strength in themselves. The photo of the Mg sandwiched film seems to show similar features as the transition metals. The middle layer of Mg, however, is not as clearly defined as those of Cr and Mo. From EDX analysis, Mg is found to uniformly diffuse into the Al films to make an Al-Mg solid-solution. The high power durability of the Mg sandwiched film is expected to derive its hardening effect from the formation of the Al-Mg solid-solution. As Al-Mg can not be etched by RIE because of the high boiling point of $MgCl_2$, annealing at low temperature for diffusion of Mg must be carried out after RIE etching.

3. Application to High Power SAW Devices

3.1. Introduction

The application of ladder type SAW filters as antenna duplexer of cellular phones is described in this section. The antenna duplexer is one of the most important RF parts for FDD (Frequency Division Duplex) type cellular phone systems widely used around the world. An antenna duplexer is placed in the front-end part of the RF circuit as shown in Fig.43 in order to separate the transmitting signal and the receiving signal from each other at the antenna port. As the high power of 1~1.2W level is applied to the antenna duplexer, high power durability, low insertion-loss, high-level isolation between the transmitting and receiving bands, and high level intercept point are needed for this device. Large sized antenna duplexers made of dielectric filters are currently used for cellular phones because it has been very difficult for the conventional transversal-type SAW filters to achieve satisfactory performance. The first attempt to developing a SAW antenna duplexer was reported by Hikita et al.[4] using a ladder type SAW filter with a band elimination characteristic as the transmitting filter and a transversal type SAW filter (IIDT type) with a band-pass characteristic as the receiving filter. Since then, several papers on SAW duplexers have been published [33, 36-40]. In the early stage of the development, some characteristics such as the rejection level of the receiving band, size,

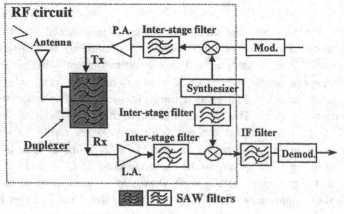

Fig.43. A Block diagram of RF circuit of a cellular phone.

cost, and other characteristics were not satisfactory. As an example of success in practical use, recent developments for AMPS specification reported by the authors [33, 36] are described in this section. The development of a ladder type SAW filter and high power durability of IDTs described in the former two sections are the basic technologies in our development on SAW duplexers. In addition, the package design of the antenna duplexer is also a core technology to realize small size and a high characteristic performance.

3.2. General design of antenna duplexer

First of all, the structure of the antenna duplexer and the electrical conditions required for the two filters composing a duplexer are described. Fig.44 shows the basic structure of an antenna duplexer. It has two filters, a transmitting filter and a receiving filter, and three ports, a transmitting port, a receiving port, and an antenna port. The output port of the transmitting filter and the input port of the receiving filter are connected to each other at the antenna port. In the frequency band of the transmitting state (Bt), the RF signal flows from the transmitting port to the antenna port. In the frequency band of the receiving state (Br), the RF signal flows from the antenna port to the receiving port. The shaded arrows show these flows. In the transmitting band (Bt), the following relations must be satisfied:

$$Z_2 = \infty \ \Omega, \quad Z_1 = Z_0 = \sim 50 \ \Omega \tag{15}$$

Here, Z_1 and Z_2 are the impedance of the transmitting filter and the receiving filter, respectively, viewed from the antenna port, and Z_0 is the total impedance connecting Z_1 and Z_2 in parallel. Eq. (15) means that the output signal from the transmitting filter passes through to the antenna port without going to the receiving filter. On the other hand, in the receiving band (Br), the following relations must be satisfied:

$$Z_1 = \infty \ \Omega, \quad Z_2 = Z_0 = \sim 50 \ \Omega \tag{16}$$

In the same manner, Eq. (16) means that the input signal from the antenna port makes a one way pass towards the receiving filter. As a result, the transmitting and the receiving signals are completely separated. In order to satisfy Eq. (15) and Eq. (16), the S_{11} curve

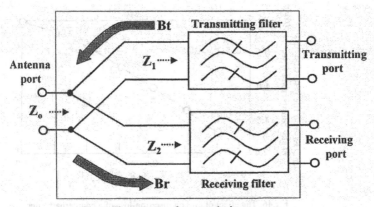

Fig.44. The structure of antenna duplexer.

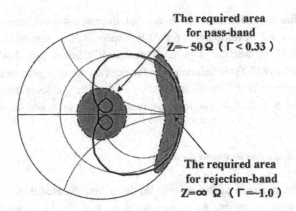

Fig.45. Optimum area of S_{11} required for each filter.

each filter viewed from the antenna port must agree with the shape indicated in Fig. 45. The impedance of its pass-band must be around 50 Ω, other words, the absolute value of S_{11} must be less than 0.33 (SWR<2.0). For the rejection band, whichever it is the higher frequency side or the lower frequency side, the impedance must be infinity or sufficiently larger than 50 Ω. The actual filter usually does not have the optimum characteristic of Fig.45. In such a case, we can easily optimize it by rotating the phase angle if the absolute value of S_{11} in the rejection band is sufficiently large. The absolute value of S_{11} in the rejection band is very a important factor for making the insertion-loss of the counterpart filter in duplexer low. Fig 46 shows the effect of the absolute value of S_{11} on the insertion-loss of the counterpart filter. This relationship is derived from

$$\text{Loss (dB)} = 20 \log | 2(1+|S_{11}|)/(3-|S_{11}|) | \tag{17}$$

If the increase of the insertion-loss by coupling two filters is permitted to be 0.5 dB maximum, it is noticed from Fig.46 that the $|S_{11}|$ value must be more than 0.8.

Investigating S_{11} of a ladder type SAW filter, it is found that the S_{11} curve greatly depends on the resonator configuration. When the outermost arm of the filter is composed of a series arm SAW resonator, the S_{11} curve of the higher rejection band

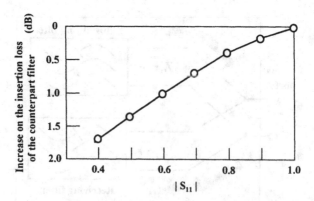

Fig.46. The effect of $|S_{11}|$ on insertion-loss of a counterpart filter.

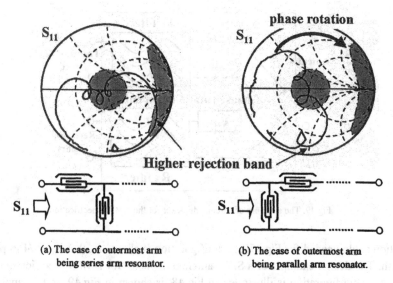

(a) The case of outermost arm being series arm resonator.

(b) The case of outermost arm being parallel arm resonator.

Fig. 47. S_{11} of a ladder type SAW filter depending on its configuration.

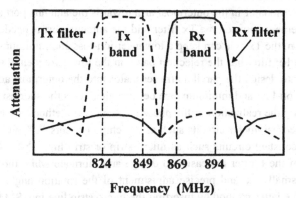

Fig. 48. Frequency specification of AMPS.

enters the required area as shown in Fig.47(a). On the contrary, in the case of the outermost arm being a parallel arm resonator like Fig.47(b), the S_{11} curve of both the lower rejection band and the higher rejection band takes a path far from the required area for the rejection band. Then, a ladder type SAW filter having the counterpart filter's pass band in its higher rejection band as in Fig.47(a), such as the transmitting filter (Tx) having the receiving filter (Rx) in its higher rejection band according to the AMPS specification (see Fig.48), can be used as it is without any additive elements. In the reverse case that has the counterpart filter in the lower rejection band just like the Rx filter of the AMPS specification, however, some additive element to rotate the phase angle is needed in order to achieve the optimum condition shown in Fig.45. In this case, the SAW resonator's configuration of Fig.47(b) is better than that of Fig.47(a), because the necessary length of the phase rotation element becomes much shorter. The actual

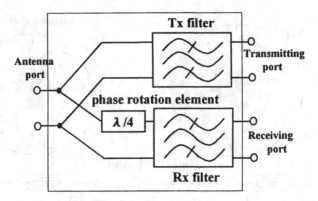

Fig.49. The structure of antenna duplexer for the AMPS specification.

rotation angle needed for Fig.47(b)'s configuration is about π, that is, $\lambda/4$ in physical length. The design example of a SAW antenna duplexer for AMPS specification whose frequency configuration is illustrated in Fig.48, is shown in Fig.49. In this application, the Tx filter that has the rejection band on the higher frequency side is designed to have a series arm resonator in the outermost arm towards the antenna port and hence no extra electrical element between the Tx filter and the antenna port is needed. Because of this configuration, the Tx side can keep a low insertion-loss and minimum size. On the other hand, as the Rx filter has the rejection band on the lower frequency side, it is better for the Rx filter to design the parallel arm resonator into the outermost arm in order to keep a low insertion-loss and minimum size because that gives the shortest phase rotation.

Regarding the realization of the phase rotation, two methods are well known. One is to use lumped parameter circuits such as LC chip elements [39, 40], and the other is to use distributed constant circuits such as micro strip or strip lines [34, 36]. Generally speaking, the merits of the former are easier to design and fabricate while those of the latter are capable of small size and precise adjustment of the rotation angle. The authors have developed the latter method by inserting the micro strip line in a SMT package as small as possible, as described in the next section.

3.3. Package design of antenna duplexer

The package design has an important role in the case of the antenna duplexer. In order to keep the package size small and the insertion-loss low, the package was designed as follows.
a) The package has two cavities for mounting the two SAW filters and has a micro strip line for the phase rotation built in itself, and it is made of multi-layered aluminum ceramics.
b) The micro strip line is mounted on the backside of the package in a folded shape. The surface area of micro strip line is a little recessed so as to have an air gap between the top of micro strip line and the bottom surface of package.

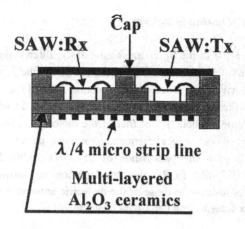

Fig.50. Cross sectional view of duplexer package.

λ /4 micro strip line

Fig.51. The outward aspect of SAW duplexer. The package size is 9.5×7.5×2.0 mm.

c) The micro strip line is gold plated to decrease the insertion-loss of the micro strip line made of tungsten.

Fig.50 schematically shows the cross sectional view of package designed for AMPS specification. Fig.51 shows the photograph of the package. On the back plane, the folded micro strip line plated by gold appears.

In addition to the above technologies, it is important to utilize the parasitic inductance of the package effectively. The package has three levels of electrical ground, the ground for each of the two chips laid under them, the ground for the interconnection of these, and the true ground for the complete duplexer circuit. These are connected by via holes or side dented metalized strips that have very small parasitic inductance, the order of 0.01nH. It has been found by the authors that these small values of parasitic inductance affect the isolation among the three ports, especially the through isolation between the Rx and the Txt. The effect is described in Ref. 36 in detail.

3.4. Characteristics of antenna duplexer

Fig.52 shows an example of the S_{21} characteristic of the antenna duplexer developed for the AMPS specification. For both the Tx and the Rx filters, the design of the filter chips uses a ladder type SAW filter as described in section 3.2 and the substrate is 42Y-Xcut LiTaO$_3$ [19]. The two filters are mounted in the package of Fig. 51 whose size is 9.5 x 7.5 x 2mm and the volume is about 1/8 of that of the dielectric antenna duplexer that has the same specification. The typical frequency response performances are 2.0/3.6 dB (Tx/Rx) insertion loss in the pass-bands of 25 MHz width, 53/45 dB stop-band attenuation, and 55/52 dB Tx-Rx through isolation as shown in Fig.52. These characteristics are comparable to those of the dielectric antenna duplexer except for the superiority of the Tx filter having band-pass characteristics.

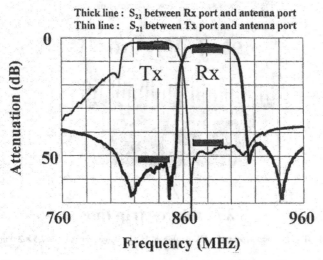

(a) The characteristics of S_{21} between Rx port and antenna port and between Tx port and antenna port.

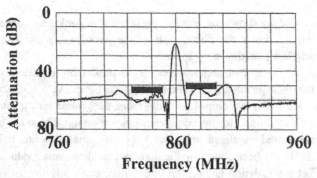

(b) The characteristic of S_{21} between Tx port and Rx port.

Fig.52. The characteristics of S_{21} between each ports.

4. Conclusion

The principle, the simulation, the design rule, and the characteristics for cellular phone applications are described for the ladder type SAW filter. The features of the ladder type SAW filter have been shown to be a low insertion loss, a wide band-width, and suitable for high power applications. The small sized RF filters for cellular phones with the center frequency ranging from 800 MHz to 2.5GHz are realized using the design of this filter. In addition, the power durability of the ladder type SAW filter is improved by investigating the structure and materials of the IDT. The Cu sandwiched film and the Mg sandwiched film are concluded to have sufficient power durability for the application to front-end filters of cellular phones. Using these technologies including the package technology, we have developed the small antenna duplexer for cellular phones. These small SAW devices utilizing the ladder type SAW filter are widely used in the world, and contribute the miniaturization and the high performance of cellular phones.

References

1. S.C.C Tseng, and G. W. Lynch, "SAW planar network", *IEEE Proceedings of Ultrasonics Symposium*, pp282-285, 1974.
2. G. L. Matthaei, E. B. Savage, and F. Barman, "Synthesis of Acoustic-Surface-Wave-Resonator Filters Using Any of Various Coupling Mechanisms", *IEEE Tras. SU*, vol. SU-25, p72, 1978.
3. M. Tanaka, T. Morita, K. Ono, and Y. Nakazawa, "Narrow band-pass double mode SAW filter", *IEEE Proceedings of 38th Frequency Control Symposium*, pp286-293, 1984.
4. M. Hikita, Y. Ishida, T. Tabuchi, and K. Kurosawa, "Miniature SAW antenna duplexer for 800-MHz portable telephone used in cellular radio system", *IEEE trans. Microwave Theory Tech.*, vol. MTT-36, p1047-1056, 1988.
5. Y. Satoh, O. Ikata, T. Matsuda, T. Nishihara, and T. Miyashita, "Resonator-type low-loss filters", Proc. Int. Symp. SAW Devices for Mobile Comm., pp.179-185, 1992.
6. O. Ikata, T.Miyashita, T. Matsuda, T. Nishihara, and Y. Satoh, "Development of low-loss band-pass filters using SAW resonators for portable telephones", *IEEE Ultrasonics Symposium*, pp111-115, 1992.
7. M. Hikita, T. Tabuchi, N. Shibagaki, and T. Akagi, "SAW filters applicable to high-power", Proc. Int. Symp. SAW Devices for Mobile Comm., pp.105-112, 1992.
8. M. Hikita, N. Shibagaki, T. Akagi, and K. Sakiyama, "Design methodology and synthesis techniques for ladder type SAW resonator coupled ffilters", *IEEE Ultrasonics Symposium*, pp15-24, 1993
9. J. Heighway, S.N. Kondratiev, and Victor P. Plessky, "Balanced bridge SAW impedance element filters", *IEEE Ultrasonics Symposium*, pp27-30, 1994.
10. E. J. Staples, "UHF Surface acoustic wave resonators", *Proc. of 28th Annual Symposium on Frequency Control*, pp280-285, 1974.
11. W. R. Smith, H. M. Gerard, and W. R. Jones, "Analysis and design of dispersive

interdigital surface-wave transducers", *IEEE Trans. Microwave Theory & Tech.* vol. MTT-20, No.7, pp458-471, 1972.

12. T. Kojima and K. Shibayama, " An analysis of reflection characteristics of the surface-acoustic-wave reflector by an equivalent circuit model", Jpn. J. Appl. Physi., 26, Suppl. 26-1, pp117-119, 1987.

13. H. A. Haus, and P. V. Wright, "The analysis of grating structures by coupling-of-modes theory", *IEEE Ultrasonics Symposium Proceedings*, pp277-281, 1980.

14. K. Hashimoto and M. Yamaguchi, "General-purpose simulation for leaky surface acoustic wave devices based on coupling-of-modes theory", *Proc. IEEE Ultrasonics Symposium*, pp117-122, 1996.

15. K. Yamanouchi, "Generation, Propagation, and Attenuation of 10 GHz-Range SAW in $LiNbO_3$", *Proc. IEEE Ultrasonics Symposium*, pp57-62, 1998.

16. A. Yuhara, T. Mizutani, N. Hosaka, J. Yamada and S. Kobayashi, "Dry process technology for high frequency SAW devices", *Proc. IEEE Ultrasonics Symposium*, pp343-349, 1989.

17. O Ikata, Y. Satoh, T. Miyashita, T. Matsuda, and Y. Fujiwara, "Development of 800 MHz band SAW filters using weighting for the number of finger pairs", *Proc. IEEE Ultrasonics Symposium*, pp83-86, 1990.

18. T. Morita, Y. Watanabe, M. Tanaka, and Y. Nakazawa, "Wide band low-loss double mode SAW filters", *Proc. IEEE Ultrasonics Symposium*, pp95-104, 1992.

19. O. Kawachi, G. Endoh, M. Ueda, O. Ikata, K. Hashimoto and M. Yamaguchi, "Optimum Cut of LiTaO3 for high performance leaky surface acoustic wave filters", *Proc. IEEE Ultrasonics Symposium*, pp71-76, 1996.

20. H. Yatsuda et. al., "Miniaturized SAW filters using a flip-chip technique", *IEEE Trans., UFFC*, vol.43, pp125-130, 1996.

21. J.I.Latham, W.R.Shreve, N.J.Tolar and P.B.Ghate, "Improved metallization for SAW devices", Thin Solid Films, 64, pp9-15, 1979.

22. Y.Ebata and S.Morishita, "Metal-migration of aluminum film on SAW resonator", Trans. IECE Japan Pt. C, J67-C, pp278-285, 1984(in Japanese).

23. A.Yuhara, H.Watanabe and J.Yamada, "Sputtered Al-Ti electrodes for high frequency and high power SAW devices", Jpn. J. Appl. Phys. 26 Sppl. 26-1 p135, 1987.

24. A.Yuhara, N.Hosaka, H.Watanabe, J.Yamada, "Al electrodes fabrication technology for high frequency and high power durable SAW devices", *Proc. IEEE Ultrasonics Sympium*, pp493-496, 1990.

25. N.Kimura, M.Nakano and K.Sato, "Power durability of Al-Ta alloy film electrodes using in SAW filters", Trans. IECE Japan Pt. C-II, J80-CII, pp356-357, 1997(in Japanese).

26. T.Tabuchi et al., IECE Technical Report, US87-18, pp23-28, 1987 (in Japanese).

27. T. Tabuchi et al., IECE Spring National Convention Record of Japan, A-234, p236,1988 (in Japanese).

28. H.Ieki and A.Sakurai, "SAW resonators using epitaxially Al electrodes", Trans. IECE Japan Pt. A, J76-A, pp145-152, 1993 (in Japanese).

29. N.Matsukura, A.Kamijo, E.Ootsuka, Y.Takahashi, N.Sakairi and Y.Yamamoto,

"Power durability of highly textured Al electrode in SAW devices",Jpn. J. Appl. Phys.35, pp2983-2986, 1996.

30. T.Matsuda, H.Uchishiba, O.Lkata, T.Nishihara and Y.Satoh, "L and S band low-loss filters using SAW resonators", *Proc. IEEE Ultrasonics Sympium,*.pp.163-167, 1994

31. T.Nishihara, H.Uchishiba, O.Ikata and Y. Satoh, "Improved power durability of SAW filters for an antenna duplexers", Jpn. J. Appl. Phys. Vol.34, pp2688-2692, 1995.

32. J.P. Laine, V.P.Plessky, and M.Salomaa, "Investigations on the power tolerance of ladder impedance-element SAW filters", *Proc. IEEE Ultrasonics Sympium.* pp15-18, 1996.

33. J.Yamada, N.Hosaka, A.Yuhara and A.Iwama, "Sputtered Al-Ti electrodes for high power durable SAW devices", *Proc. IEEE Ultrasonics Sympium,* pp285-290, 1988.

34. O.Ikata, Y.Satoh, H.Uchishiba, H.Taniguchi, N.Hirasawa, K.Hashimoto and H.Ohmori, "Development of small antenna duplexer using SAW filters for handheld phones", *Proc. IEEE Ultrasonics Sympium,* pp111-114, 1993.

35. J.K.Howard, J.F.White, "Intermetallic compounds of Al and transitions metals: Effect of electromigration in 1-2-um wide lines", J. Appl. Phys. 49(7), pp4083-4093, 1978.

36. O.Ikata, N.nishihara, Y.Satoh, H.Fukushima, and N.Hirasawa, "A design of antenna duplexer using ladder type SAW filters", *Proc. IEEE Ultrasonics Sympium,* pp1-4, 1998.

37. M. Hikita, N. Shibagaki, T. Akagi, and K. Sakiyama,"Design methodology and synthesis techniques for ladder-type SAW resonator coupled filters", *Proc. of IEEE Ultrasonics Symposium,* pp15-24, 1993.

38. N. Shibagaki, T. Akagi, K. Hasegawa, K Sakiyama, M. Hikita, "New design procedures and experimental results of SAW filters for duplexers considering wide temperature range", *Proc. of IEEE Ultrasonics Symposium,* pp129-134, 1994.

39. M. Hikita, N. Shibagaki, K. Asai, K. Sakiyama, and A. Sumioka, "New miniature SAW antenna duplexer used in GHz-band digital mobile cellular radio", *Proc. of IEEE Ultrasonics Symposium,* pp33-38, 1995.

40. N Shibagaki, K. Sakiyama, M. Hikita, "Precise design technique for a SAW resonator coupled filter on 36 YX-LiTaO3 for use in a GSM SAW duplexer module for satisfying all GSM system specifications", *Proc. of IEEE Ultrasonics Symposium,* pp19-24, 1996.

41. G. Endoh, M. Ueda, O. Kawachi, and Y. Fujiwara, "High performance balanced type SAW filters in the range of 900MHz and 1.9GHz", *Proc. of IEEE Ultrasonics Symposium,* pp41-44, 1997.